PROJECT MANAGER'S MANAGER'S PORTABLE HANDBOOK

**David I. Cleland
Lewis R. Ireland**

Second Edition

McGRAW-HILL

New York   Chicago   San Francisco   Lisbon   London
Madrid   Mexico City   Milan   New Delhi   San Juan
Seoul   Singapore   Sydney   Toronto

The McGraw·Hill Companies

Cataloging-in-Publication Data is on file with the Library of Congress

1 2 3 4 5 6 7 8 9 0   DOC/DOC   0 9 8 7 6 5 4

ISBN 0-07-143774-6

Coventry University

*The sponsoring editor for this book was Larry Hager, the production supervisor
was Pamela A. Pelton, and the art director for the cover was Margaret Webster-
Shapiro. It set in Times Roman by ProImage Corporation.*

Printed and bound by RR Donnelley.

This book is printed on acid-free paper.

CONTENTS

Section 8. The Project Culture   399

Section 9. Project Communications   443

Section 10. Improving Project Management   465

PREFACE

According to *Fortune* magazine, "Project Management is going to be huge in the next decade . . . project management is the wave of the future." Truly, project management is an idea whose time has come, and its future is most promising. The interest and emphasis on project management as the management system of choice is growing throughout many industries and businesses.

The continuing growth and acceptance of project management as the system of choice for managing change dictates that people have ready access to current information on the concepts and practices. One manager voiced his satisfaction with this Handbook by announcing at a meeting, *"Don't leave home without it!"* The authors take this one step further and say, *"Don't go to work without it!"*

Project management has grown over the past 50 years through the dedicated efforts of practitioners and academicians. Practitioners have worked through the processes and developed on an empirical basis many of the practices that are in use today. Academicians have conducted research to find better methods for project performance. Both the practitioner and the academician have contributed to the thousands of articles and hundreds of books that describe project management theory and practice.

Many professional associations have been started to promote project management and provide a forum for the interchange of project information. These associations have provided a means of publishing and distributing project information as well as providing other services. The largest and best-known project management professional association is the Project Management Institute of Newtown Square, Pennsylvania.

Today, the abundance of literature on project management has eclipsed the ability of busy people to find a single source to keep them abreast of the fundamental knowledge, skills and attitudes required to manage projects. These individuals need a relevant, concise and pragmatic source of information on how to manage projects. This Handbook is dedicated to that purpose.

As the theory and practice of project management continue to expand, new topic material appears. These new topic areas have been integrated into this second edition of the *Portable Handbook.*

This Handbook provides a single source of key summary information on the current theory and practices of project management. Users of this Handbook would most likely be managers of project managers, project managers, project management practitioners, and teachers and students of project management. Users may range from the seasoned project manager to the newly assigned practitioner.

This Handbook is designed to be easy to use, with readily available information. As a reference source, it will provide practical information in summary form on how to define, design, develop, and produce project results. Its size provides an easy-to-carry ready source of pertinent project management information. This Handbook is an everyday working document to provide practitioners key information on project management.

Contemporary theory and practice of major project topic areas provide the building blocks on knowledge. Coverage of topic areas is in summary form, buttressed by two references that provide in-depth information on each topic area. The summary content of this Handbook avoids the fluff and expanded narrative used to describe topics in the existing project management books.

The use of tables and models to show information further contributes to the condensed form of information. Key user questions are presented in each topic area to enhance the user's understanding of the material presented and the concepts discussed. Considerable effort has been made to distill the information to the most compact form.

ACKNOWLEDGMENTS

The conceptualization, definition, and preparation of this book became a key project for the authors. As is typical of all projects, this project took on a life cycle of its own and required the cooperative effort of many people who served as advocates, adversaries, and consultants. Our deep appreciation goes to all of these book project "stakeholders," without whom this project could not have been imagined and completed.

Our deep appreciation to all of our students, clients, and associates who have provided the opportunities for us to engage in discussions and practices of project management. Our students and associates in the project management community have been particularly influential in their support. We have tried to capture, in summary fashion, the best of the concepts, processes, and techniques of project management theory and practice.

Our thanks to Dr. Boyapa Bidanda, chairman of the Industrial Engineering Department, and Dr. Gerald D. Holder, Dean of the School of Engineering at the University of Pittsburgh, who provided an intellectual environment in which a project such as this Handbook could be pursued.

We also recognize and thank Claire Zubritzky and Lisa Bopp of the Industrial Engineering Department, who provided outstanding administrative support to this project.

We acknowledge and recognize the support given by Ouida F. Ireland, who reviewed drafts and commented on areas that were unclear or confusing for both content and format. She has encouraged the writing of this Handbook and supported the authors' efforts to complete this project on time.

INTRODUCTION

Project managers and other professionals working on projects depend upon four competencies. First is knowledge, the understanding of project management theory, concepts, and practices. Second is skill, the capability to use techniques and tools of the profession to obtain the proper results. Third is ability, the capability to integrate and use knowledge and skills in an effective manner. Fourth is motivation, the capability to develop and maintain the proper values, attitudes, and aspirations that helps all project stakeholders to work together for the betterment of the project.

Project managers and other professionals will find that this book provides valuable summary information on the building and enhancement of their competencies in working in the project management environment. The format of the book is adaptable to the user looking up a topic area of project management and finding fundamental summary guidance on how individual competencies can be developed and enhanced. If the user wishes to gain further insight into a topic area, the references at the end of each topic area provide additional information.

The format of the book provides for an organization of the topic areas by major heading. Users who want basic summary information in these topic areas can use whatever order best suits their purposes.

Section 1 recounts information on The Discipline of Project Management to include such topics as Project Success and Failure, Project Management Competency, A Project Management Philosophy, Ethics in Project Management, and The Project Management Process. This section provides a framework for understanding the discipline of project management.

Section 2, Project Organizational Design, provides standards for how an appropriate organizational design can be constructed for the management of projects. Authority-Responsibility-Accountability and the basic concepts of how to organize a project are explored in this section. A user who wants some models on how to get the project stakeholders organized will find valuable information in this section.

Section 3, Alternative Project Applications, deals with a topic of growing interest in the field of project management, namely, how project management can be used to deal with non-traditional projects as elements in the strategic management of the organization.

Section 4, The Strategic Context of Projects, puts forth the notion of projects as building blocks in the design and execution of organizational strategies. In this section, enterprise senior managers are charged with the responsibility of maintaining oversight over the selection and use of projects and project portfolios, to facilitate the development of new and improved products, services, and organizational processes. The use of program management as an umbrella for related projects in organizational strategy is also provided.

Section 5, Project Leadership, suggests standards and processes for project managers and others to use in providing a sense of direction and commitment for managing the people associated with a project. The key message is that a competent person can lead and can manage a project team and should know the difference between leadership and managership. A general description of overall team leadership is also presented.

Section 6, Project Initiation and Execution, addresses the key considerations to be evaluated when selecting a project to support organizational purposes, how to prepare winning proposals, and how to start up a project. Further topics discussed include Contract Negotiations and Administration, Managing Quality in Projects, and Project Termination.

Section 7, Project Planning and Control, presents the basic means and methodology on how to plan for projects and how to maintain monitoring, evaluation, and control over the resources used on a project. How to establish project portfolios and use of earned value management to monitor and control projects are also presented. The many subtopics in this section are building blocks in the management functions of planning and control.

Section 8, The Project Culture, provides insight into how the cultural ambience of a project and its stakeholders can influence the outcome of the project. Motivation, Project Manager Capabilities, and the Positive and Negative Aspects of Teams are presented.

Section 9, Project Communications, describes the concept and use of a Project Management Information System, Communication in Project Meetings, Negotiations, and other important considerations involved in the exchange of information during the management of a project are also provided in this section.

Section 10, Improving Project Management, provides a prescription for how project stakeholders can manage their project responsibilities during some of the major challenges that arise in the management of projects. The topic area of New Managers is provided to give the project stakeholders insight into some of the major forces that are shaping the role of those managers associated with a project.

PROJECT MANAGER'S PORTABLE HANDBOOK

SECTION 1
THE DISCIPLINE OF PROJECT MANAGEMENT

1.1   PROJECT SUCCESS OR FAILURE

1.1.1   Introduction

It is important to be able to anticipate whether a project will be either a success or failure. Success or failure is determined by the measures applied for evaluating the project during its life cycle—and when it is complete. Looking ahead and determining those actions that contribute to success or failure can often avoid adverse outcomes.

1.1.2   Project Success or Failure

1.1.2.1   Success and Failure are in the Eye of the Beholder.   The words "success" and "failure," like the word "beauty," are in the eyes of the beholder. In the project management context, the word "success" is used in the context of achieving something desirable, planned, or attempted—that is, the delivery of the project results on time, within budget,

1

and having an operational or strategic fit with the enterprise's mission, objectives, and goals.

The word "failure" describes the condition or fact of not achieving the expected end results. Project failure is a condition that exists when the project results have not been delivered as was expected. However, if the project results are acceptable to the user, then overrun of costs and schedule may be tolerated.

Determining success or failure requires that performance standards be developed on the project, which can be compared to the results that are being produced. It is important to remember that a key performance factor is whether or not the results of the project can be effectively used by the customer.

Product success or failure may be perceived differently by different project stakeholders:

- A project that has overrun cost and schedule goals but provides the user with the results that had been expected may be judged a success by the user.
- A project team member who gains valuable experience on the project team may consider the project a success.
- A supplier who has provided substantial resources to the project may consider the project a success.
- A contractor who lost a bid to do work for the project may consider the project a failure.

Because of a project's temporary nature, a determination of relative success or failure may be difficult. The objective nature of a determination of project success or failure makes it challenging to develop objective measures of performance. The meaning of project success or failure may vary depending on the period in the life cycle when the determination is made.

Nevertheless, there are some general standards that can be used in judging whether a project is a success or a failure.

1.1.3   Project Success—Factors

- The project results have been delivered to the customer, who sees the project as having an appropriate fit with the mission, objectives, and goals of the enterprise.
- The project work packages have been accomplished on time and within budget.
- The overall project results have been accomplished on time and within budget.

- The project stakeholders are happy with the way the project was managed and the results that have been produced.
- The project team members believe that serving on the project team was a valuable experience for them.
- A profit has been realized on the work accomplished on the project.
- The project work has resulted in some technological breakthroughs that promise to give the enterprise a competitive edge.
- Effective teamwork has been carried out on the project.
- The project has opened up new business improvements or opportunities for the project customer.
- Contemporary project management theory and practice have been carried out on the project.

1.1.4 Project Failures—Factors

- The project has overrun costs and schedules.
- The project does not have an appropriate fit with the customer's mission, objectives, and goals.
- The project has failed to meet its technical performance expectations.
- The project was permitted to run beyond the point where its results were needed to support the customer's expectations.
- Inadequate management processes were carried out on the project.
- A faulty design of the project's technical performance standards was conducted.
- The project stakeholders were unhappy with the progress on the project and/or the results that were obtained.
- Top management failed to review and support the project.
- Unqualified people served on the project team.
- The project met the initial requirements, but did not solve the longer-range business need.

Given the general standards for determining project success or failure, the contributing factors are shown in Table 1.1 and Table 1.2.

Given some of the likely causes of project success or failure, we must recognize that both are end results that can be affected by many forces and factors. These forces and factors are not considered all-inclusive, as each project tends to be unique and there might be additional reasons for success or failure. It is important that the reader recognize that a determination of success or failure is dependent on many matters. Awareness

TABLE 1.1   Factors Likely to Contribute to Failure

- Inadequate status/progress reports
- Insufficient senior management support and oversight
- Inadequate competencies of project manager regarding understanding of technology; administrative skills; interpersonal skills; communication skills; ability to make decisions; and limited vision—does not see big picture
- Poor relations with project stakeholders
- Poor customer relationships
- Lack of project team participating in the making of decisions
- Lack of team spirit on project
- Inadequate resources
- Insufficient planning and control
- Inadequate engineering change management
- Unrealistic schedules
- Underestimated cost, leading to underfunding
- Unfavorable public opinion
- Untimely planned project termination
- Inefficiencies in use of resources
- Poor definition of authority and responsibility for project team
- Lack of commitment on part of team members

TABLE 1.2   Factors Contributing to Project Success

- Adequate senior management oversight and support
- Early effective planning
- Appropriate organizational design
- Delegated authority and responsibility
- Efficient system for monitoring, evaluating, and controlling the use of resources on the project
- Effective contingency planning
- Strong team member participation in the making and execution of decisions on the project
- Realistic cost and schedule objectives
- Customer commitment to the project
- Adequate and continuing customer oversight
- Project manager's commitment to established technical performance objectives; budgets; schedules; and use of state-of-the-art management concepts and processes
- Adequate management information system
- An effective and efficient management system for the project

of these factors and forces will improve the chances that the project will be more successful and less likely to fail.

1.1.5 Key User Questions

1. Have performance standards been developed for the project that can be used as criteria for determining project success or failure?
2. Do the project team, general managers, and senior management understand what are the likely factors of success or failure?
3. Are the project accomplishments reviewed on a regular basis to determine whether or not success or failure factors are present?
4. Are provisions in place to conduct major progress reviews to determine if the project results are likely to have an appropriate operational or strategic fit?
5. Does the culture of the organization support the management of projects, taking into consideration both success and failure factors?

1.1.6 Summary

In this section, some ideas concerning project success and failure were presented. Included in the textual material were some examples of the forces and factors that can contribute to project success or failure. Finally, various key user questions suggested some criteria that can be used to determine the relative success or failure of a project.

1.1.7 Annotated Bibliography

1. DeLucia, Al and Jackie, *Recipes for Project Success* (Newtown Square, PA: Project Management Institute, 2001). This book offers ten key tips for project success, including tips on planning, running a project, and controlling the work through analogy to cooking. Basic project management information is provided for the newly appointed project manager who needs a quick, simple explanation of techniques and principles. This book gives good tips on what to do and what issues to avoid that place the project at risk. The principles given are proven through experience and good, sound logic.
2. Pinto, Jeffrey K., "The Elements of Project Success," in David I. Cleland, *Field Guide to Project Management,* 2nd ed., (New York, NY: John Wiley & Sons, 2004). The author makes a key point that the

process for analyzing and predicting the likelihood of success or failure of a project is by no means a simple one. He provides several examples to exemplify the challenges of determining project success. In addition, a Ten-Factor Critical Success Model in project implementation is offered to the reader.

1.2   PROJECT MANAGEMENT: A DISTINCT AND CHANGING DISCIPLINE

1.2.1   Introduction

Formal project management has been around for more than 50 years and is practiced in many different industries. The history of project management practices dates back hundreds of years. These processes and practices have been documented and disseminated over the past few decades to contribute to the evolving discipline. A subset of the management discipline is emerging.

1.2.2   Project Management: Profiles of Change

The practices of project management can be traced to antiquity, as evidenced in past major construction projects such as the Great Pyramids, canals, bridges, cathedrals, and other infrastructure projects. Today, project management is an idea whose time has come. A description of the evolution of project management as suggested in the literature follows:

The article "The Project Manager," published in the *Harvard Business Review* in May-June 1959, set forth some essential notions about the discipline:

- Projects are organized by tasks requiring an integration across the traditional functional structure of the enterprise.
- Unique authority-responsibility-accountability relationships arise when a project is managed across the traditional elements of the organization.
- A project team is a unique organizational unit dedicated to delivery of project results on time, within budget, and within predetermined technical specifications.

In 1961, Gerald Fish wrote in the *Harvard Business Review* about the growing obsolescence of the line-staff concept, and described the growing trend in contemporary organizations toward a "functional teamwork" approach to organizational design.

An important contribution to the project management literature appeared in the form of the "matrix" organization, first described by Professor John F. Mee of Indiana University in a 1964 article in *Business Horizons*. The contributions that this article made included the first description of the nature of the evolving "matrix" organization to include the "web of relationships" that replaced the line and staff relationship of work performance.

1.2.2.1 Role of Senior Management in Projects.   In 1968, a landmark study of the practices of senior management in leading industrial corporations noted the responsibilities of senior managers for project management. The study noted high-level committees (such as the board of directors) were widely used as a valuable organizational design to:

- Establish broad policies.
- Coordinate line and technical management.
- Render collective judgments on the evaluation of corporate undertakings.
- Conduct periodic reviews and monitoring of ongoing programs and projects.

1.2.3 The Emergence of Project Management

Distinctive factors that stand out in the emergence of project management, include:

- Demonstration of effectiveness through such noteworthy initiatives as the Manhattan Project and the Polaris Submarine.
- Development of specialized techniques for scheduling project activities such as PERT, CPM, and cost-schedule control systems.
- An early definition of a project as "any undertaking that has definite, final objectives representing specified values to be used in the satisfaction of some need or desire." (Ralph Currier Davis, *The Fundamentals of Top Management* (New York, NY: Harper and Brothers, 1951), p. 268.)
- The emergence of concepts which support the growing field of project management to include:

- A distinct life cycle
- Cost considerations
- Schedule factors
- A technical performance capability
- An assessment of the operational or strategic fit of the project results into the project owner's organization.

Some of the unique characteristics of project management coming forth in its evolution include:

- Projects are ad hoc endeavors, which have a defined life cycle.
- Projects are building blocks in the design and execution of organizational strategies.
- Projects are the leading edge of new and improved organizational products, services, and organizational processes.
- Projects provide a philosophy and strategy for the management of change in the organization.
- The management of projects entails the crossing of functional and organizational boundaries.
- The management of a project requires that an inter-functional and inter-organizational focal point be established in the organization.
- The traditional management functions of planning, organizing, motivation, directing, and control are carried out in the management of a project.
- Both leadership and managerial capabilities are required for the successful completion of a project.
- The principal outcomes of a project are the accomplishment of technical performance, cost, and schedule objectives.

Projects are terminated upon successful completion of the cost, schedule, and technical performance objectives—or earlier in their life cycle when the project results no longer promise to have an operational or strategic fit in the organization's future.

1.2.3.1 Authority-Responsibility. In 1967, Cleland, writing in *Business Horizons* magazine, described the difference between de facto (earned) authority, and de jure (legal) authority.

De jure, or legal authority comes from the organizational position that an individual holds and is reflected in project documentation such as a letter of appointment, position descriptions, policy documents, and related documentation.

De facto authority is that which comes from the individual's knowledge, expertise, interpersonal skills, experience, and demonstrated experience to work with people.

1.2.3.2 A Project Management System.   In 1977, Cleland published a short article in the *Project Management Quarterly* which described a Project Management System. This system is described in Section 7 of this Handbook.

1.2.4   Contribution of Project Management

Project management during its evolution contributed to a theory and practice in its own right as it matured as a discipline. A summary of the major changes that have come about in project management since its emergence includes:

* Recognition that project management is a discipline in its own right, as a branch of knowledge and skills
* Discovery and establishment of the legitimacy of the "matrix" organizational design as a means for delegating authority, responsibility, and accountability for the management of project resources
* Stimulated and propagated the growth of professional associations in the field
* Developed and disseminated the concept of a Project Management System as a performance standard for the management of project resources
* Provided the "strategic pathway" for the emergence and use of alternative teams in the operational and strategic management of the organization
* Became the principal means for the management of ad hoc activities in organizations
* Tested and established the legitimacy of the "horizontal dimension" in contemporary organizations
* Defined the concept of the influence of project stakeholders, and the importance of being able to manage project stakeholders
* Created and defined a new career path for managers and professionals

1.2.5   A Contemporary Model of Project Management

An article in *Fortune* magazine captured the essence of today's state of project management. The key messages of this article are shown in Table 1.3.

TABLE 1.3   Key Messages Regarding Project Management

- Mid-level management positions are being cut.
- Project managers are a new class of managers to fill the niche formerly held by middle managers.
- Project management is the wave of the future.
- Project management is spreading out of its traditional uses.
- Managing projects is managing change.
- Expertise in project management is a source of power for middle managers.
- Job security is elusive in project management, as each project has a beginning and an end.
- Project leadership is what project managers do.

Source: Thomas H. Stewart, "The Corporate Jungle Spawns a New Species," *Fortune,* July 10, 1995, pp. 179–180.

1.2.6   Key User Questions

1. Do the people in the organization who are involved in project management have a sense of how and why this discipline emerged?

2. In the existing organization, are there any historical projects that have impacted the management of change in either the operational or strategic elements of that organization?

3. Do the people understand that the key questions to be dealt with in the management of a project include its cost, schedule, technical performance deliverables, and operational and strategic fit in the organization?

4. Do the key members of the organization accept the idea that project management is an idea whose time has come, and has progressed from being a special case to an important element in the management of an organization?

5. How do senior managers in the enterprise see their role in the use of project, service, and process development through the use of project management processes?

1.2.7   Summary

In this section, a brief summary of some of the key characteristics of the evolution of project management was presented.

The origins of project management appeared in antiquity, and are represented in the relics of historical periods. Today project management is viewed as an idea whose time has come. The continued evolution of project management has created a distinct philosophy reflected in management literature of the discipline.

1.2.8 Annotated Bibliography

1. Cleland, David I., "Project Management: Profile of Changes," *Proceedings of the 29th Annual Project Management Institute 1998 Seminars & Symposium,* Long Beach, CA, October, 1998 pp. 1206–1210. This reference provides a brief overview of the evolution of project management expressed in the major changes that have occurred in this discipline over the past several years.

2. Drucker, Peter F., "The Coming of the New Organization," *Harvard Business Review*, January-February 1988, pp. 45–53. This article describes some of the alternative designs of contemporary organizations to include that characteristic of today's project-driven matrix organization.

1.3 PROJECT MANAGEMENT COMPETENCY

1.3.1 Introduction

Project management competency is essential for organizations' projects as building blocks to their future growth and profitability. In the past, knowledge was considered the key to success, but the concept of individual competency within project management is recognized as a more powerful approach to business.

Webster's II New College Dictionary defines competence as being properly qualified and capable, adequate for the stipulated purpose. *Roget's International Thesaurus* offers many synonyms for competence, such as ability, ableness, enablement, capability, capableness, capacity, efficiency efficacy, sufficiency, adequacy, "the stuff," and so forth.

Competency Has Many Synonyms			
Ability	Ableness	Enablement	Capability
Capableness	Capacity	Efficiency	Efficacy
Sufficiency	Adequacy		The Stuff

Competency depends on the personal characteristics of an individual, reflected in his or her knowledge, skills, and attitudes. Knowledge consists of suitable familiarity, awareness, and comprehension acquired by study and experience. Skill is the ability to apply knowledge and attitude is a state of mind or feeling.

Foundations of Competency		
Knowledge	**Skills**	**Attitudes**

1.3.2   Knowledge

The project manager must possess several basic elements of knowledge:

- An understanding of the technology of the project. A project is usually carried out within the context of a relevant technology. These technologies vary widely. For example, the technologies involved in a highway construction project will be different than those involved in a concurrent engineering project, or a project to develop an information system for an organization.

- A general understanding of the concept and process of strategic management as applied to the organization involved. Strategic management deals with the development of future initiatives for the organization as if that future mattered. It is particularly important that the project managers understand how their project is a building block in the design and execution of organizational strategies of the organization.

- A solid understanding of the theory, process, and practice of project management in the management of ad hoc initiatives supporting organizational purposes. In particular, the competent project manager should understand how the management of projects differs from the management of traditional work or activity within that organization.

- An understanding of and familiarity with the management processes involved in the management of a project. These processes, usually defined in the context of project planning, organizing, motivation, direction, and control, constitute the framework from which a project is managed within the context of the project stakeholders.

- A conceptual and working understanding of the relevance of a *project management system (PMS)* to the management of a project. A PMS consists of the following subsystems:

  a. A *facilitative organizational subsystem (FOS)* is the organizational arrangement that is used to superimpose the project teams on the functional structure of the enterprise. The resulting *matrix* portrays the formal authority and responsibility patterns and reporting relationships of members of the project team. Two complementary organizational units tend to emerge in such an organizational context: the project team and the functional units.

b. The *project planning subsystem (PPS)*, which starts with the development of a *work breakdown structure* (*WBS*). Under the project planning activity the objectives, goals, and strategies are prepared. The project cost, schedule, and technical performance parameters are defined, along with strategies on what project resources are required and how these resources will be managed during the life of the project.

c. The *project management information subsystem* (*PMIS*) contains the information required to manage the project. The PMIS plays a key role in the monitoring, evaluation, and control of the resources used on the project. The PMIS may be formal or informal and is needed to determine the status of the project and how effectively the project resources are being utilized, and to gain an understanding of how soon—and how well—the project will be completed.

d. The *project control subsystem (PCS)* provides for the selection of performance standards for the project technical performance specifications, schedules, and cost objectives. This subsystem is used to compare actual progress with planned progress, with guidance on what might be required to get the project back on the correct trajectory. The rational for a control subsystem arises out of the need to monitor the various organizations that are performing work on the project. The PCS and the PMIS work together to provide the intelligence to determine project status and how likely the project results will be to contribute to the goals and objectives of the organization.

e. The *cultural ambience subsystem* (*CAS*) portrays how the project is carried out within the human and social context of the enterprise. The emotional patterns of the social groups, their perceptions, attitudes, prejudices, assumptions, experiences and values, all go to develop the organization's cultural ambience for project management. This ambience influences how people act and react, how they think and feel, and what they say in the organizations, all of which ultimately determines what is taken for socially acceptable behavior in the organization.

f. The *human subsystem* (*HS*) involves just about everything associated with human element. An understanding of the human subsystem requires some knowledge of sociology, psychology, anthropology, communications, philosophy, leadership, and so on. Motivation is an important consideration in the management of the project team. Project managers must find ways of putting themselves into the context of human subsystem of the project so that the project team members trust and are loyal in supporting project purposes. The artful management style that project managers develop and encourage within the project team may well determine the success or failure of the project.

1.3.3 Skill

Skill is the proficiency, ability, or dexterity required to carry out responsibilities required for a profession and the performance of a role in a social setting. An individual performing a project management role requires a unique combination of skills to effectively function in a project environment. These skills include:

- Interpersonal skills—the ability to work and through people in the accomplishment of the project goals and objectives. They involve the ability to understand people, to empathize with them, to provide them with rewards for their work, and to have them believe and trust your major considerations in working successfully with people. Lack of interpersonal skills is recognized as one of the major causes of project failures in organizations.

- Communication skills—the ability to effectively transmit information between individuals through a common system of symbols, signs, or behavior. Peter Drucker, probably the best-known author in the field of management, believes that the ability to communicate heads the list of criteria for success. He also believes that one's effectiveness depends on the ability to reach others through the spoken or written word when working in organizations. This ability to communicate is perhaps the most important of all the skills an individual can possess.

- Ability to sense and see the "systems" applications of the work that the project manager and the project team members do is an important skill. This means that one who sees the system's application perceives the larger context in which the project exists. Such a larger context would place the project itself as a building block in the strategic management of the enterprise. Also, in the management of the project, consideration would be given to the multiple stakeholders who have—or believe that they have—a stake or claim in the management of the project and in the results that the project produces.

- Political sensitivity is an important skill for the project manager to have. Political issues are likely to emerge during the conceptual phase of the project as well as during the production (construction) of the project results. Then, when the project user employs the results of the project into the operational or strategic elements of the organization, additional political considerations will emerge. Politics matter—an effective project manager knows the politics, particularly in the management of the project stakeholders. The effective management of stakeholders, taking into consideration their needs, can spell the difference between success or failure of the project.

- Management style is the manner or mode in which the project manager, as well as other managers and individuals, carry out their responsibilities. How people conduct themselves is often a reflection of their knowledge, skills, and attitudes. A project manager who maintains an interest in the well-being of the team members and who can clearly communicate with these people will likely have a successful project. Sensitivity to the effects of decisions on everyone involved, being a good listener, and treating team members as human beings, rather than just individuals who happen to be working on the project, are important management style considerations.

- Building conceptual models of the project and the manner in which it should be managed is an important skill. Models serve as conceptual performance guides that can guide the making and execution of decisions in the allocation of resources on the project. A conceptual model for the organizational design of the project, how the project will be planned, the management style to be followed, how the project results will be monitored, evaluated, and controlled, are all places where conceptual models play a role in guiding the thinking and action of the project team members.

1.3.4 Attitudes

Attitudes—states of mind or feeling about the project and how it is to be managed—are important components of one's ability. The point of view, the position, and the opinion held and expressed by the people on the project will influence their behavior in working with the team members. The attitude that a person holds will influence how that person manages the project team. How a manager—or a professional—comes across during the many face-to-face relationships will influence how that person is perceived. Attitude plays an important role in working with and through people.

Douglas McGregor's concept of Theory Y and Theory X within an individual's needs has considerable merit. Theory X sees people as having an inherent dislike of work and avoiding it when possible. Because of this characteristic people have to be controlled, directed, and threatened with punishment to get them to do their best work. In contrast, Theory Y states that physical and mental effort in work is as natural as play or rest. The average individual learns under proper conditions to seek responsibilities.

Theory Y places the responsibility for motivation and attitudes of people squarely on the manager.

A. H. Maslow put forth the idea of a "hierarchy of needs," which places the various needs of individual in relationship to one another. He believes that an individual has five basic levels of needs:

- Sustaining the body, such as by satisfaction of hunger, thirst, and sleep
- Adequate protection to ensure safe and secure lives
- Satisfactory associations with other people
- Esteem: self-respect and the respect and approval of others
- The opportunity to achieve one's potential for maximum self-development, leading to personal success

Maslow believes that the needs of an individual can only be attained if the lower level of needs has first been satisfied.

1.3.5   Authority and Responsibility

The prudent and caring project manager should be aware of these needs and, as far as possible, contribute to the cultural ambience of the organization to facilitate the attainment of these needs. How the project managers and other individuals perceive their role and responsibility in the management of the project is important. Authority consists of the authority that is granted to the organizational position that the individual occupies, and the influence that an individual exercises because of his or her knowledge, skills, and attitudes. What this means is that one cannot depend on just the delegated grant of authority, but must develop the personal attributes that can be used to influence people working on the project. This is particularly important when dealing with the project stakeholders, many of whom are not under the delegated authority from the organization.

1.3.6   Emotional Intelligence

The notion of *emotional intelligence* (*EI*) is explained in a book by Daniel Goleman of that title. EI has its roots in the idea of social intelligence, proposed by E. L. Thorndike in 1920. He defined social intelligence as "the ability to understand and manage men and women, boys and girls— to act wisely in human relations."

Goleman states that "Emotional intelligence comprises the skills that help people harmonize, and should become increasingly valued as a workplace asset in the years to come." Emotional intelligence provides extraordinary insight into the development and use of interpersonal skills in a social setting, such as in the project management environment that managers and professionals face every day. These individuals will be well rewarded in developing an understanding of EI and applying it in the social setting of project management.

1.3.7 Individual Competency

Competency is expressed in many ways within an organization, and different levels of positions within an organization may apply different definitions. An individual competency model is one way of promoting understanding and appreciation for top performers.

Individual Competency Model		
Knowledge + Skills + Attitude = Competency		
Knowledge (Familiarity, awareness, or comprehension acquired by study or experience)	Skills (The ability to apply knowledge)	Attitude (A state of mind or feeling)
• Project "technology" • Strategic management • Project management theory and practice • Project management processes • Project management systems model	• Interpersonal skills • Communication skills • "Systems" applications • Political sensitivity • Building conceptual models	• Maslow's hierarchy of needs • McGregor's Theory X and Theory Y • Authority and responsibility • Emotional intelligence

The collective sum of individual competency has a significant influence on the capability of an organization to perform and grow at a competitive pace.

1.3.8 Key User Questions

1. How well do the managers and professionals in the organization understand and practice the basics of project management competency?

2. Do the people participating in project management understand the role that emotional intelligence plays in the social context of the management of projects in the organization?
3. Do the project managers manage their projects from a "project management system perspective?" If not, why not?
4. What confidence would the managers and professionals in the project management environment have in the idea of emotional intelligence?
5. How qualified are the principals in the project management environment to perform their roles with due regard for the human and social elements of their work?

1.3.9   Summary

Project management competency includes all of the understanding and insight developed over the last several decades in the research and practice of the human subsystem in organizations. As the reader ponders the basic thoughts expressed in this topic area of project management, he or she should gain an awareness of the extraordinary role that people play in the management of projects as well as some of the important considerations to be followed in the management of people.

Project managers should be aware of, and live by the basic equation:

$$\textbf{Knowledge + Skills + Attitude = Competency}$$

1.3.10   Annotated Bibliography

1. Cleland, David I., ed., *Field Guide To Project Management,* 2nd ed. (New York, NY: John Wiley & Sons, 2004). This book provides fundamental blueprints for successful project planning and execution. The basics of project management are presented. This is a must-own volume for project managers, product developers, team leaders, and executive personnel. This book is directed to those people who are involved in the day-to-day challenges of managing projects.
2. Culp, Gordon, and Anne Smith, *Managing People (Including Yourself) for Project Success* (New York, NY: Van Nostrand Reinhold, 1992). This practical guidebook clearly shows how productive communication, motivation, and leadership skills and self-management techniques can make a dramatic difference in meeting project objectives and goals. By using real-life project examples and enjoyable hands-on exercises, the book shows you how to build on the basic principles of managing projects.

1.4   A PROJECT MANAGEMENT PHILOSOPHY

1.4.1   A Project Management Philosophy

This book is about project management, a field of practice and study that has evolved since antiquity. In the last 50 years, there has been a dramatic increase in both book and periodical literature, and this Handbook presents the key elements to be found in this discipline. In this section, a philosophy of project management will be presented as an overview of this important discipline, and as a basing point from which people who work in project management can use to gain a conceptual understanding of the major elements of this discipline. Figure 1.1 presents an overview of the

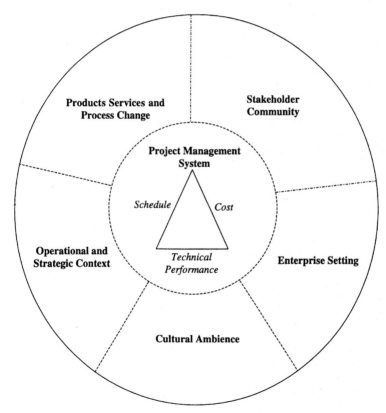

FIGURE 1.1   Major considerations—project management.

major considerations involved in the management of a project. These considerations are covered in this Project Management Portable Handbook, along with many other issues with which the project team should be conversant.

1.4.2   A Philosophy

A philosophy is an outlook about something, such as a field of thought and practice. Other meanings of this word include a "way of thinking" about the field, such as an inquiry into the nature of project management to include its conceptual framework, processes, techniques, and framework of principles. Those who are involved in project management will find that the development of a philosophy or way of thinking about this discipline is valuable.

A philosophy is a way of thinking

1.4.3   The Conceptual Framework of Project Management

A project consists of a combination of organizational resources pulled together to create something that did not previously exist, and that will provide an enhanced performance capability in the design and execution of organizational strategies. Other characteristics of a project include:

- Projects are the principal means by which the organization deals with change.
- Changes in organizational products, services, or organizational processes are brought about through the use of projects.
- Each project has specific objectives regarding its cost, schedule, and technical performance capability.
- Each project, when completed, should add to the operational and/or strategic capability of the organization.
- Projects have a distinct life cycle, starting with the emergence of an idea, through to the delivery of the project results to a user or "customer."
- Projects change the organizational design and culture of an organization, primarily through the workings of the "matrix organization."

- A focal point is designated in the organization through which the resources supporting a project are planned, integrated, and utilized to deliver the project results.
- Specialized planning, organization, motivation, leadership, and control techniques have emerged to support the management of a project's resources.
- Today, project management is recognized as having a rightful place in the continued emergence of the management discipline.
- Although project management emerged in the construction industry, today it is practiced in all industries, in military entities, educational entities, ecclesiastical organizations, in the social field, and in the political domain.
- The management of the project stakeholders poses a major challenge for the team.
- A body of knowledge has developed to describe the art and science of project management.
- This body of knowledge is changing the way that contemporary organizations are managed.
- Project management is the wave of the future—the role played by project management in the years ahead will be challenging, exciting, and crucial.

1.4.4 The Project Management Functions

Project management is carried out through a management process consisting of the core functions of management to utilize resources to accomplish project ends. These core functions are shown in Fig. 1.2 and are briefly discussed below.

Planning—Development of the objectives, goals, and strategies to provide for the commitment of resources to support the project.

Organizing—Identification of the human and non-human resources required, providing a suitable layout for these resources, and the establishment of individual and collective roles of the members of the project team, who serve as a focal point for the use of resources to support the project.

Motivating—The process of establishing a cultural system which brings out the best of people in their project work.

Directing—Providing for the leadership competency necessary to ensure the making and execution of decisions involving the project.

FIGURE 1.2   The core functions of project management.

Control—Monitoring, evaluating, and controlling the use of resources on the project consistent with project and organizational plans.

Summary material will be presented on two of the core functions of project management. Elsewhere in this book additional information will be provided on these functions as well as on the three additional functions, namely Motivation, Direction, and Control.

1.4.5   The Essential Work Package Elements of Project Planning

Project planning is the process of thinking through, and making explicit, the objectives, goals, and strategies necessary to bring the project through its life cycle. The work packages involved in project planning include:

• Establish the strategic fit of the project. Ensure that the project is truly a building block in the design and execution of organizational strategies, and that it provides the project owner with an operational capability not currently existing or improves an existing capability.

- Identify strategic issues likely to affect the project.
- Develop the project technical performance objective. Describe the project deliverable end product(s) that satisfies a customer's needs in terms of capability, capacity, quality, quantity, reliability, efficiency, etc.
- Describe the project through the development of the project WBS. Develop a product-oriented family tree division of hardware, software, services, and other tasks to organize, define, and graphically display the product to be produced, as well as the work to be accomplished to achieve the specified product.
- Identify and make provisions for the assignment of the functional work packages. Decide which work packages will be done in-house, obtain the commitment of the responsible functional work managers, and plan for the allocation of appropriate funds through the organizational work authorization system.
- Identify project work packages that will be sub-contracted. Develop procurement specifications and other desired contractual terms for the delivery of the goods and services to be provided by outside vendors.
- Develop the master and work package schedules. Use the appropriate scheduling techniques to determine the time dimensions of the project through a collaborative effort of the project team.
- Develop the logic networks and relationships of the project work packages. Determine how the project parts can fit together in a logical relationship.
- Identify the strategic issues that the project is likely to face. Develop a strategy for how to deal with these issues.
- Estimate the project costs. Determine what it will cost to design, develop, and manufacture (construct) the project, including an assessment of the probability of staying within the estimated costs.
- Perform risk analysis. Establish the degree or probability of suffering a setback in the project's schedule, cost, or technical performance parameters.
- Develop the project budgets, funding plans, and other resource plans. Establish how the project funds should be utilized, and develop the necessary information to monitor and control the use of funds on the project.
- Ensure the development of organizational cost accounting system interfaces. Since the project management information system is tied in closely with cost accounting, establish the appropriate interfaces with that function.

- Select the organizational design. Provide the basis for getting the project team organized, including delineation of authority, responsibility, and accountability. At a minimum, establish the legal authority of the organizational board of directors, senior management, and project and functional managers, as well as the work package managers and project professionals. Use the linear responsibility chart (LRC) to determine individual and collective roles on the project team.

- Provide for the project management information system. An information system is essential to monitor, evaluate, and control the use of resources on the project. Accordingly, develop such a system as part of the project plan.

- Assess the organizational cultural ambience. Project management works best where a supportive culture exists. Project documentation, management style, training, and attitudes all work together to make up the culture in which project management is found. Determine what project management training would be required. What cultural fine-tuning is required?

- Develop project control concepts, processes, and techniques. How will the project's status be judged through a review process? On what basis? How often? By whom? How? Ask and answer these questions prospectively during the planning phase.

- Develop the project team. Establish a strategy for creating and maintaining effective project team operations.

- Integrate contemporaneous state-of-the-art project management philosophies, concepts, and techniques. The art and science of project management continues to evolve. Take care to keep project management approaches up-to-date.

- Design project administration policies, procedures, and methodologies. Administrative considerations often are overlooked. Take care of them during early project planning, and do not leave them to chance.

- Plan for the nature and timing of the project audits. Determine the type of audit best suited to get an independent evaluation of where the project stands at critical junctures.

- Determine who the project stakeholders are and plan for the management of these stakeholders. Think through how these stakeholders might change through the life cycle of the project.

1.4.6 Organizing for Project Management

Project organizing deals with the determination of the individual and collective roles that people in the organizational play in supporting project objectives, goals, and strategies. The major considerations involved in such organizing include:

- The project-driven matrix organization has a distinct structure, which at first glimpse appears to violate traditional organization principles.
- The matrix organizational design is a blend of the project and functional organizational units in which there is a sharing of authority responsibility, and accountability.
- The functional managers and the project managers in the enterprise carry out complementary roles with regards to the project.

In its most elementary form, the interface between the project effort and the functional effort constitutes the key of the matrix organization carried out through the project work packages.

Extraordinary authority-responsibility-accountability relationships exist in the matrix organization.

A Linear Responsibility Chart (LRC) can be used to establish the authority and responsibility relationships involved in the individual and collective roles with the organization.

Authority is defined as the legal right to act; responsibility is the obligation to act. The ambiguities of the matrix organization provide ample opportunity for the ability of the project manager to exercise legal authority as well as the influence that arises from that individual's competencies.

Care must be taken to prescribe the legal authority that the manager, and the members of the project team have.

In the final matters of managing the project, the knowledge, skills, and attitudes of the project manager will be the principal determinant in the success or failure of the project.

1.4.7 Key User Questions

1. Do the members of the project teams recognize what is meant by a "project management philosophy"—and what implications such a philosophy has in the theory and practice of project management in the organization?
2. Has the idea of a conceptual framework of project management been communicated to the members of the project teams in the organization, and do these members understand the meaning of such a framework as a guide to their thought and actions?
3. Do the people in the organization understand the fundamentals involved in the project management process?

4. Have the fundamentals of organizational design been discussed with the project team members?
5. In Fig. 1.1 the major considerations that are involved in the management of project are presented. Do the key project participants understand these considerations?

1.4.8 Summary

This section presented the idea of a *Project Management Philosophy* as a way of thinking about the conceptual framework, processes, techniques, and principles that are involved in the management of projects. A philosophy was described as a way of thinking about the management of projects. Two figures were presented which identified the major considerations and the use of key management functions in the project management process.

1.4.9 Annotated Bibliography

1. Cleland, David I., *Project Management: Strategic Design and Implementation,* 4th ed. (New York, NY: McGraw-Hill, 2002). The collective chapters of this book provide the basic information on which to build a project management philosophy. Other book publications on project management also provide excellent information with which to develop a "way of thinking" about the subject.
2. Milosevic, Dragan Z., *Project Management Tool Box* (New York, NY: John Wiley & Sons, 2003). This book provides detailed descriptions of the basic tools used in the management of projects. The author offers more than 50 tools and techniques that ensure a seamless performance of orderly project activities. Based on the industry-standard Project Management Institute's Body of Knowledge, the book gives information on how to plan and control projects, build effective teams and evaluate risks, and align a project management toolbox with corporate strategy.

1.5 BENEFITS OF PROJECT MANAGEMENT

1.5.1 Introduction

Project management is a profession that bridges many industries and is a process that delivers unique benefits. Project management is a discipline that has significant advantages over other processes as well as being adaptable to fit the unique needs of different industries. Project management can be tailored to fit many different situations around the world and can be designed to accommodate various levels of sophistication.

> **Project management is the principal change agent**

1.5.2 Background

Organizations and people gain significantly from improved processes that provide the optimum solution to business requirements. Project management has the potential, when fully implemented, to provide the most effective means of developing and delivering new product services and organizational processes. The project management process is a streamlined process to focus specifically on the end result and delivery to customers.

The process of project management as a means of dealing with organizational change is successfully used in many industries today. The number of different applications of the process in the management of change is also expanding. The basic concepts and processes of project management permit flexibility to meet the needs for an effective change strategy that realizes the most benefits.

Benefits are derived from processes when there is a minimum use of resources and maximum stakeholder satisfaction. Waste in any area is a negative benefit, or an opportunity, for a project to improve. Waste is defined as "any consumption of resources (materials, people's time, energy, talent, and money) in excess of the amount required to do the job.

Some of the most frequent opportunities for improving productivity and avoiding waste are shown in Figure 1.3.

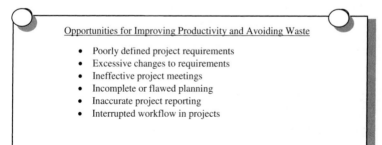

Opportunities for Improving Productivity and Avoiding Waste

- Poorly defined project requirements
- Excessive changes to requirements
- Ineffective project meetings
- Incomplete or flawed planning
- Inaccurate project reporting
- Interrupted workflow in projects

FIGURE 1.3   Opportunities for improving productivity and avoiding waste.

1.5.3 Project Management and Benefits

The number of benefits realized is directly related to the effectiveness of the implementation of project management. Well-developed project management processes that are tailored to the organization provide the most benefits. These processes must also be closely followed to ensure the practices align with the intended outcomes.

An example of following the project management process is in the area of planning. Planning the project is critical to successfully defining the path to project completion and product delivery. Weak or flawed planning will allow the project to drift off course and waste effort. Positive benefits of improved productivity and effectiveness are derived when the planning is adequate to guide the project team to the completion of all project work.

1.5.4 Benefits of Project Management

Benefits of project management encompass several areas. Identifying the different groups of stakeholders in projects is helpful in determining the type of benefit derived by each. These benefits are best listed as improvements or enhancements to the group. The most likely stakeholders are:

- The organization as a business, either for-profit or not-for-profit;
- Senior management of the organization, which includes the manager of project managers, to the president of the organization, and through to the board of directors for major projects;
- Project leaders and project team members, or those working the project; and
- Customers of projects, the consumer, user, owner, and financier.

FIGURE 1.4

Benefits that are derived from project management as compared to other methods are listed below. Benefits are generally considered within the different groups in Fig. 1.4.

1.5.5 Benefits to the Organization

- Improved productivity through providing the most direct path to the solution of a problem
- Improved profits through reducing wasted time and energy on the wrong solutions
- Improved employee morale through greater job satisfaction
- Improved competitive position within industry through bringing faster results to situations
- Improved project process and workflow definition
- Improved capability and maturity in business solutions
- More success and fewer failures through the dedicated focus on work
- Better decision making on continuation/termination of work efforts
- Improved reward system for senior managers, project leaders, and project team members
- Smoother integration of project results into the organization
- Improved product, service, or organizational process development and implementation

1.5.5.1 Benefits for Senior Managers

- Confidence in outcomes for work efforts through better predictability
- Reduced number of changes to work effort during execution
- Faster delivery of products and services that meet customer requirements
- Better information for leadership decision-making
- Improved communications with stakeholders
- Confidence in the organization's business capability
- Improved approval process for new work initiation through better requirement definition
- Better facilitation of organization mission, objectives, and goals

1.5.5.2 Benefits for Project Leaders and Team Members

- Improved job satisfaction through better performance
- Reduced hassle from changing requirements

- Pride in workmanship
- Confidence in ability to manage/work towards solution
- Less effort (hours worked) with better results
- Improved communications with senior managers and customers
- Confidence in ability to complete the work
- Better work tracking and control through better information
- Improved professional development

1.5.5.3 Benefits for Customers

- Confidence in senior management, project leader, and project team
- Confidence in delivery of required product and services
- Confidence in delivery on time within price
- Improved visibility into work planning and execution process
- Improved satisfaction with the product or service
- Improved product definition and communication of own requirements
- Improved working relationship with the project team

Benefits derived from project management are typically viewed as a difference between what is provided under current planning and execution practices and what will be done in the future. In the project environment, benefits are most often viewed as

- Technical—the result of the project's product or service
- Customer satisfaction—the customer's feeling as to the value of the end product, service, and other factors
- Delivery time—meeting the date when the product or service is needed
- Price/cost—the price or cost has not exceeded the value delivered

The change between the level of benefits delivered before and after new project management practices must be measured to have a quantifiable meaning. Without solid project plans, it is often difficult to measure the quantitative advances objectively. Unfortunately, too many measures of effectiveness are developed and implemented from a subjective basis.

1.5.6 Measures of Success

Measures of success are used for two primary purposes. First, a measure of success tells the project team when the work is complete. Second, if the work cannot be completed, it establishes a means for measuring the degree of success in meeting the requirements.

Measures of success permit gauging progress as well as identifying a benchmark for subsequent improvement efforts. Starting a task or group of tasks in a project should begin with the desired outcome or desired results. This can be communicated to the performing party and sets expectations for the completed work.

Incomplete work, or failure to meet the desired measure of success, establishes a point from which improvements can be made. When a task or group of tasks fail to meet the agreed upon measure of success, it is from this point that an evaluation can be made for future improvements. Future improvements may result from process improvement or training of the workforce or both.

Setting measures of success should stretch the capability of the project team. The stretch measure of success will often provide better results with a failure than an easy goal. Additional benefits will be derived from stretch measures of success. Project teams given stretch goals must be informed that goals may not be attainable, but their efforts will provide the best results, although the progress may be less than goals.

1.5.7 Key User Questions

1. Do senior managers understand the correlation between proper project planning and the benefits derived from the results of a project?
2. Do you see opportunities within your organization to improve processes and gain additional benefits for the organization?
3. Do you see opportunities to make the project work less stressful and more productive?
4. Do members of the project team understand the benefits of following work processes and the benefits derived for the organization?
5. What measures of success are used in your organization to identify process and work improvements?

1.5.8 Summary

Project management processes have the capability to bring numerous benefits to an organization and the people working the projects. The list of benefits derived from using proven project management methodology, techniques, standards, tools, and practices have been identified within categories as to who is the beneficiary. Each category may have a different perceptions of the benefits derived from project management based on their organizational position.

Benefits derived from improved processes include more confidence in the outcome of the project, less stress on the performing project team, higher productivity rates, less waste of valuable resources, reduced costs for projects, and faster time to market. The intangible benefits include an improved organization image as a company with a core competency in project management. An organization may derive other benefits from using the best practices of project management and may significantly improve their relative position in the industry.

1.5.9   Annotated Bibliography

1. Damelio, Robert, *The Basics of Process Mapping* (New York, NY: Productivity Press, 1996). This book provides a hands-on team resource, training supplement, and reference for process mapping. The author presents techniques for gaining an accurate assessment of work flow and customer/supplier relationships. The book provides strategies for project improvements by utilizing important relationship mapping and cross-functional process mapping data.

2. Imai, Masaaki, *Kaizen: The Key to Japan's Competitive Success* (New York, NY: McGraw-Hill, 1986), chap. 1, 2. This book discusses the concept of improving processes to reduce waste and improve productivity. "Kaizen" means improvement and the Japanese method of establishing a method for continual improvement for any process.

1.6   ETHICS IN PROJECT MANAGEMENT

1.6.1   Introduction

Ethics in project management encompasses many areas of personal and professional conduct because of the range of project locations. Project personnel are required to work in different countries that have unique cultural aspects and different value systems. In some countries, bribes, a conflict in the US environment, are expected and often demanded to ensure continuity of work on a project.

Ethics in all environments are not universally understood or agreed upon. Ethics can often be situational and loosely applied or they can be rigorous and demanding on project personnel. The lack of a code of ethics or training in ethics can also affect how personnel respond to challenging situations. Random or unequal application of a code of ethics can also affect practiced ethics.

A code of ethics is required for all recognized professions. It is expected that professionals will freely state their values and live to those statements of ethical conduct. The code of ethics for a professional group must be advertised and enforced to be effective.

> **Practiced ethics do not ensure project success, but the lack of ethics leads to project failure.**

1.6.2   Ethical Obligations of a Professional

To be considered professional, individuals must clearly state their code of ethics by which they can be expected to abide. The code of ethics must consider the legitimate needs of others as well as identifying all obligations that the professional assumes. This code of ethics sets the expectation of those being served and clearly states the obligations.

Project managers may have a code of conduct within their organization to guide them in the manner in which they work with others. The best known code of conduct for project managers is the *Project Management Professional Code of Ethics*. This code was developed in 1983 by the Project Management Institute as a part of its Project Management Professional (PMP®) Certification Program. Individuals being certified to be a Project Management Professional must subscribe to and support the PMP® Code of Ethics.

When a code of ethics is not adopted by an individual or a group of individuals, the bounds of ethical conduct are not defined. Individuals will vary in the practice and enforcement of ethical conduct to the extent that there is no consistent practice. Random and situational ethics do not support professionalism or build confidence in others that the collective group can or will abide by any rules of conduct.

1.6.3   Code of Ethics

The code of ethics of PMI® for project managers is provided as a model for all professional project management practitioners to follow.

Code of Ethics for the Project Manager

Preamble: Project Managers, in the pursuit of the profession, affect the quality of life for all people in our society. Therefore, it is vital that Project Managers conduct their work in an ethical manner to earn and maintain the confidence of team members, colleagues, employees, employers, clients and the public.

Article I: Project Managers shall maintain high standards of personal and professional conduct and:

A: Accept responsibility for their actions.

B: Undertake projects and accept responsibility only if qualified by training or experience, or after full disclosure to their employers or clients of pertinent qualifications.

C: Maintain their professional skills at the state of art and recognize the importance of continued personal development and education.

D: Advance the integrity and prestige of the profession by practicing in a dignified manner.

E: Support this code and encourage colleagues and co-workers to act in accordance with this code.

F: Support the professional society by actively participating and encouraging colleagues and co-workers to participate.

G: Obey the laws of the country in which work is being performed.

Article II: Project Managers shall, in their work:

A: Provide necessary project leadership to promote maximum productivity while striving to minimize cost.

B: Apply state of the art project management tools and techniques to ensure quality, cost and time objectives, as set forth in the project plan, are met.

C: Treat fairly all project team members, colleagues and co-workers, regardless of race, religion, sex, age or national origin.

D: Protect project team members from physical and mental harm.

E: Provide suitable working conditions and opportunities for project team members.

F: Seek, accept and offer honest criticism of work, and properly credit the contribution of others.

G: Assist project team members, colleagues and co-workers in their professional development.

Article III: Project Managers shall, in their relations with their employers and clients:

A: Act as faithful agents or trustees for their employers and clients in professional or business matters.

B: Keep information on the business affairs or technical processes of an employer or client in confidence while employed, and later, until such information is properly released.

C: Inform their employers, clients, professional societies or public agencies of which they are members or to which they may make any presentations, of any circumstance that could lead to a conflict of interest.

D: Neither give nor accept, directly or indirectly, any gift, payment or service of more than nominal value to or from those having business relationships with their employers or clients.

E: Be honest and realistic in reporting project quality, cost and time.
Article IV: Project Managers shall, in fulfilling their responsibilities to the community:
A: Protect the safety, health and welfare of the public and speak out against abuses in these areas affecting the public interest.
B: Seek to extend public knowledge and appreciation of the project management profession and its achievements.

1.6.4   Ethics for the Project Management Practitioner

The above code of ethics is an example and was derived from an extensive study of ethics. The study addressed all the areas of ethical obligations that engineers must exhibit to be professionals. For project management professionals, the obligations are depicted below.

Obligations of Project Management Professionals to Others

- Society in general or the public's interest
- Clients or customers of project work
- Employers of the project management professional
- Employees working for the project management professional
- Project team members
- Professional colleagues in project management and related fields

---

The above list is a general approach to whom there is owed an ethical obligation. If a person is a project manager, for example, there is a greater obligation to the team members than if the project manager is a team member. The greater the responsibility within project management, the greater the ethical obligation to serve those in subordinate positions.

The greater the professional responsibility, the greater the ethical obligation. Project managers must demonstrate ethical behavior to team members.

Honesty and integrity are essential elements of any profession. Honesty and integrity are the foundation for trust and confidence in the professional. Individuals deviating from the truth and fabricating progress results on a project will soon have both team members and superiors questioning their ability to effectively manage a project.

Willingness to accept responsibility and be held accountable for one's actions is another key element of ethical behavior. A project manager is responsible for all the activities that occur or fail to occur on a project. It is unethical and unfair to place the blame on others when the project manager is clearly at fault.

1.6.5 Enforcement of the Code of Ethics

A code of ethics is only effective for professionals when the professional abides by all aspects of the obligation. Deviations from the code of ethics must be enforced to ensure credibility of the profession. A system is required to investigate reported abuses and implement corrective actions.

A self-correcting system for abuses is best. However, the violator of the code of ethics will typically deny any wrongdoing. When violations are reported, it is necessary for the professional group to conduct a fair and impartial investigation to determine the facts. The facts must be reviewed and a finding of a violation or no violation must be made. The finding is then referred to a body authorized to either dismiss the allegation or initiate some form of punishment.

Punishment for violations of a code of ethics may range from an oral or written warning if the offense is minor to removal of credentials if the person has committed a serious breach of ethics. Because professional groups only have authority over the code of ethics in a voluntary agreement with the individual, punishment can only encompass removing what the professional group may grant. Where the violation of the code of ethics is also a violation of law, the individual may be punished under the law by the body having jurisdiction.

Punishment for violations of the code of ethics is a harsh and disruptive process. It is best to avoid violations of the code of ethics through an active program of education and reinforcement of the need to abide by these rules. When an erosion of the code of ethics occurs through small deviations, the corrective measures must be enforced to prevent further degradation.

1.6.6 User Questions

1. Are there obligations that exist to set expectations for acceptable behavior in your organization?

2. What elements of the code of ethics would be most difficult to enforce within your organization?

3. When there is a violation of the code of ethics, does this equate to a less than honest approach to working relationships?

4. Which ethical obligation is the most difficult to meet, and is it realistic for your organization?

5. How does a code of ethics contribute to the project team's performance and what are some examples?

1.6.7 Summary

Ethical behavior for all true project management practitioners is guided by a code of ethics. If the code of ethics is missing, then there is no consistent professional obligation to others and the individual cannot be considered a professional.

Codes of ethics identify obligations to different parties. These obligations are formally documented and made available to others as a means of establishing the expectations of others. These expectations, like a contract, provide a basis for the professional's conduct and the expected outcomes from the professional.

The extent to which a professional follows the code of ethics determines the confidence level that the other party assigns. Individuals operating on the boundaries of ethical conduct will find an erosion of confidence in their professional competence. Honesty and integrity are fundamental to building and maintaining this confidence as a professional.

1.6.8 Annotated Bibliography

1. Ireland, Lewis R., Joann Schrock, and Walter Pike, "Ethics for the Project Manager," *Project Management Quarterly,* Project Management Institute, Drexel Hill, PA, August 1983. This article deals specifically with the ethical obligations of a project manager to several parties. It is the foundation for the current Project Management Institute's Certification Program. These fundamentals have not changed, but there have been changes to some words to accommodate the shift from a project manager, the leader, to anyone working in a project.

2. *IEEE Study on Ethics and Ethical Conduct for Engineers,* McLean, VA, c. 1980. This study of ethics and ethical conduct describes the requirements for engineers to be considered professionals. There are extensive examples of ethical behavior and misbehavior.

1.7   PROJECT LIFE CYCLE

1.7.1   Introduction

New products, services, and organizational processes have their genesis in ideas evolving with the enterprise. Typically such ideas go through a distinct life cycle—a natural and pervasive order of thought and action. In each phase of this life cycle different levels of thought and activities are required within the enterprise to assess the value of the emerging idea as it evolves during its life cycle. The representative phases of a life cycle usually include those phases depicted in Fig. 1.5.

The life cycle of a project can last from just a few weeks or months to ten or more years, such as in the pharmaceutical industry, or a major construction project such as the Channel Tunnel. These phases, and what an analysis of what such phases will do for the project are discussed below.

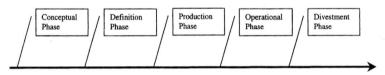

FIGURE 1.5   Generic model of project life cycle.

1.7.2   The Conceptual Phase

During this phase the environment is examined, forecasts are prepared, objectives and alternatives are evaluated, and the initial examination of the technical performance, cost and schedule aspects of the idea's development are examined. Other activities undertaken during the conceptual phase are cited in Table 1.4.

There should be a high mortality rate of potential projects during the conceptual phases. Rightly so, as it is during this phase that sufficient study should be carried out about the project's expectations that a decision can be made regarding the potential usefulness and survivability of the project.

1.7.3   The Definition Phase

The purpose of the definition phase is to determine the cost, schedule, technical performance expectations, resource requirements, and likely operational and strategic fit of the probable project results. Issues to be resolved during the definition phase include these listed in Table 1.5.

TABLE 1.4   Conceptual Phase Activities

- Initial assessment of the resources required.
- Development of preliminary insight into the operational or strategic value of the project in complementing existing enterprise purposes.
- Determination if the expected project results are needed. ✓
- Establishment of preliminary objectives and goals for the project. ⌣
- Organization of a team to manage the project.
- Selling the organization on the project approach.
- Preparation of a preliminary project plan to include a proposal if required by the final project user.
- Determination of existing needs or potential deficiencies of existing products, services, or organizational processes, as appropriate.
- Determination of initial technical, environmental, and economic feasibility, and the practicability of the project's expected outcome.
- Selection and preparation of an initial design for the expected outcome.
- Initial determination of expected stakeholder interfaces.
- Preliminary determination of how the project results will be integrated into existing enterprise strategies.

TABLE 1.5   Definition Phase Activities

- Full assessment of the project outcomes before major resources are committed to continue development of the project.
- Identify need for further study and development for the project.
- Confirm the decision to continue development, create a "prototype", and assess the full impact of the project for production or installation.
- Firm identification of the human and non-human resources that will be required for the continued development and deployment of the project results.
- Preparation of final system performance requirements. ✓
- Preparation of plans to support the project results.
- Identification of areas of the project where high risk and uncertainty dictate further assessment.
- Definition of system inter- and intra-system interfaces.
- Development of a preliminary logistic support, technical documentation, and after-sale plan.
- Preparation of suitable documentation required to support the system to include policies, procedures, job descriptions, budget and funding protocol, and other documentation necessary to track and report on the progress being made.
- Development of protocols on how the project will be monitored, evaluated, and controlled.

1.7.4 The Production (Construction) Phase

During this period the project results are produced (constructed) and delivered as an effective, economical and supportable product, service, or organizational process. The plans and strategies conceived and defined during the proceeding phases are updated to support production (construction) initiatives. Other major elements of work carried out during this phase are reflected in Table 1.6.

1.7.5 The Operational Phase

Entry into this phase indicates that the project results have been proven economical, feasible, and practicable and are worthy of being implemented by the user to support their operational or strategic initiatives. Other major elements of work carried out during this phase include those shown in Table 1.7.

1.7.6 The Divestment Phase

In this phase, the enterprise "gets out of the business" which the project results provided. The "getting out of the business" may be caused by loss

TABLE 1.6 Production Phase Activities

- Identification and integration of the resources required to facilitate the production processes such as raw materials inventory, vendor parts, supplies, labor, and funds.
- Verification of system production specifications.
- Actual production, construction, and installation.
- Final development and approval of after-sale logistic support to include after-sale services.
- Performance of final testing to determine adequacy of the project results to do the things it is intended to do.
- Development of technical manuals and affiliated documentation describing how the project results are intended to operate.
- Development and finalization of plans to support the project results during its operational phase.
- Build and test tooling.
- Develop production process strategies to include equipment specification, tooling support, and labor force indoctrination and training.
- Process engineering changes as needed.

TABLE 1.7 Operational Phase Activities

- Operating the project results along the intended lines.
- Integration of the project's results into existing organizational systems.
- User field evaluation of the technical, cost, schedule, and economic sufficiency of the project results to meet actual operating conditions.
- Provide feedback to enterprise planners concerned with the development of new products, services, or organizational processes.
- Evaluation of adequacy of supporting systems in the organization to complement the project's operational results.

---

of customer demand, emergence of new products, services, or processes—all of which have a finite lifetime. Other major activities during this period include:

- Phase down in the use of the project results
- Development of plans for and the transfer project resources to other elements of the organization
- Evaluation of problems and opportunities associated with the use of the project results
- Recommendations for the management of future projects and programs
- Identification and evaluation for new or improved management techniques

Of course, each project has its unique life cycle. The material that was presented in this section was provided to present a "generic" set of issues involving most projects. The reader should be able to take these generic issues and find out how each of the elements in the five life cycle phases suggested can be used, and also provide a basing point for the identification of the other issues appropriate to a particular project.

1.7.7 Key User Questions

1. Have "boiler plate" criteria been developed for how products, services, and organizational processes and their associated activities will be managed, giving due consideration for the issues likely to arise in the life cycle of these activities?
2. Do the project team members, and other project stakeholders understand the meaning and implications of managing projects on a "life-cycle" basis?

3. Have key points been established to facilitate the making and execution of decisions involving the major issues in the management of projects—in particular those points at which the use of resources on a project should be continued, or a decision made to terminate the project?

4. Do the project monitoring, evaluation, and control processes take into consideration the key issues or elements involved in the life cycle of each project?

5. Does the culture of the enterprise support the concept of a project life cycle, which can be used as one of the templates for the management of the project?

1.7.8 Summary

In this section, the concept and processes involved in a project's life cycle were presented as a useful protocol for the management of a project. A generic life cycle approach was used to illustrate how the use of such an approach can bring about added order and improved protocol for the management of a project. The reader and user of the life cycle approach suggested in this section, should find that the guidance put forth in the chapter can improve the efficiency and effectiveness with which product, service, or process development is conceptualized and carried out.

1.7.9 Annotated Bibliography

1. Cleland, David I., and Lewis R. Ireland, *Project Management: Strategic Design and Implementation,* 4th ed. (New York, NY: McGraw-Hill, 2002). Chapter 2, "The Project Management Process," provides a comprehensive description of these management functions woven into the project life cycle woven into the description. Such topics as managing the life cycle, and risk and uncertainty in a life cycle, as well as figures and tables reinforce the role of a life-cycle in the overall process of managing a project.

2. Belanger, Thomas C., "Innovative Life Cycles," in David I. Cleland, ed., *Field Guide to Project Management,* 2nd ed. (New York, NY: John Wiley & Sons, 2004). This chapter provides an overview of the key considerations in choosing an innovative project life cycle. The advantages of using a life cycle approach, the benefits likely to accrue to the user, and several models are presented, which provide an overall mes-

sage of the importance of choosing and using a project life cycle strategy.

1.8 PROJECT MANAGEMENT BODIES OF KNOWLEDGE AND PM CERTIFICATION

1.8.1 Introduction

As project management receives more recognition as a profession with its body of knowledge and certification of professionals within the project management community, there is a continuing need to revise and define contemporary theory and practice. This section gives a top-level view of the bodies of knowledge currently in use for training and certification.

Bodies of knowledge circumscribe the knowledge areas that professionals are expected to master and apply to their profession. The project management bodies of knowledge have reached several plateaus over the past 30 years, and continue to grow. National cultures have influenced how the bodies of knowledge have been defined and applied, and there are ongoing initiatives that will lead to new and better definitions and compositions.

A natural result of the recognition of project management as a profession is the establishment of new certification programs. Certification of individuals is accomplished by project management professional societies around the world and at different levels. Individuals may be certified in knowledge of the profession or performance-based competency.

Certification is valued by individuals and organizations because of the confidence placed in an individual's ability to perform on the job.

1.8.2 Project Management Bodies of Knowledge

The project management profession has been described in the literature for more than 50 years. Bodies of knowledge have been codified into formal documents since 1983 and continue to evolve with the profession. Several professional organizations have developed different versions of a project management body of knowledge.

The Project Management Institute (PMI©) in Newtown Square, Pennsylvania, has been the leader in developing a project management body of knowledge. Other organizations, such as the International Project Management Association in Switzerland, the Association for Project Management in England, the American Society for the Advancement of Project

Management in the United States, and the Australian Institute for Project Management in Australia, have developed bodies of knowledge. All organizations refer to their respective bodies of knowledge as standards.

The Project Management Institute's project management body of knowledge is the most widely recognized and used. It is officially referred to as *A Guide to the Project Management Body of Knowledge,* or PMBOK® Guide. The 2002 edition defines the body of knowledge in nine areas.

Nine Knowledge Areas
Project Integration Management
Project Scope Management
Project Time Management
Project Cost Management
Project Quality Management
Project Human Resource Management
Project Communications Management
Project Risk Management
Project Procurement Management

The International Project Management Association (IPMA) has developed a body of knowledge called a competency baseline, or ICB, that defines various functions within project management. IPMA has defined its body of knowledge or ICB in 42 categories that provide a scope for the body of knowledge and experience suggested by the association.

The ICB has 28 core elements that are considered essential for a comprehensive understanding of the project management profession.

ICB Core Elements
1 Projects and Project Management
2 Project Management Implementation
3 Management by Projects
4 System Approach and Integration
5 Project Context
6 Project Phases and Life Cycle
7 Project Development and Appraisal
8 Project Objectives and Strategies
9 Project Success and Failure Criteria

ICB Core Elements (*Continued*)
10 Project Start Up
11 Project Close Out
12 Project Structures
13 Content, Scope
14 Time Schedules
15 Resources
16 Project Cost and Finance
17 Configurations and Changes
18 Project Risks
19 Performance Measurement
20 Project Controlling
21 Information, Documentation, Reporting
22 Project Organisation
23 Teamwork
24 Leadership
25 Communication
26 Conflicts and Crises
27 Procurement, Contracts
28 Project Quality

Supplementing the core elements are 14 additional areas that are general categories to enhance understanding of the profession.

Supplemental Elements
29 Informatics in Projects
30 Standards and Regulations
31 Problem Solving
32 Negotiations, Meetings
33 Permanent Organization
34 Business Processes
35 Personnel Development
36 Organizational Learning
37 Management of Change
38 Marketing, Product Management
39 System Management
40 Safety, Health, and Environment
41 Legal Aspects
42 Finance and Accounting

IPMA is an umbrella association of national associations. These national associations use the 42 elements to develop their respective project management bodies of knowledge in their native languages. Several language translations and refinement for these 42 elements include English (both American and British), German, French, and Spanish. Visit the IPMA website at *www.ipma.ch* for a listing of associations and their respective programs.

The Australian Institute of Project Management (AIPM) has developed its body of knowledge around the nine PMI knowledge areas. See *www.aipm.com.au* for details.

The Association for Project Management (APM) developed its body of knowledge around the 42 elements of IPMA's ICB. See *www.apm.org.uk* for details.

The American Society for the Advancement of Project Management (*asapm*) developed its body of knowledge around the 42 elements of the IPMA's ICB. See *www.asapm.org* for details.

1.8.3   Certification in Project Management

1.8.3.1   The Value of Project Management Certification.   The value of a professional certification may be subject to challenge by those believing the program is flawed or that certification is not required. Many ask the question "Is certification a valid practice or is it just someone's idea of what a person should know?" Others may ask whether certification is a differentiating factor that truly sets apart those going through a certification process. One must understand the basic definitions to address such questions.

First, we must agree to similar definitions of *value* and *certification*. Value of any product or service is determined by the consumer or customer—not by the provider. Value, or the worth of something, is determined by the amount that a person is willing to pay or exchange for certification.

> *Value* is defined in *The American Heritage Dictionary, Second College Edition* as "An amount considered to be a suitable equivalent for something else; a fair price or return for goods or services."

> *Certification,* **or the authentication of one's qualifications, is defined in** *The American Heritage Dictionary* **as the act of certifying. "To guarantee as meeting a standard."**

The worth of certification can then be described as the fair price one would pay to become qualified through meeting established standards within a given profession. In this discussion, we view project management certification as meeting established standards at any defined level of the project management profession.

1.8.3.2 Certifying Organization.   In the context of certifying that one meets established standards, an organization, acting independently of outside influence, authenticates through rigorous testing, examining, and reviewing of a candidate's knowledge, attitude, experiences, and skills that he or she meets the established standards of the program. The certifying organization sets the standards by which individuals qualify to receive a specific knowledge or competency professional designation.

The match between what the consumer needs and what an organization offers in certification provides one means of measuring value. A close match between a customer's requirements for project management knowledge or competency and the qualifications of the professional designation provides significant value that can contribute to organizational effectiveness. On the other hand, artificial standards or requirements unrelated to individual performance in a certification program detract from the value of certification.

1.8.3.3 Beneficiaries of Certification.   To view certification in a holistic manner, let's consider who the beneficiaries of project management certification are. The simple answer is, an individual whose qualifications have been authenticated by an independent process that is directly related to the profession. Perhaps a better answer is, anyone or organization that receives value from project management certification. This includes, as a minimum, the following:

- Individual—one who has met the requirements of the certification process and has exhibited those qualities, characteristics, traits, and knowledge necessary to be designated a project manager or other component of the project management process.

- Organization—a company or an agency that employs or hires an individual meeting the certification standards and whose performance exceeds that of individuals not meeting the certification standards.
- Customer—consumer of products or services who has a greater degree of confidence in the performance of the individual certified as a project manager or other designation and receives greater value for contracted products or services.
- Professional society—association or agency through public recognition as an authoritative, certifying body that serves the project management community.
- Project management community—body of project management practitioners that takes pride in having an established profession that promotes its value to others.
- Public at large—people in general because they receive greater value through products and services when project management is properly designed and implemented in a work environment.

Many *individuals* place value on certification because it is a career-enhancing move to build on one's professional capability. Certification provides goals for individuals to achieve and obtain recognition within the business arena. Some items that may provide value to an individual are:

- Being a stakeholder in the project management profession through peer recognition as a supporter of one's chosen discipline
- Improving job opportunities through employer recognition of superior qualifications authenticated by an independent agency
- Increasing pay for being certified within one's profession that may boost salaries from 5 to 20 percent, depending upon the recognized value by an organization
- Positioning within an organization for advancement and increased salary opportunities
- Demonstrating superior knowledge and competence to prospective employers
- Be within the project management profession as meeting professional standards.

Organizations seeking improvements in their project management processes and practices need a means of assessing individual professional qualifications that relate to national practices. Certification provides an independent means for assessing these qualifications. Some of the items providing value to an organization are:

- Having confidence that employees have a consistent understanding of the profession and are operating from one store of knowledge
- Being able to build customer confidence through showing that employees meet certification requirements and are consistent in their practices
- Using certification as a differentiator from competitors in sales and corporate image marketing
- Establishing certification as a part of career development for individuals
- Using certification as a criterion for hiring, promotion, and downsizing
- Setting standards by which project management performance can be measured

Customers obtain value from certification in several ways. Customers can rely on certified individuals to perform in a standardized manner that gives greater benefits from the delivered product or service. Some of these items of value are:

- Being able to confirm qualifications of an individual being considered for consulting work
- Having confidence that a service will be provided within a standard established by the certifying agency
- Having confidence that the performing individual has the capability to produce goods and services that meet the requirements

Professional societies derive value from conducting a professional certification program. These societies are recognized as leaders in the profession and builders of the project management profession. Because the project management profession continues to grow in its application to different industries, societies are challenged to maintain currency in the state-of-the-art practices. Some items of value to professional societies are:

- Recognition as a leader in the profession for the continual evolving of the project management profession
- Recognition as a provider of valuable services to individuals, organizations, the public at large, and the project management community

The *project management community* has a need for opportunities for individual growth and professional improvement. A project management certification program within the community enhances the stature of the profession and supports expansion of project management education programs. Some items of value to the project management community are:

- Professional standards are available for application to individual performance.
- Goals may be established to support professional growth.
- Consistency in project management practices builds upon the community.

The *public at large* continually demands improvements in products and services. Project management has the capability to support those changes necessary to deliver faster, better, and less expensive products and services—when individuals and organizations work together to design project management systems and to implement those systems in a consistent manner. Some of the value items that project management can deliver to the public at large are:

- Having less expensive products and services
- Having a better grade of products and services
- Having confidence that certified individuals will perform in a consistent ethical manner
- Achieving advances in the project management systems, including processes and practices, that lead to improved delivery of products and services

Certification leads to improved products and services at a minimum of cost while delivering benefits to many. Project management growth in design and implementation continues to advance when such efforts as certification are established.

**Two types of project management certification:
Knowledge and Performance Competency**

1.8.3.4   Value of Knowledge-Based versus Competency-Based Certification.   The basic difference between the two certification approaches is that knowledge is measured against a knowledge-based standard and competency is measured against a competency-based standard. Knowledge is typically defined in a body of knowledge and the standard is applied through a test. Competency, however, is a broader-based standard that encompasses knowledge of the subject, skills or the ability to apply the knowledge, and attitude. Competency testing may include a sample of a candidate's experience.

Compared to knowledge-based certification, competency-based certification gives everyone a greater degree of assurance that a certified individual will be able to perform at a significantly higher performance standard. Knowledge examination measures a candidate's understanding of the vocabulary, the body of knowledge, and perhaps some practices. The ability to perform to an acceptable level of proficiency is assumed by many customers. Competency-based certification, however, measures a candidate's ability to perform project management tasks at a given level.

One can see that the relative value of competency-based certification is significantly greater than knowledge-based certification. Competency includes knowledge and examines the areas of skills and attitude.

1.8.4 Project Management Certification

Certification in project management is unique to the different professional organizations and continues to evolve to meet the needs of individuals. Current information on certification can be obtained from the professional association. English language certification in project management is currently offered by the following organizations:

- American Society for the Advancement of Project Management (USA)—multilevel certification program based on competency. See *www.asapm.org*.
- Australian Institute for Project Management (Australia)—multilevel certification program based on competency. See *www.aipm.com.au*.
- Association of Project Management (United Kingdom)—multilevel certification program based on competency. See *www.apm.org.uk*.
- Project Management Institute (USA and serving other countries)—single level certification based on knowledge. See *www.pmi.org*.

1.8.5 User Questions

1. Why is it necessary to have a body of knowledge for a profession, and what is its significance?

2. Why is project management considered a profession when it overlaps many industries around the world and requires industry-specific knowledge?

3. Why do some countries permit professional associations to develop standards while other countries require that the standards be registered with the government?

4. What is the difference between an individual who has a professional designation in a project management area and one who doesn't have the designation?

5. What is the basis for testing a person's project management knowledge?

1.8.6   Summary

Bodies of knowledge for project management vary based on the perceived needs of the national culture and its application of the discipline. The Project Management Institute, headquartered in Newtown Square, Pennsylvania, has the most widely accepted project management body of knowledge that focuses on single-project application.

Certification in a professional designation validates an individual's qualifications and makes that person more valuable to the employer or business partner. There are several professional associations serving the interests of the project management profession: to include the American Society for the Advancement of Project Management, the Association for Project Management in the United Kingdom, the Australian Institute for Project Management in Australia, the International Project Management Association in Switzerland, and the Project Management Institute in the United States.

Each professional association has its unique certification process based on an established body of knowledge and experience, education, and service criteria. Each has unique designations that result from its respective processes.

The future of project management certification by all professional associations is good. Each association serves the needs of members and nonmembers in their respective national borders and international clients through certification. The trend shows more interest by individuals in developing nations and a growth rate of nearly 30 percent for the PMP designation.

1.8.7   Annotated Bibliography

1. Project Management Institute, *A Guide to the Project Management Body of Knowledge* (Newtown Square, PA: Project Management Institute, 2002). This is the latest refinement to the project management body of knowledge and represents current thought on the project management profession for single-project management. The processes in this document are helpful in planning and organizing efforts for projects.

2. Ireland, Lewis R., "The Value of Project Management Certification," *asapm* website (*www.asapm.org*), April 2003. This article reviews benefits for different individuals and organizations participating in project management certification programs. It describes the various benefits that are derived by elements and explores a wide range of beneficiaries.

1.9 PROJECT MANAGEMENT PROCESS*

1.9.1 Introduction

A process is defined as a protocol for dealing with activities in the design, development, and production (construction) of something—such as a project. A project management process provides a paradigm for how the management functions of planning, organizing, motivating, directing, and control can be carried out. Figure 1.6 provides a basic model of the functions involved in the management of a project.

FIGURE 1.6 Basic functions of project management.

A concise way of describing the project management processes through its major functions are indicated below:

- *Planning:* What are we aiming for and why? In carrying out the planning for a project the organization's mission is used as the basing point for the determination of the projects objectives, goals, and strategies. During the planning process, the policies, procedures, techniques, and documentation need to flesh out the anticipated use of resources to accomplish the project purposes are established.

- *Organizing:* What's involved and why? In carrying out this function the needed human and non-human resources are determined, and the desired patterns of authority, responsibility, and accountability are established.

---

*Note: This Project Management Process is presented in a slightly different context in Section 1.4.4, The Project Management Functions.

- *Motivating:* What brings out the best performance of the project team members and other people who support the project?

- *Directing:* Who decides what and when? In the discharge of this function, the project managers and other managers provide for the face-to-face leadership in the making and oversight of the execution of decisions involved in the commitment of resources on the project.

- *Controlling:* Who judges results and by what standards? In this function the project manager, team members, and other managers carry out the monitoring, evaluation, and control of the use of resources supporting the project.

1.9.2 Further Information on the Project Management Processes

In Table 1.8, a more detailed description of representative functions in the project management process are portrayed.

Each of the activities noted under these functions are only representative. The management functions used in the management of a project are the principal foci around which the making and implementing of decisions about the project are carried out.

A series of questions that can help in planning and reviewing how well these functions are carried out is presented in Table 1.9.

The management functions are key to understanding how the project resources can be managed. Each of these functions—planning, organizing, motivating, direction, and control—is a "work package" in itself. Taken together, the functions provide for the overall "work package" for how the entire project is to be carried out.

1.9.3 Key User Questions Regarding the Project Management Functions

1. Do the managers in the organization understand the major functions that are involved in the management of projects in the enterprise?

2. Have the major outputs for each of the function been identified?

3. Do the project team members understand how the planning and control functions are linked in the management of a project?

4. Have relevant policies, procedures, and protocols been established to flesh out the planning elements in the management of projects?

5. Does management consider the needs of individual team members in developing an ambience for enhancing the motivation of people associated with the project?

TABLE 1.8 Representative Functions/Process of Project Management

<hr>

Planning: What Are We Aiming for and Why?

<hr>

Develop project objectives, goals, and strategies.
Develop project work breakdown structure.
Develop precedence diagrams to establish logical relationship of project activities and milestones.
Develop time-based schedule for the project based on the time precedence diagram.
Plan for the resource support of the project.

<hr>

Organizing: What's Involved and Why?

<hr>

Establish organizational design for the team.
Identify and assign project roles to members of the project team.
Define project management policies, procedures, and techniques.
Prepare project management charter and other delegation instruments.
Establish standards for the authority, responsibility, and accountability of the project team.

<hr>

Motivating: What Motivates People to Do Their Best Work?

<hr>

Determine project team member needs.
Assess factors that motivate these people to do their best work.
Provide appropriate counseling and mentoring as required.
Establish rewards program for project team members.
Conduct initial study of impact of motivation on productivity.

<hr>

Directing: Who Decides What and When?

<hr>

Establish "limits" of authority for decision making for the allocation of project resources.
Develop leadership style.
Enhance interpersonal skills.
Prepare plan for increasing participative management techniques in managing the project team.
Develop consensus decision-making techniques for the project team.

<hr>

Controlling: Who Judges Results and by What Standards?

<hr>

Establish cost, schedule, and technical performance standards for the project.
Prepare plans for the means to evaluate project progress.
Establish a project management information system for the project.
Prepare project review strategy.
Evaluate project progress.

TABLE 1.9 Representative Questions of the Team Management Functions

---

Team Planning

What is the mission or "business" of the team?
What are the team's principal objectives?
What team goals must be attained in order to reach team objectives?
What is the strategy that will be used by the team to accomplish its purposes?
What resources are available for the team's use in accomplishing its mission?

Team Organization

What is the basic organizational design of the team?
What are the individual and collective roles on the team that must be identified, defined, and negotiated?
Will the team members understand and accept the authority, responsibility, and accountability that is assigned to them as individuals and as a team?
Do the team members understand their authority and responsibility to make decisions?
How can the team effort be coordinated so that the members will work in harmony, not against one another?

Team Motivation

What motivates the team members to do their best work?
Does the team manager provide a leadership style acceptable to the members of the team?
Is the team "productive"? If not, why not?
What can be done to increase the satisfaction and productivity of the team members?
Are the team meetings conducted in such a manner that people attending are encouraged or discouraged?

Team Direction

Is the team leader qualified to lead the team?
Is the team leader's style acceptable to the members of the team?
Do the individual members of the team assume leadership in the areas where they are expected to lead?
Is there anything that the team leader can do to increase the satisfaction of the team members?
Does the team leader inspire confidence, trust, loyalty, and commitment among the team members?

TABLE 1.9   Representative Questions of the Team Management Functions
(*Continued*)

Team Control

Have performance standards been established for the team? For the individual
    members?
What feedback on the team's performance does the manager have who appointed
    the team?
How often does the team get together to formally review its progress?
Has the team attained its objectives and goals in an effective and efficient
    manner?
Do the team members understand the nature of control in the operation of the
    team?

Adapted from M. H. Mescon et al., *Management* (New York, NY: Harper & Row, 1981),
p. 167.

1.9.4 Summary

In this section, the management functions were presented as the major
building blocks in the process of managing projects. Planning, organizing,
motivation, direction, and control were presented as these major building
blocks. Examples are given of why and how the use of such functions
were provided during the process of managing projects.

1.9.5 Annotated Bibliography

1. Cleland, David I., and Lewis R. Ireland, *Project Management: Strategic
    Design & Implementation*, 4th ed. (New York, NY: McGraw-Hill,
    2002), chap. 2, "The Project Management Process." This chapter de-
    scribes how the management functions are the major elements found
    in the management of projects. In addition, the chapter material pro-
    vides suggestions of how these management functions can be estab-
    lished and improved in the management of projects.

SECTION 2
PROJECT ORGANIZATIONAL DESIGN

2.1   ORGANIZING FOR PROJECT MANAGEMENT

2.1.1   Introduction

This section provides an examination of the organizational design used for the management of projects. It addresses shortcomings in the traditional organization designs as well as the strengths of the project organization.

2.1.2   Shortcomings of the Traditional Organizational Design

- Traditional organizational hierarchies tend to be slow, inflexible, and fail to provide for an organizational focus of project activities.

- Barriers commonly exist in the traditional organization, which stifles the horizontal flow of activities required when projects are undertaken.

- Inadequate delegation of authority and responsibility to support project activities is a common problem in the traditional organizations.

A modification of traditional organizational design is required to support project activities.

> **The project organization is an integration of the project stakeholders.**

2.1.3   The Project Organization

- The project organization is a temporary design used to denote an inter-organizational team pulled together for the management of the project.
- Personnel in the project organization are drawn from the supporting functional elements of the enterprise.

When a project team is assembled and superimposed on the existing traditional structure, a matrix organization is formed. Figure 2.1 portrays a basic project management matrix organizational design.

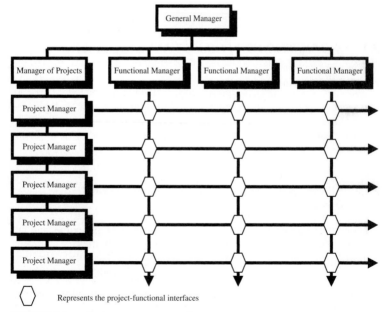

⬡   Represents the project-functional interfaces

FIGURE 2.1   A basic management matrix.

2.1.4 Alternative Forms of the Project Organization

Functional Organization—The project is divided and assigned to functional entities with project coordination being carried out informally by functional and higher level managers.

The Functional Matrix—The project manager is assigned with the authority to manage the project across the functions of the enterprise.

The Balanced Matrix—A design where the project manager shares the authority and responsibility for the project with the functional managers.

The Traditional Matrix—The project manager and the functional manager share explicit complementary authority and responsibility for the management of the project.

2.1.5 Traditional Departmentalization

The most commonly used traditional means used to decentralize authority, responsibility, and accountability include the following:

- Functional Departmentalization, where the decentralized organizational units are based on common specialties, such as finance, engineering, and manufacturing
- Product Departmentalization, in which organizational units are responsible for a product or product lines
- Customer Departmentalization, where the decentralized units are designated around customer groups, such as the Department of Defense
- Territorial Departmentalization, where the organizational units are based on geographic lines; for example, Southwestern Pennsylvania marketing area
- Process Departmentalization, where the human and other resources are based on a flow of work such as an oil refinery

2.1.6 The Matrix Organization

In the matrix organization, there is a sharing of authority, responsibility, and accountability among the project team and the supporting functional units of the organization. The matrix organizational unit also takes into consideration outside stakeholder organizations that have vested interests in the project. The matrix organization is characterized by specific delineation of individual and collective roles in the management of the project.

2.1.7   The Project-Functional Interface in the Matrix Organization

It is at this interface that the relative and complementary authority-responsibility-accountability roles of the project manager and the functional manager come into focus. Table 2.1 suggests a boilerplate model that can be used as a guide to understand the interface of relative authority, responsibility, and accountability within the matrix organization.

TABLE 2.1   The Project-Functional Interface

Project manager	Functional manager
• What is to be done? • When will the task be done? • Why will the task be done? • How much money is available to do the task? • How well has the total project been done?	• How will the task be done? • Where will the task be done? • Who will do the task? • How well has the functional input been integrated into the project?

Source: David I. Cleland and Lewis R. Ireland, *Project Management: Strategic Design and Implementation,* 4th ed. (New York, NY: McGraw-Hill, 2002), p. 233.

2.1.8   Basic Form of the Matrix

In its most basic form, a matrix organization looks like the model in Fig. 2.2, where the project and functional interface comes about through the project work package. Each work package is a "bundle of skills" for which an individual or individuals have responsibility to carry out in supporting the project.

2.1.9   Roles of the Project Manager

Many roles are carried out by the project manager. A few of these key roles are indicated below:

• A strategist in developing a sense of direction for the use of project resources
• A negotiator in obtaining resources to support the project
• An organizer to pull together a team to act as a focal point for the management of the project
• A leader to recruit and provide oversight over the planning and execution of resources to support the project

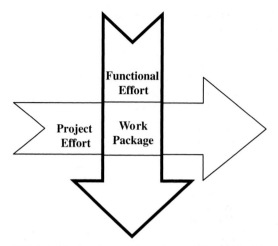

FIGURE 2.2 Interfaces of the project and functional effort around the project work package. (*Source: David I. Cleland and Lewis R. Ireland, Project Management: Strategic Design and Implementation,* 4th ed. (New York, NY: McGraw-Hill, 2002), p. 237.

- A mentor in providing counseling and consultation to members of the project team
- A motivator in creating an environment for the project team that brings out the best performance of the team
- A controller who maintains oversight over the efficacy with which resources are being used to support project objectives
- Finally, a diplomat who builds and maintains alliances with the project stakeholders to gain their support of the project purposes

2.1.10   A Controversial Design

The matrix organizational design has been praised and condemned; it has had its value, problems, and abuses. The review that follows examines some of the characteristics of a *weak* and a *strong* matrix.

1. Weak matrix:

- Failure to understand the individual and collective roles of the participants in the matrix

- An inherent suspicion of an organizational design that departs from the traditional model of the organization
- A failure on the part of senior management to stipulate in writing the relative roles to be carried out in the matrix organization
- Lack of trust, integrity, loyalty, and commitment on the part of the members
- Failure to develop the project team

2. Strong matrix:

- Individual and collective roles have been defined in terms of authority-responsibility-accountability.
- The project manager delegates authority as required to strengthen the team members.
- Team members respect the prerogatives of the functional managers, and the roles of other stakeholders on the project.
- Conflict over territorial issues are promptly resolved.
- Continuous team development is carried out to define and strengthen the roles of the team members and other stakeholders.

2.1.11   Project Manager–Customer Interface

The interactions between a project manager and the customer is depicted in Figure 2.3. The reciprocal interrelationships portrayed in this figure are

FIGURE 2.3   Contractor—project manager relationships. (***Source:***   Adapted from David I. Cleland, "Project Management—An Innovation in Management Thought and Theory," *Air University Review,* January-February 1965, p. 19.)

only representative of the myriad of interaction that occurs. Key decisions and factors likely to have impact on the project should be channeled through the respective project managers.

2.1.12 Key User Questions

1. Do the participants on the project understand their individual and collective roles on the project?
2. Has appropriate documentation been created and distributed that stipulates the relevant authority-responsibility-accountability of the project participants?
3. Is the type of project organization that is being used appropriate for the projects that are being managed?
4. Is there an effective means for the resolution of conflict in the organization?
5. Are there any changes needed in the organizational design for the projects underway in the enterprise?

2.1.13 Summary

In this section, alternative designs for the project organization were presented. In such organizations a clear definition of the individual and collective roles of the project participants is required. Several models of organizational design were presented to include the commonly used "matrix organization." The characteristics of both a "weak" and a "strong" design were presented, along with a brief description of the interface that project managers have with customers.

2.1.14 Annotated Bibliography

1. Cleland, David I., and Lewis R. Ireland, *Project Management: Strategic Design and Implementation,* 4th ed. (New York, NY: McGraw-Hill, 2002), chap. 8, "Organizing for Project Management." This chapter examines the project-driven organization form to include alternative means for the design of a project organization. Included in the examination is a description of how authority-responsibility-accountability can be delegated to project participants.

2. Middleton, C. J. "How to Set Up a Project Organization," *Harvard Business Review*, April 1967. This is an early and classic article that describes a strategy for establishing the project organization. The author reviews the need for a balance of power in the matrix organization between the project manager and the functional managers. He describes the relative roles of the project manager vis-à-vis the functional managers, and the responsibility that the sponsoring general managers have for the solution of conflict between these two managers.

2.2   PROJECT ORGANIZATION CHARTING

2.2.1   Introduction

This section of the Handbook describes Project Organization Charting, or the alignment of the organization within the context of authority and responsibility. The Linear Responsibility Chart (LRC) is a primary tool for this.

2.2.2   The Traditional Organization Chart

At best, the typical pyramidal organization chart is an oversimplification of the organization because of its general nature. Although it describes the framework of the organization, and can be used to acquaint people with the nature of the organizational structure—how the work in the organization is generally broken down—it lacks specificity regarding individual and collective roles. It does little to clarify the myriad reciprocal relationships among members of the project stakeholders. As a static model of the enterprise, the traditional chart does little to clarify how people are supposed to operate together in doing the work of the enterprise. An alternative is needed—the LRC.

The LRC portrays who works with whom

2.2.3   Linear Responsibility Chart

An LRC displays the coupling of the project work packages with people in the organization. These couplings are the outcome of bringing the key elements of an LRC together. These key elements include:

- An organizational position
- An element of work—the work package
- An organizational interface point
- A legend to describe a relationship
- A procedure for creating the LRC
- A commitment to make the LRC work

Figure 2.4 shows the basic nature of an LRC with the "P" indicating Primary responsibility for the work package.

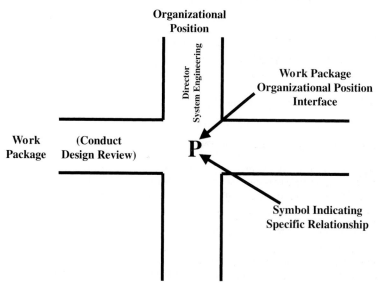

FIGURE 2.4 Essential structure of a linear responsibility chart. (*Source:* David I. Cleland and Lewis R. Ireland, *Project Management: Strategic Design and Implementation,* 4th ed. (New York, NY: McGraw-Hill, 2002), p. 272.)

2.2.4 Work Package—Organizational Position Interfaces

A work package is a single, discrete unit of work that has a singular identity that can be assigned to one individual, and to other individuals who are involved in doing the work on the work package. Work packages are directly related to specific organizational positions. These positions and the responsibilities assigned to them in carrying out the use of resources on the work package requirements constitute the basis for the LRC. The authority and responsibility regarding a work package and an organizational position are depicted in a suitable legend, a sample of which follows:

1. Actual responsibility
2. General supervision
3. Must be consulted
4. May be consulted
5. Must be notified
6. Approval authority

Figure 2.5 shows an LRC for the authority-responsibility-accountability relationships within a matrix organization.

2.2.5   Development of the LRC

The development of an LRC should be done cooperatively with the project team members. Once completed, an LRC can become a "living document" to be used for the following:

- Portray formal, expected authority-responsibility-accountability roles
- Acquaint all stakeholders with the specifics of how the work packages are divided up on the project
- Contribute to the commitment and motivation of team members since they can see what is expected of them
- Provide a standard for the role that the project manager and the team members can monitor regarding what people are doing on the project

2.2.6   Key User Questions

1. Do the project team members understand the usefulness and inadequacies of the traditional organizational chart?
2. Do you see the opportunity to use the LRC to enhance the understanding of roles in the project?
3. Have the team members participated in the development of an LRC for the project?

Activity	General Manager	Manager of Projects	Project Manager	Functional Manager
Establish department Policies and objectives	1	3	3	3
Integration of projects	2	1	3	3
Project direction	4	2	1	3
Project charter	6	2	1	5
Project planning	4	2	1	3
Project—functional conflict resolution	1	3	3	3
Functional planning	2	4	3	1
Functional direction	2	4	5	1
Project budget	4	6	1	3
Project WBS	4	6	1	3
Project control	4	2	1	3
Functional control	2	4	3	1
Overhead management	2	4	3	1
Strategic programs	6	3	4	1

Legend:

1   Actual responsibility

2   General supervision

3   Must be consulted

4   May be consulted

5   Must be notified

6   Approval authority

FIGURE 2.5   A linear responsibility chart of project management relationships.

4. Do the members of the project team understand their individual and collective roles on the project as a result of having participated in the development of the LRC?

5. Has the LRC for the project organization been used as an aid to further the development of the project team?

2.2.7   Summary

In this section, the use of an LRC has been described as a means of better understanding the individual and collective roles on the project team. An LRC can serve to facilitate the coordinated performance of the team—since everyone would better understand their individual and collective role on the project team.

2.2.8   Annotated Bibliography

1. Cleland, David I., and Lewis R. Ireland, *Project Management: Strategic Design and Implementation,* 4th ed. (New York, NY: McGraw-Hill, 2002), chap. 9, "Project Authority."

2. Worley, Christopher G., and Charles J. Teplitz, "The Use of 'Expert' Power as an Emerging Influence Style within Successful U.S. Matrix Organizations," *Project Management Journal,* February 1993.

2.3   AUTHORITY—RESPONSIBILITY—ACCOUNTABILITY

2.3.1   Introduction

In this section, a description is offered concerning the matter of how authority, responsibility, and accountability are provided to complement the organizational design used for management of projects.

2.3.2   Defined Authority

Authority is defined as a legal or rightful power to command or act. As applied to the manager, authority is the power to command others to act or not to act. There are two basic kinds of authority:

- *De jure* authority is the legal or rightful power to command or act in the management of a project. This authority is usually expressed in the form of a policy, position description, appointment letter, or other form of documentation. De jure authority attaches to an organizational position.
- *De facto* authority is that influence brought to the management of a project by reason of a person's knowledge, skill, interpersonal abilities, competency, expertise, and so forth.

Power is the possession of an organizational role augmented with sufficient knowledge, expertise, interpersonal skills, dedication, networks, alliances, and so forth, that gives an individual extraordinary influence over other people.

Authority, Responsibility, and Accountability are unifying forces in organizations

---

2.3.3 Authority-Responsibility-Accountability (A-R-A) Failures

Most failures of A-R-A in the project organizational design can be attributed to one or more of the following factors:

- Failure to define the legal A-R of the major participants in the project organizational design, such as the matrix form
- Unwillingness of project people to share A-R in the project affairs
- Lack of understanding of the theoretical construction of the matrix
- Existence of a cultural ambience that reinforces the "command and control" mentality of the traditional organizational design

2.3.4 Documenting Authority, Responsibility, and Accountability (A-R-A)

Suitable documentation should be published concerning the A-R of the major project participants. Figure 2.6 is one example of the nature of such documentation.

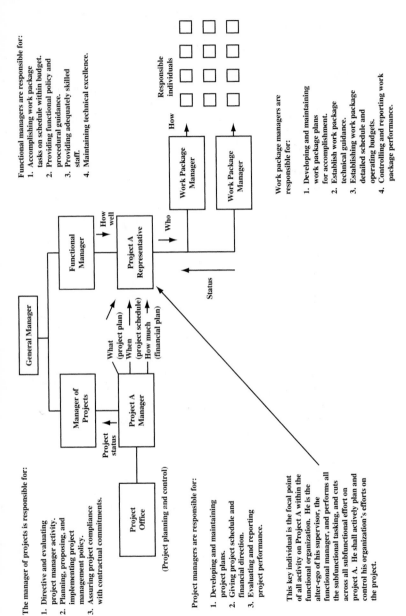

The manager of projects is responsible for:

1. Directive and evaluating project manager activity.
2. Planning, proposing, and implementing project management policy.
3. Assuring project compliance with contractual commitments.

(Project planning and control)

Project managers are responsible for:

1. Developing and maintaining project plans.
2. Giving project schedule and financial direction.
3. Evaluating and reporting project performance.

This key individual is the focal point of all activity on Project A within the functional organization. He is the alter-ego of his supervisor, the functional manager, and performs all the subfunctional tasking, and cuts across all subfunctional effort on project A. He shall actively plan and control his organization's efforts on the project.

Functional managers are responsible for:

1. Accomplishing work package tasks on schedule within budget.
2. Providing functional policy and procedural guidance.
3. Providing adequately skilled staff.
4. Maintaining technical excellence.

Work package managers are responsible for:

1. Developing and maintaining work package plans for accomplishment.
2. Establish work package technical guidance.
3. Establishing work package detailed schedule and operating budgets.
4. Controlling and reporting work package performance.

Figure 2.6 Project-functional organizational interface. (*Source:* David I. Cleland and William R. King,

2.3.5   Defining Responsibility

Responsibility, a corollary of authority, is a state, quality, or fact of being answerable for the use of resources on a project and the realization of the project objectives.

2.3.6   Defining Accountability

Accountability is the state of assuming liability for something of value, whether through a contract or because of one's position of responsibility.

Figure 2.7 is a summary of Fig. 2.6 and is one way of portraying these forces. Cost, schedule, and technical performance parameters are elements around which authority-responsibility-accountability force flow. Each organizational level in this figure provides an ambience for both individual and collective roles.

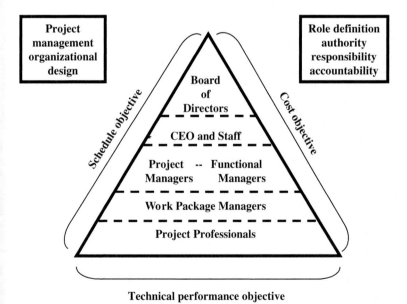

FIGURE 2.7   Project management organizational design.

2.3.7 Key User Questions

1. As you consider the existence and flow of A-R-A in your organization, do the project people understand their individual and collective roles?

2. Has specific documentation been developed that adequately delegates A-R-A to the project participants?

3. Do the project people understand that their delegations of *De Jure* authority has to be reinforced with *De Facto* authority?

4. Are there any barriers to the adequate exercise of authority in the enterprise?

5. Do the people associated with the project understand the role of power and influence in the management of the project, as well as the support provided by the functional elements of the enterprise?

2.3.8 Summary

In this section, the concepts of authority-responsibility-accountability have been presented. Authority was described as having two major elements: *De Jure* and *De Facto*. Responsibility was described as the state of being held answerable for the use of resources and results. Accountability is the state of being liable for the use of resources in the creation of value. Great care should be taken to develop and disseminate organizational documentation throughout the enterprise that describes and delegates the elements of authority, responsibility, and accountability needed for people to perform on the project.

2.3.9 Annotated Bibliography

1. Cleland, David I., and Lewis R. Ireland, *Project Management: Strategic Design and Implementation,* 4th ed. (New York, NY: McGraw-Hill, 2002), chap. 9, "Project Authority." This reference describes the nature and use of authority in the process of managing a project, paying attention to how authority best works with elements of responsibility and accountability in the affairs of a project. Several key suggestions and models are suggested to enhance an understanding of the individual and collective roles in the management of a project.

2. Fisher, Kimball, *Leading Self-Directed Work Teams,* (New York, NY: McGraw-Hill, 1993). There are many similarities between a project team and a "self-directed" team. This book explains how team leadership skills such as coaching, facilitating, group dynamics, and many more of the issues likely to confront the project manager—and the members of the project team—can be managed. The author profiles the most innovative team leader practices from known and successful industrial organizations. The reader should be able to recognize how the knowledge expressed in this book can be applied to the management of a project team.

2.4 PROJECT MANAGEMENT TRAINING

2.4.1 Introduction

Training is an important aspect of raising knowledge levels and changing behavior of individuals on projects. Training gives the individuals greater capability to perform at higher levels of productivity and make greater contributions to project success. Interpersonal skills training is as important as knowledge training on project management functions because of the team challenges in managing projects.

Skill training also is important for the tools of project management. These tools fulfill the requirements for time planning (schedule tool), reporting (graphics and correspondence), cost tracking and computing (spreadsheet), and managing data (database tool). Tool training, however, does not replace the knowledge requirements for project management functions.

All personnel do not require the same type or level of training within knowledge or skill areas. Some training is for familiarization, or broad general knowledge, while other training is for detailed working knowledge. The position and responsibilities of the person dictate the level of knowledge required and training requirements.

Training also has to consider whether the individuals being trained possess knowledge that is inappropriate and must be unlearned prior to new knowledge being learned or whether the training is the first exposure. There is also training to reinforce concepts and knowledge as well as advanced training that builds on prior training.

We are trained and ready to do the job!

2.4.2 Project Management Knowledge and Skills

Senior leaders have been promoted within an organization because they are experts in traditional management concepts. Few senior leaders have been involved in project management as a discipline. Those few who have been in project management have not been able to keep pace with the maturing of the methodologies and technologies.

Senior leaders have also been omitted from the training in current project management discipline through failure to develop appropriate courses of instruction. Senior leaders do not need the same level of knowledge and do not need to know the details.

General knowledge of concepts and principles as well as an ability to interpret the information generated by project reporting is more appropriate. Senior leaders must be able to use that information to determine how to support projects, how to link strategic goals to projects, and how to take corrective action when the information gives early signs of failure.

Project leaders must know the overall status of the project at any time. There is a need to be able to interpret the information generated by the project to recognize early signs of slippage in any area. Therefore, the knowledge required is a working level understanding of the project and being able to visualize the convergence of the technical solution. Project leaders must be human resource managers as well as negotiators, communicators, and technically qualified in their industry or discipline.

Project leaders, depending upon the size of the project and the authority delegated, may have need for more detailed working knowledge to be a "working project leader" or more general knowledge if a "managing" project leader.

Project planners are the principle planners for projects and must have the detailed knowledge of the concepts, principles, techniques, and tools of planning. Planning knowledge and skills are not taught in the US system of higher education. Therefore, many of the planners have learned planning in the apprenticeship method with much of the work being trial and error. Planning, as the design of how the work will be performed, is a critical aspect of project management.

Project team members must be familiar with the methodology and flow of project management before being assigned to a project for the most effective and productive effort. Understanding the guidelines and checkpoints in a project supports the work. They also must understand their role in measuring project progress, such as how to report progress against a schedule.

2.4.3   General Categories of People Involved in Projects

There are generally four levels of individuals who need to have some knowledge of project management. This includes the capability of project

TABLE 2.2   Responsibility—Role—Knowledge Matrix

Responsibility	Role	Knowledge of PM
Strategic direction	Senior leaders	Capabilities of project management to support strategic purposes.
Allocates resources	Manager/Sponsor of project leaders	Knowledge of organization's strategic objectives and goals and project linkages.
Applies resources	Project leader	Knowledge of project and processes.
Uses resource	Project team member	Detailed knowledge of components of project work.

management to deliver against strategic goals, approval process and allocation of resources to projects while tracking their progress, direction of project planning, execution and close-out, and details of the project work and work processes.

Table 2.2 summarizes the categories of people involved in projects and the level of knowledge required to effectively meet their responsibilities.

In evaluating the training requirements, it is helpful to match responsibilities, knowledge, and skill requirements to the position. A detailed analysis of an organization's requirements is helpful to understand the type and level of training needed to improve individuals' capabilities.

A general list of knowledge and skill areas is shown in Table 2.3. It matches the training requirement to the position. This is a start point and assumes that all persons need training in the identified area.

One of the major challenges in projects is for the senior leaders to understand project capabilities and the role of senior leadership in linking projects to strategic goals. Typically, projects are initiated based on near-term requirements that may or may not contribute to the organization's strategic position in the future. Only senior leadership can bridge the strategic to the operational areas.

Tool training is often viewed as the solution to weak project management. Training in a scheduling tool may be viewed as the only need for project leaders. However, the list in the table above clearly demonstrates the need for training in the methodology, techniques, standards, principles, fundamentals, and interpersonal skills. An effective project leader needs both knowledge and skills.

TABLE 2.3 Knowledge and Skill Areas for Project Participants

Knowledge/Skill	SL	PS	PL	PP	PC	TM
Strategic planning and plan	X	X				
Organization's strategic goals	X	X	X			
Project decision-making	X	X	X			
Project information (understanding and interpreting project data)	X	X	X	X	X	X
Project leadership skills (coaching, conflict resolution, facilitation, motivation, negotiation)			X			
Oral and written communication	X	X	X	X	X	X
Project planning		X	X	X		
Project meeting management			X			X
Organizational project methodology	X	X	X	X	X	X
Project best practices		X	X	X	X	X
Project scope management			X	X	X	X
Project schedule/time planning and management			X	X	X	X
Project quality management			X	X	X	
Project cost/budget planning and management			X	X	X	
Project risk planning and management			X	X	X	
HR planning and management			X	X	X	
Project procurement planning and management			X	X	X	
Team role and responsibilities			X	X	X	X
Project tools (scheduling, correspondence, spreadsheet, graphics)			X	X	X	X

Legend:
SL = Senior Leadership
PS = Project Sponsor
PL = Project Leader
PP = Project Planner
PC = Project Controller
TM = Team Member

2.4.4 Developing Training Programs

Organizations need a plan for the selecting and designing training for individuals having responsibilities for project successes. Training should begin with senior leaders and cascade down to the working level of the project. This training may be accomplished in parallel rather than series.

TABLE 2.4 Role and Training Matrix

Role/Position	Training time	Advanced training
Senior leaders	6 hours intensive program	One or two hours to refine the knowledge, as appropriate.
Project leaders	32 hours of fundamentals and as many as 32 hours in interpersonal skills.	24 hours of advanced training in techniques and best practices of leadership.
Project planner and controller	32 hours in fundamentals and 16 hours in planning concepts and practices.	Refresher training as required in methodology and standards.
	24 hours of tools training.	Tool training as required for new tools.
Project team member	32 hours in fundamentals	Tool training as required by work position.

It is important to understand that the actions of the senior leaders affect the project planning, execution, and control. Therefore, senior leaders need to understand their roles and implement the practices for linking strategic goals to projects. This prevents training project team members in their procedures and then changing the procedures to meet new guidance from the senior leadership.

Referring back to the general requirements in Table 2.3, the time allocated to the training of personnel is depicted in Table 2.4.

This general outline of training provides a point of departure. An audit should be conducted to identify training requirements for each individual. What are the knowledge and skills needs for each position within an organization and what do the individuals possess? What existing knowledge is inappropriate for the organization and what retraining is required? These basic questions should guide a person in making an audit training of needs.

2.4.5 Key User Questions

1. What training in project management concepts at a broad general level is helpful for the senior leadership to understand this discipline?
2. The fundamentals of project management are needed by the project team to understand how the project is designed. Of the participants in

the project management process, who does not need the fundamental training?

3. If project planning is weak for projects, what is the primary reason and what can be done about it?

4. What type of training does the project planner require and how is this different from other members of the project team?

5. What unique skills does the project leader require that is not an absolute requirement for other members of the project team?

2.4.6 Summary

Project management training has a broader scope than just the project team. The range of training includes senior leaders and project sponsors to link their roles in the project management process with the work being accomplished. Senior leadership needs to understand the general area of projects and the indicators of progress to fulfill their strategic and operational responsibilities.

The project team, to include the project leader, needs training for basic concepts, processes and principles of project management. The training on the organization's methodology and best practices establishes the foundation for advanced and individual needs. Selected individuals in the project team will need advanced training.

2.4.7 Annotated Bibliography

1. Elbeik, Sam, and Mark Thomas, *Project Skills* (*New Skills Portfolio*), (Butterworth-Heinemann, January 1999), 200 pp. This book on project skills required for success will round out the understanding for training project team members.

2. Various sources: Training and education is rapidly changing in the number and quality of providers. One should search the Internet for the most recent training providers to identify the courses and compare them with one's needs. Distance learning on the Internet is moving to the forefront and may be the right solution for busy project managers seeking knowledge.

2.5 *INTERNATIONAL PROJECTS*

2.5.1 Introduction

International projects are defined as projects that bridge one or more national borders. International projects may be driven by one organization or they may be a partnership or consortium. The organizational relationships of the project owners drive the implementation.

International projects bridge more than national borders; there are cultural differences, time zone considerations, language differences, and monetary differences. Differences among the project participants increase the chance of miscommunication and misunderstandings. There are advantages to international projects that override any challenges to project planning and execution.

**Large international projects
are the wave of the future!**

2.5.2   International Project Rationale

Cooperative efforts between parties of different nations have significant benefits with only a few difficulties. At the national level, two countries may decide to develop cooperative research and development when it would be too costly for one nation to invest in the effort. Sharing the investment cost makes the project affordable.

Technology may drive the international project. When one country has advanced technology and another country needs that technology, joint co-operative efforts through projects will benefit both parties through lower cost and shorter time to obtain advanced technology.

Difference in labor costs is another reason for international projects. Where a project is labor intensive and the cost of labor is significantly less in another country, an organization may seek the less expensive resources. For example, some computer programming is one area that has been accomplished in India rather than the United States because of a significant difference in labor costs.

International projects can be cooperative efforts between organizations in different nations to build products for sale in both nations. Organizations work as a team to build different parts of a product, and both organizations market the product in their respective nations. This lowers the cost of the product and capitalizes on the capabilities of each organization.

Consortia with several different national organizations cooperating to build a product can also be an international project. The Concorde aircraft is an example of the cooperative efforts of nations to build a supersonic passenger plane that serves more than one national interest.

2.5.3   Types of International Projects

The different types of international projects may be defined by their characteristics. The general types will provide an understanding of the arrangements that may be used and of the existing thoughts on cooperation across borders. Figure 2.8 shows the more common types of international projects.

Descriptions of the types of international projects are below. These descriptions are representative of international projects at the top level, and the projects are typically tailored to meet the business needs of the stakeholders.

- Multi-National Projects—the cooperative efforts of two or more nations to build a product that serves the interests of both nations. The interested nations prepare and sign a cooperative agreement that defines the area and extent of cooperation. Typical business arrangements will be:

Types of International Projects

Multi-National Projects

One Organization with Project Elements in Different Nations

One Organization with Product Components built in Several Nations

Consortium of Organizations for One or More Projects

FIGURE 2.8   Types of international projects.

- Division of work between nations, i.e., contracts in countries
- Steering Committee or Management Structure of the Project
- Customs and duties waived
- Monetary unit that will be used
- Transfer of funds between national entities
- Intellectual property rights

- One Organization with Project Elements in Different Nations—the efforts, controlled by one organization, across one or more national borders to achieve a specific purpose. This is used by major corporations when it is in their best interests to have a presence in one or more countries. Components of the project work will be performed in various countries under the control of the parent organization. The end product of the project affects all the organization's elements in the participating countries.

- One Organization with Project (product) Components Built in Several Nations—the combined efforts of several different nations' businesses to build a single project. The different national entities may be subcontractors or partners. The end product of the project is typically for one or more organizations, but not necessarily the participating organizations.

- A Consortium of Organizations—the efforts of several organizations in different nations, functioning as one for a single purpose, and having one or more projects. This is used when there is a need for several countries to develop a product or products that will serve their common interests, combining the capabilities of several organizations to meet a threat or interest. The end product of the project will typically serve all participants, but may be for profit.

There is an unlimited number of arrangements that can be made for international projects. The more common type are simple and straight forward. When projects span national boundaries, there are inherent difficulties such as taxes, differences in monetary stability, and differences in quality practices. Keeping the projects simple and uncomplicated will facilitate completion.

2.5.4   Advantages of International Projects

International projects have many advantages that set the scene for more in the future. Figure 2.9 summarizes some of these advantages.

Detailed discussion of the advantages is shown below.

FIGURE 2.9   Advantages of international projects.

- Financial—richer and more developed nations can create work opportunities in emerging nations through international projects. Having the emerging nation, with low-cost labor, perform labor intensive functions is profitable. This also provides the emerging nation a source of revenue that builds on its economy.

- Hard currency transfer—many nations have high inflation rates that exceed 1000 percent a year. When the project work is paid in "hard currency," or currency that has a low inflation rate, the receiving organization has stability and leverage within its national boundaries. The stable currency provides a growth for the performing organization that cannot be achieved through projects internal to the borders.

- Technology transfer—technology from one nation can be transferred to another nation through cooperative efforts of international projects. Technology transfer may be through providing information to another nation or through reverse engineering of the product. The net result is that a nation's technology capital is increased through the transfer.

- Capitalize on existing technology—another country may be significantly ahead in selected technology. Transfer of technology may take time and the need can be fulfilled by the owning nation. Therefore, the owning national organization can obtain the technology through contractual relationships.

2.5.5   Disadvantages of International Projects

International projects, in comparison with domestic projects, have disadvantages or additional challenges that an organization must address. These challenges can be overcome at a price and with knowledge that the advantages must outweigh the disadvantages. Figure 2.10 summarizes some of the disadvantages of international projects.

Detailed discussion of the disadvantages is shown below.

• Technology transfer—while technology transfer can be an advantage, some nations place restrictions on the transfer of technology. Most recently, computer technology has been the subject of illegal transfers. From an organization's perspective, it may not want to share its trade secrets in manufacturing and other proprietary information.

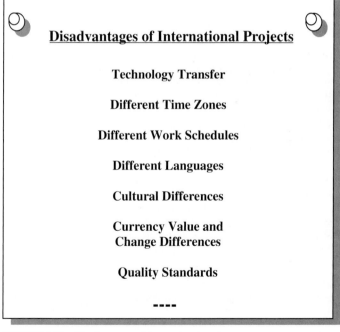

Disadvantages of International Projects

Technology Transfer

Different Time Zones

Different Work Schedules

Different Languages

Cultural Differences

**Currency Value and
Change Differences**

Quality Standards

FIGURE 2.10   Disadvantages of international projects.

- Different time zones—time zones may inhibit communication between the parent organization and the partners around the world. Communication windows may be limited to a few working hours a day or at night.

- Different work schedules—national holidays and work hours can limit communication windows and other opportunities. National holidays can also be celebrated at different times. Take for example Thanksgiving in the United States and Canada, celebrated in November and October respectively.

- Different languages—languages can make it difficult to bridge the nuances and shoptalk of the technical field. American English, for example, can be significantly different from the English language spoken in Canada, England, Ireland, Scotland, Australia, New Zealand and former British colonies.

- Difference in meeting agendas because of cultural differences—meetings between Europeans, North Americans, and Asians can be significantly different in format. The silent time between an American ending his/her point and the time that the Asian responds can be uncomfortable for the American. However, this is a matter of respect by the Asian to wait until the other person is finished speaking.

- Measuring projects in own nation currency—long-term projects using the local currency as measures of success may have difficulty when one currency is devalued through inflation or another currency increases in value. The project cost can vary significantly with changes to currency values.

- Quality standards—quality of work is different across national borders. The materials available may be inferior and the workmanship can be at a low level of sophistication. These imbalances in quality can adversely affect the end product and its utility.

Depending upon the agreements between the different parties of an international project, there will be other disadvantages identified. Any organization should explore the differences between the cultures prior to initiating a new project. These differences, once identified, may be avoidable through seeking a common ground.

2.5.6   Developing an International Project Plan

Anticipating the advantages and disadvantages will give an organization important aspects an international project may encounter. Understanding the culture, the capability of another country, the materials available to perform the work, and the trained resources to perform the work is just a start. National laws, customs and duties, and transportation systems may be either favorable or unfavorable.

International projects between governments typically use an agreement that has approximately 20 paragraphs. These follow a standard procedure to ensure all areas are addressed. The agreements are morally binding, but not legally binding because of the lack of a court that has jurisdiction over the two or more nations. Organizations, however, should designate the country that has legal jurisdiction should negotiation fail.

Governments specify which language the agreement will be interpreted in and the dictionary that is used for all words. This provides a means of improving the communication as well as ensuring disputes are easier to settle. This does not always prevent misunderstandings because of the use of same words in different ways.

Planning for international projects must consider how the business will be conducted and the cultural aspects of all nationalities involved. It is interesting to note that many countries have regional cultures that vary within a country. Planning must account for the regional differences.

A checklist of items to address in the project plan would include the following:

- Technology and capacity of the people involved to complete tasks
- Skills and knowledge in the technical and managerial areas required
- Unique aspects of the culture that must be considered
- Transportation systems for shipping or delivery of the products
- Manufacturing materials in grade and quantity
- Quality assurance and quality control procedures
- Communication and reporting means, both tools and skills
- Political stability in the countries involved
- Laws regarding labor, taxes, duties and fees
- Government indemnification of loss for all reasons other than force majeure
- Currency used for payments and monetary unit used for accounting
- Time zone differences and the affect on project reporting
- Holidays and on-work periods of all countries
- Interfaces between and among participants of the project

Project planning must be accomplished in one language to assure the best understanding. Even among English-speaking countries, there will be differences in understanding the plan. The use of graphics and illustrations will promote understanding more than the use of words or semantic notations.

2.5.7 Key User Questions

1. When are projects considered "international projects"?
2. What advantage is there for two governments to cooperate through international projects?
3. When disputes occur in a contract between the parent organization and a contractor in another country, where would the court be located for a legal decision?
4. What is perhaps the most challenging area for planning and executing international contracts and why?
5. When invoking quality standards, international projects must consider what aspects of quality?

2.5.8 Summary

Both governments and many private organizations seek international projects to leverage the capabilities of another country. This leveraging provides benefits to both countries as well as providing specific benefits to the participating organizations. Typically, the leveraging is for financial savings or gain.

International projects have both advantages and disadvantages for an organization starting and executing a project that spans national borders. These advantages and disadvantages need to be considered when the business case is being developed.

Attempting to transplant one's culture into another nation will be difficult and long-term at best. Work ethic varies widely, from hard working to relaxed and lethargic. The manner in which the different national cultures approach business and interpersonal relationships varies from very formal to informal. Customs of respect for others vary widely also.

Communication may be the most challenging area for the project. Even within the English-speaking nations, there is a wide variation in the use of words and terms. When a language is translated, the nuances of such areas as technical language or shop talk can be troublesome.

2.5.9 Annotated Bibliography

1. Zeitoun, Dr. Alaa A., "Managing Projects across Multi-National Cultures, A Unique Experience," *Proceedings of the 29th Annual Project Management Institute 1998 Seminars and Symposium*, Long Beach, CA, October 1998. This paper presents some unique aspects of managing international projects, including the qualifications for a project manager. Cultural aspects for Africa, Latin America, Middle East, and United States are described.

2. Kothari, Dhanu, and Romeo Mitchell, "Achieving Global Competitiveness through Project Management," *Proceedings of the 29th Annual Project Management Institute 1998 Seminars and Symposium,* Long Beach, CA, October 1998. This paper provides information on the trends in the global market arena. It addresses the project management core competencies needed for operating around the world.

2.6 WORKING IN PROJECTS

2.6.1 Introduction

Project work is unlike much of the production work, and is unique in its execution. It usually is not repetitive, but has some difference from prior projects. Project work is stimulating and challenging to most project practitioners.

> **Working in projects is working with change!**

2.6.2 Understanding Project Work for the Project Practitioner

Project practitioners, whether assigned full or part-time to a project, benefit from understanding the basic characteristics of the project as they relate to their responsibilities. Figure 2.11 depicts project characteristics to be understood.

Characteristics of a project that affect practitioners in meeting their obligations are:

• Projects focus on goals centered on time, cost, and quality. In literature, these goals are equal. In reality, whether stated or not, these goals are ranked. Time is the most important goal, i.e., product delivery. Quality of the project's product is second, i.e., the functionality and attributes that result from building to the design. Cost is the last goal to be considered.

   Project practitioners will receive tasks that are time constrained. Each task is dependent upon receiving something and providing something to

Project Practitioners Responsibilities

Understand Project Time, Cost, & Quality Goals

Recognize Skills Available & that Skills Assigned to a Project May be Different

Duration of Assignment to Project is Temporary

Understand Specific Work Assignment

Participate in Project Planning

Report Progress of Work

FIGURE 2.11   Characteristics to be understood by project practitioners.

another task. This input-output process synchronizes the task with other tasks.

• Project practitioners will be assigned based on the need for skills. Often the skill requirements are not an exact match with the project practitioner. There may be greater or lesser skills. This mismatch is not a mistake; it is the way the process works and the best method available for assigning people.

When assigned to a project, recognize that there will be some difference between the skill requested and your skills. One should only raise an objection when the task cannot be completed because of a lack of knowledge. However, it is appropriate to identify the mismatch for future assignments and quality of work.

• Projects are temporary in nature. Duration of the assignment to a project can be predicted based on the need for a particular skill set. The assignment may be to accomplish one task or a series of tasks.

• Project practitioners should anticipate when they would be released from the project. Release may be to a resource pool or to another project.

The goal should be to complete the current assignment as quickly and effectively as possible so as to contribute in other projects.

- Understanding the work requirement is critical to a project practitioner's successful contribution to the project. The fast paced nature of some projects may cause miscommunication of requirements and misinterpretation of requirements.

- Project practitioners must understand the specific assignment and can confirm that by giving feedback to the project manager. Such statements as "I understand the requirement to be this" is one way of providing this feedback loop. It may be important to understand some of the background on the project. This may be obtained in a similar fashion by stating "I need some more baseline information on the project. Where may I get this information?" Ensure the task is understood prior to beginning work for the best solution.

- Project practitioners participate in project planning meetings and project status meetings. Everyone has a responsibility to understand the information being discussed and to contribute to the success of the project. There is a tendency not to raise issues if it does not directly affect the person's individual work.

- Project practitioners have unique insight into the process and can make valuable contributions to planning and status meetings. When something is missing or going wrong, a person has a responsibility to identify the issue. This may be a missing interface specification or a leap over critical work in the project.

- Project participants will make reports of progress and status on the assigned tasks. This reporting of work accomplished is important to the project manager's ability to accurately portray the project to customers and senior managers.

- Reporting progress with accuracy and honesty is a basic requirement. The project procedures should define the process for collecting information and reporting that information to a project control person. Some projects need progress to be measured in 10 percent increments and amount of work to complete the task in hours. Other projects require more precision such as a computed percentage by a scheduling system. When asked for future projections, ensure they are realistic.

There are other unique aspects to projects. A project participant, however, should always ensure he or she meets the assigned responsibilities by understanding the requirement, ensuring their own skill sets are capable of performing the tasks, honest and accurate reporting of progress, and closure of the assigned task. Project practitioners should anticipate when his or her assignment to the project will be completed and plan for transition to other work.

2.6.3 Understanding the Project Manager

Project managers are different in management and leadership styles. The temporary nature of a project practitioner's assignment may limit the exposure to the project manager and may not show the project manager's style. Longer-term assignments means understanding what the project manager wants and how he or she expects delivery.

Being assigned as a member of the project team on a full-time basis requires that the project practitioner quickly becomes aware of the project manager's style. Understanding this style will allow rapid assimilation into the project environment and make the project practitioner more effective. Some means of understanding the project manager's style are:

- Observe how the project manager interfaces with others. Is it friendly or more formal? Is assigning tasks a request or an order, written or formal?
- How are meetings conducted? Rigid, free-flowing, formal, informal, tightly controlled, other?
- How enthusiastic is the project manager over the project? High energy and enthusiastic, or other?
- What communication style is exhibited by the project manager? Open, formal, informal, clear and concise, minimum information, respectful, and other? Are there mission-type instructions or detailed instructions?
- How demanding is the project manager? Expects perfection, expects timely delivery of performance and information, expects extraordinary effort on tasks, easy going and not demanding?
- What projects has this person previously managed and what were the results?
- What close associations does this person have with others, both internal and external to the project? No associations with others and considered a loner; some associations with others in a technical field; likes to associate only with people external to the organization?

Project practitioners need to know the project manager to be able to understand instructions and how to please the project manager. Often, good performance is not enough. A person needs to build relations and avoid confrontations over technical or management issues.

2.6.4 Project Practitioner's Growth

A project practitioner has an opportunity to grow both personally and professionally while working in projects. These growth opportunities are found in rapidly changing environments that continue to challenge the physical and intellectual being of a person. Meeting the challenges exercises the mind to find new and innovative ways of doing the work.

Projects provide the environment to grow. Work is non-repetitive and requires individual contribution to solve the problems. It is a team effort in the best sense, and through the team effort a person gains insights into others' thought processes. Everyone contributes to the outcome of a successful project.

Personal growth is gained through working with and respecting others' competencies. The dynamic nature of temporary assignments will expose a person to more individuals than the production type work. These contacts with others expands the number of colleagues and the network of professionals.

Many projects are on the forefront of technology, leading the industry in technological advances. Working in these projects broadens one's knowledge of a technical field and the existing state of the art for that technology. Exposure to new technology pushes the imagination to see where and how far technology may advance in the next 10 years.

In many situations, people have infrequent opportunity to communicate with different groups. In projects, project practitioners may be required to brief customers, senior managers, and others on their work. This drives a person to improve communication skills and gives the opportunity to make both information and decision briefings. A person learns better communication methods in fast moving situations.

Projects provide opportunities for project practitioners to be recognized for their work and contributions. The project team will acknowledge the work and, if exceptional, senior management will learn of the contribution. There may be plaques and certificates to recognize and publicly acknowledge a person's contributions.

Last, the knowledge and skills learned while in projects will build on a project practitioner's capability for future opportunities. The project practitioner will learn more and faster in the fast-paced environment of projects when there is total involvement in the work. Involvement in solving problems while performing the work will give a new depth to an individual's understanding of the organization's business.

2.6.5 Key User Questions

1. Would a passive person without imagination work well on a high-technology project? Why or why not?

2. What are some of the benefits for personal growth working on a project?

3. What are some of the benefits for professional growth working on a project?

4. Why is project work more broadening professionally than working in a production area or a research laboratory?

5. How may a person assigned temporarily become rapidly oriented to the project and project manager?

2.6.6 Summary

Most people working in projects are not trained in project management techniques, but are individuals who perform selected elements of the project's work. These individuals are project practitioners, i.e., those who bring skill sets other than project management and who may have little experience or understanding of project management. Getting started in the project may be difficult for the project practitioner.

Individuals assigned to projects have the expectation that their skill sets will be used. Assignments are made to be a "best fit", which does not always mean a good fit.

General skills that project practitioners will need are communication skills, progress measurement and reporting skills, and interpersonal skills. These "soft skills" may not be present in all individuals and may require some team work to help the newly assigned project practitioner.

Specific skills that may be useful in a project environment are analytical skills to assess problems, problem solving skills for finding the best solutions, and research skills to collect objective information.

2.6.7 Annotated Bibliography

1. West, Jimmie L., "Building a High-Performing Project Team," *Field Guide to Project Management* (New York, NY: Van Nostrand Reinhold, 1998), chap. 18, pp. 239–254. This chapter highlights the means to develop the project team and understand the stages of development.

2. Yourzak, Robert J. "Motivation in the Project Environment," in *Field Guide to Project Management,* 2nd ed. (New York, NY: John Wiley & Sons, 2004). This chapter highlights motivational factors in teams and lists the results of surveys regarding what motivates people. A careful perusal of this chapter, based on research carried out by the author, will provide useful insight into how a person's motivation impacts performance.

2.7   PROJECT OFFICE

2.7.1   Introduction

The project office is coming into use in many organizations to achieve the benefits of consolidation for many project management functions. Centralization of the functions permits an organization to gain consistency in practices as well as using common standards for such items as schedules and reports. Benefits derived from the project office are dependent upon the functions, structure, and resources assigned.

A project office is defined by its most common current usage.

> *The Project Office is a collection of project functions that serve project managers in the performance of their duties. It relieves project managers of routine, critical functions while establishing consistent and uniform practices in the functions performed. It may also serve as a central repository that "contracts out" to line organizations.*

A project office is typically started to reduce the cost of project management functions in an organization and improve the quality of project information provided to senior management. The actual implementation of the project office, however, achieves benefits that extend beyond this component. One of the primary benefits is that senior management receives uniform reports for decision making. Consistency for the organization, as well as all projects, provides a greater project management capability for the organization.

A project office is a support group that provides services to project managers, senior managers, and functional managers working on projects. The project office is usually not a decision making body that replaces either senior managers or project managers. It does prepare information and reports, however, that support the decision making process used by senior and project managers.

A project office may also be called by several different names. Some of the more common names are

- Project Support Office
- Program Support Office
- Project Management Office
- Project Management Support Office
- Program Office

2.7.2 What is a Project Office?

A project office is what an organization wants it to be. It can be as simple as a few people preparing and maintaining schedules to several people performing planning, reporting, quality assurance, collecting performance information, and functioning as the communication center for several projects. The project office is defined by the business needs for the organization and grows with those needs.

Functions that a project office performs are shown in Table 2.5.

TABLE 2.5 Functions Performed by the Project Office

Work area	Services rendered
Project planning support	Maintain methodology and variances from standard practices
	Store and update templates for planning
	Store and retrieve lessons learned
	Maintain progress metrics
	Provide cost and time estimating consultation
Project audit	Process checklist for each milestone
	Support project intervention for deficiencies
	Maintain corrective action log
Project control support	Maintain change control log and follow-up
	Maintain change control actions and closure items
	Validate timesheet entries and follow-up
	Conduct trend analysis on progress
	Support development of status reports
	Conduct distillation and summary of all projects
Project team support	Participate in teambuilding exercises
	Mentor and coach in project management techniques
Project management skills development	Conduct skills assessment for future projects
	Participate in project performance evaluations
	Support continuous learning by project teams
Project management process maintenance	Maintain project methodology baseline and changes
	Identify general training requirements for the process
	Maintain policies, procedures, and practices for project management
	Institutionalize project management
Project management tools	Conduct tool needs assessment for projects and organization
	Evaluate current tools for adequacy and compatibility with projects
	Coordinate tool training for project teams
	Provide technical expertise on tools

TABLE 2.5 Functions Performed by the Project Office (*Continued*)

Work area	Services rendered
Project executive support	Recommend priorities for new projects Recommend cross-project resource allocation Review project performance evaluations Serve as the project management consultant to executives
Project reports	Collect and validate information on periodic or continuous basis Prepare and distribute reports Prepare reports for senior management
Issues	Establish a log and track issues for the project manager Close issues after resolution Maintain history of issues for reference
Risk	Conduct risk assessment, quantification, and mitigation Track risks and closure of risk events Prepare contingency plans
Action items	Establish log and track action items Close action items after completion Maintain history of actions
Communication	Prepare communication plan Update communication plan as needed Distribute reports to stakeholders Maintain record copies of communications
Schedules	Prepare schedules in an automated system Maintain schedule status based on reported progress Produce schedules as required
Cost	Prepare budget Maintain budget based on expenditures Report budget status
Quality	Prepare Quality Assurance and Quality Control Plans Maintain QA/QC Plans Prepare test and demonstration plans Maintain records of tests
Internal project management consulting	Provide project management expertise to projects in all phases to improve plans, recover projects, advise on techniques, and advise on successes.

There are other functions that may be included in the project office. Determining whether the new items can be more effectively managed by the project manager is the criterion. One should not include items in the project office that do not directly support the projects.

2.7.3 Examples of Project Offices

Two project offices started within major organizations, one an energy company and the other a software company, reflect the differences in organizational needs. The energy company had a need to establish a project office to manage new and maintenance projects. The project management capability at the start was low because of the lack of formalized training for the project managers, few procedures because the company had recently reorganized, few techniques and practices because the former organization was distributed across several states, and weaknesses in organiation practices that negatively impacted projects. A consultant was hired to form the core of the project office and provide scheduling support.

Over time, the project office assumed the role of coordinator with other business functions such as marketing, contracting, finance, and warehousing. Senior management saw the need for expanding the project office functions as depicted in Table 2.6.

This project office functioned very well in an environment that was initially at a low level of project management maturity. The willingness of the senior managers, project managers, and functional department managers to change to better management practices was probably the greatest factor in the successful implementation of a project office. The skills and knowledge of the consultant who established the project office was also a major contributor to gaining the confidence of those being supported.

The software company's project office was less ambitious and the scope of the project office was less than the energy company's efforts. The primary difference being that the energy company started from little formal project management activity and the software company had an ongoing more formal project management capability. The software company, however, was not consistent in its practices and used project managers in technical and scheduling roles. The software company was attempting to change from random practices to uniform practices. Senior management did not understand their need for project management.

The software company established a project office with four consultants with the goal of transferring the work to internal employees. Functions planned for the project office are listed in Table 2.7.

This project office was challenged for several reasons. Senior management did not understand the functions of a project office and did not support changes to current practices by project managers. Several project

TABLE 2.6 Project Office at an Energy Company

Project Office at an Energy Company

Function	Description
Scheduling	Prepare schedule templates for large and small projects. Maintain schedules prepared by project managers Support project managers with scheduling problems
Costing	Support project managers' cost estimating work
Reports	Prepare electronic reports and maintain same in the Intranet Advise Senior Management on the meaning of electronic reports
Briefings	Prepare and present information briefings on the project management system/process Prepare and present project information for all projects (combined project summaries)
Coordination/Liaison	Conduct coordination with other company functions to ensure consistency and compatibility of functions Coordinate changes to the project management practices with other functions
Documentation	Assemble and maintain core project management practices Prepare and maintain project management best practices Prepare and maintain project management procedures
Project management knowledge center	Function as the center for project management knowledge and the primary source of current project management practices

managers viewed the project office as a hindrance to their practices as well as a threat to their jobs. Most of the project managers were well qualified in technical skills, but were familiar with only basic scheduling skills.

Resistance to change resulted in the project office being reduced from four consultants to three, then to two and finally one. The last functions performed by consultants were primarily administrative in nature while other functions were performed by one employee. The project office provided minimum services to the organization and in the end served more as an administrative office than a source of project management services.

TABLE 2.7 Project Office at a Software Company

Project Office at a Software Company	
Function	Description
Project scheduling	Develop draft schedules for project managers
	Identify interfaces between schedules and document them
	Refine schedules consistent with project management methodology
	Maintain schedules for project managers (status and progress)
Reports	Collect information for reports to senior managers
	Format monthly reports for senior managers
	Prepare briefings to inform senior managers of project status (Note: senior managers would not accept Gantt Chart formats, but required information to be formatted in tabular or narrative form.)
Project management support	Advise and support project managers on best practices for managing projects
	Provide advice on project management practices, as requested
	Provide advice to senior management, as requested
Documentation support	Establish and maintain Issues Log
	Establish and maintain Action Log
	Establish and maintain Project Interface Log
	Establish and maintain Change Control Log (cost and schedule only)
	Prepare a Risk Management Methodology for projects

2.7.4 Project Office Implementation

Starting a project office and moving it into a mature capability for an organization requires several steps. These steps are:

- Define the services to be provided by the project office. Get senior management and project manager agreement on the services. The initial project office may evolve, but it is important to have an agreed upon scope of work.

- Define the staffing skills and roles for the project office personnel. The skills and roles of the assigned personnel will determine the amount of support that can be provided.
- Establish and announce the start of the project office. Have a plan for early successes in supporting project managers and senior managers. Celebrate the early successes.
- Work closely with senior and project managers to understand their needs and meet those needs. As project managers are relieved of the routine work that the project office performs, additional requirements may emerge.
- Grow the project office's services through continually meeting business needs while providing services to the project managers.
- Refine the skills and roles of the project office as involvement continues with the customers of the project office.
- Deliver only the best products to customers.

The project office must have the support of senior management for initiation. Success of the project office will be determined by the customers. Customers are those individuals receiving products and services from the project office. The primary customers are:

- Senior managers
- Project managers or leaders
- Project team members
- Functional managers
- Stakeholders such as the recipient of the projects' products

Implementation is top down for the authority to start and maintain the project office. The continuation of the project office as a viable entity of the organization is determined by the customers. If the customers are not pleased with services, senior management's support will erode and the project office will be discontinued.

2.7.5 Key User Questions

1. Does your organization have or need a project office to consolidate routine project management functions to free project managers of some critical tasks?
2. Can your organization benefit from a project office that serves as a center of knowledge and expertise?

3. What benefits do you see delivered from a project office in your organization?
4. What skills should be included in a project office in your organization?
5. How would you select the functions that would be placed in a project office in your organization?

2.7.6 Summary

A project office serves the business needs of the organization and relieves project managers of routine tasks. Project offices also serve the needs of executives in information gathering and formatting to understand the progress being made on projects. The project office is defined by the organization to meet the business needs of the organization and to further the business goals. Extraneous functions should not be included in the project office unless they directly support the projects.

The consolidation of several functions of project management brings about consistency in practices as well as application of project management standards. This consolidation can bring about more efficiencies as well as better products to support the projects. The consolidation of functions does not take away the decisions for senior of project managers. The products of the project office support those decisions.

2.7.7 Annotated Bibliography

1. deGuzman, Melvin, *Nuts and Bolts Series 2—The Project Management Office: Gaining the Competitive Edge* (Fairfax, VA: ESI International, 1999), 70 pp. This is a guide for project stakeholders interested in starting a project management office. It focuses on the basic components of a project management office and explains how organizational and individual effectiveness can be improved.
2. Frame, J. Davidson, and Bill Christopher, eds., *The Project Office* (*Best Management Practices*) (Crisp Publications, April 1998), 88 pp. This book provides insight on the start up of a project office, with emphasis on planning for and implementing the project office.

2.8 THE PROJECT TEAM CULTURE

2.8.1 Introduction

Culture is a set of behaviors that people have and strive for in their society—whether that society is a country, an institution, a corporation, or a project team. Culture includes the totality of knowledge, belief, art, morals, law, custom, and other capabilities and attitudes expressed by people in a project. Characteristics of a project culture include:

- An interdependency with the culture of the enterprise to which the project belongs
- Because projects are the major ways in which an enterprise changes, a project culture must also change as new environmental challenges arise.
- A project culture is reflected in the manner of the people, visible or audible behavior patterns, policies, procedures, charters, plans, leadership style, and the individual and collective roles that people perform on the project team.
- A project team culture is a community, a pattern of social interaction arising out of shared interests, mutual obligations, cooperation, friendships, and work challenges.
- A project team exists as an organizational force for continuous improvement and constant change in helping to position the enterprise for its future.
- Shared interests, a lack of dominant individual egos, and the power of team trust and absolute loyalty characterizes high performing teams.
- The project team is a "body of companions" dedicated to creating something for the enterprise that does not currently exist.

> **Let each person working on a project team become a team player**

2.8.2   Strengthening the Project Team Culture

There are no magical solutions on how to build and maintain a strong supportive culture on a project. However, a few suggestions follow on what can be done:

- Keep the team members informed of the status of the project, to include both good and bad news.
- Promote the sharing of ideas, problems, opportunities, and interests among the project team, especially new members.
- Have some social activities for the project team members, but do not overdo this. Do not interfere with the personal lives of the team members.

- Cultivate an informal but disciplined working relationship to include the use of first names, networking among project stakeholders, and the cultivation of respect and dignity among all of the stakeholders.
- Avoid the use of management language and demeanor that puts a hierarchical stamp on the team and its work.
- As a team leader, advise, coach, mentor, prompt, and facilitate an environment where team members feel supported, encouraged, and rewarded.
- Keep the team informed on what the competitors are doing, particularly during the proposal submission stage of the project.
- Encourage the senior executives to visit the project, and give the team members the maximum opportunity to brief upper management.
- Be alert to the changes in attitudes that are required to deal with the expected changes in the culture.

2.8.3   The Change Issue

Unfortunately there are some people who do not like change—even on a project team—which is dealing with change in products, services, and organizational processes. People have many ways to rationalize the status quo such as indicated in Table 2.8.

TABLE 2.8   Don't Change

Don't rock the boat.
The way to get along is to go along.
Why change?
I am only a couple of years from retirement!
Things were better in the old days.
What we are doing now is good enough.
I like things the way they are.

2.8.4   End of the "Command and Control" Mindset

We are seeing the beginning of a major change in the attitudes and manner in which managers operate. These changes are causing changes in the culture of organizations, and in project teams. Table 2.9 summarizes the changes in management philosophies of the old world of "command and control" and the new world of "consensus and consent."

Project managers have to be change managers in the design and implementation of a project culture, which complements the enterprise cul-

TABLE 2.9 Changes in Management and Leadership Philosophy

The Old World Command and Control	The New World Consensus and Consent
Believes "I'm in charge."	Believes "I facilitate."
Believes "I make decisions."	Believes in maximum decentralization
Delegates authority.	of decisions.
Executes management functions.	Empowers people.
Believes leadership should be	Believes that teams execute
hierarchical.	management functions.
Practices "Theory X."	Believes that leadership should be
Exercises de jure (legal) authority.	widely dispersed.
Believes in hierarchical structure.	Believes in "Theory Y."
Believes that organizations should be	Exercises de facto (influential)
organized around function.	authority.
Follows an autocratic management style.	Believes in teams/matrix organizations.
Emphasizes individual manager's roles.	Believes that organizations should be
Believes that a manager motivates	organized around processes.
people.	Follows a participative management
Stability.	style.
Believes in single-skill tasks.	Emphasizes collective roles.
Believes "I direct."	Believes in self-motivation.
	Change.
Distrusts people.	Believes in multiple-skill tasks.
	Allows team to make decisions.
	Believes that a manager leads, as
	opposed to directs.
	Trusts people.

Source: David I. Cleland, *Strategic Management of Teams* (New York, NY: John Wiley & Sons, 1996), p. 249.

ture in which the project operates. Certain actions can be taken to help develop and maintain such a culture:

- Participate with the team members in designing and implementing a disciplined approach to planning, organizing, and controlling a project using a project management system. See Section 7.10 for a description of the Project Management System.

- Provide as much leeway as possible for the team members to do their work.

- Ensure that team members understand their authority and responsibility.

- Give team members the opportunity to actively participate in project reviews, strategy meetings, and meetings with customers.

- Make sure that the team members know what is expected of them regarding their project work package.
- Have the team members participate to the maximum in the decisions involving the project—thus facilitating their "buy in" to such decisions.
- Encourage the use of brainstorming approaches to solving problems, exploiting opportunities and challenges that face the project.
- Provide timely feedback to the team members.
- Ensure that the team members have the resources they require.
- Recognize the importance of "people-related" cultural factors that influence how people relate to the culture. These factors include:
  - Rewarding useful ideas
  - Encouraging candid expression of ideas
  - Promptly following up on team and member concerns
  - Assisting in idea development
  - Accepting different ideas—listening to that team member who is "marching to a different drummer"
  - Encouraging risk taking
  - Providing opportunities for professional growth and broadening experiences on the project
  - Encouraging interaction with the project stakeholders so that there is an appreciation by the team members of the project's breadth and depth.

2.8.5 Characteristics of a Strong Project Culture

In Table 2.10 the characteristics of a strong project culture are enumerated.

2.8.6 Cultural Features

Cultural features of an organization and a project are influenced by:

- The management leadership-and-follower style practiced by key managers and professionals
- The example set by leaders of the organization
- The attitudes displayed and communicated by key managers in their management of the organization
- The managerial and professional competencies
- The beliefs held by key managers and professionals
- The organizational plans, policies, procedures, rules, and strategies

TABLE 2.10   Project Management Culture

- There is an excitement about project management in the enterprise as the principal way of dealing with product, service, and process change.
- There is a proven track record in the organization in using projects as drivers of change in the organization.
- Appropriate project management organizational strategies, policies, procedures, and resource allocation initiatives have been developed and are being used as the hallmark of project management.
- Extraordinary efforts have been undertaken and continued in the clarification of authority, responsibility, and accountability for the project team members and other stakeholders.
- Full support of the use of project management has been recognized and fostered by the senior members of the organization.
- Unusual efforts are undertaken by all of the project stakeholders to maintain effective communication about the project and its status.
- Suitable training programs are in place to upgrade the knowledge, skills, and attitudes of team members.
- Appropriate merit evaluation and reward systems are in place which fully consider the contribution that team members are expected to make in the project.
- Project management has reached a level of maturity in the organization where it is recognized as "simply the way we do things around here."
- Experience as a project manager is a requirement for promotion to higher level management positions in the organization.

- The political, legal, social, technological, and economic systems with which the members of an organization interface
- The perceived and/or actual characteristics of the organization
- Quality and quantity of the resources (human and nonhuman) consumed in the pursuit of the organization's mission, objectives, goals, and strategies
- The knowledge, skills, and experiences of members of the organization
- Communication patterns
- Formal and informal roles

2.8.7   Key User Questions

1. Do the senior managers of the organization recognize and appreciate the importance of a supportive cultural ambience for the organization

and the project teams, which facilitate individual support and belief in the objectives and goals of the enterprise?

2. Does suitable documentation exist which stipulates the authority, responsibility, and accountability expectations of members of the project team, and guidance on how that documentation can impact the culture of the organization?

3. Has the project manager done everything that can be done to enhance the culture of the project team?

4. If a member of a project team in the organization were asked to describe the culture of the organization, would that individual be able to do so?

5. Does the culture of the organization reflect a "command and control" or a "consensus and consent" ambience?

2.8.8   Summary

In this section, a culture was defined as a set of refined behaviors that people have in their society, such as an organization or project. Furthermore, a culture is the environment of beliefs, customs, knowledge, practices, and conventionalized behavior of a particular social group—such as a project.

A few of the principal characteristics of a supportive project culture were suggested in the section. It was noted that culture of an enterprise and a project should be mutually supportive. Suggested strategies were presented to facilitate the development and propagation of a strong culture in a project. These strategies were suggested in the spirit of not being magical solutions, but if followed could help to reinforce a positive culture in the project team.

2.8.9   Annotated Bibliography

1. Cleland, David I., and Lewis R. Ireland, *Project Management: Strategic Design and Implementation,* 4th ed. (New York, NY: McGraw-Hill, 2002), chap. 20. This chapter gives a brief description of how the cultural ambience of an enterprise can impact the project team. Some of the cultural characteristics that can be supportive of a project team are provided.

2. Kotter, John P., and James L. Heskett, *Corporate Culture and Performance* (New York, NY: Free Press, 1992). This book provides an excellent overview of the concept of a culture and how that culture can impact the enterprise's performance, as well as the people in the enterprise. The book can serve as a primer for anyone who has a serious interest in how and why cultures impact an organization.

SECTION 3
ALTERNATIVE PROJECT APPLICATIONS

3.1 ALTERNATIVE PROJECT TEAMS

3.1.1 Introduction

> **Project teams are both traditional and non-traditional**

Alternative teams are becoming commonplace as a means to deal with the cross-functional and cross-organizational initiatives that enable the enterprise to deal with change.

The main focus of this Handbook has been to deal with the traditional project teams. In this section paradigm will be presented that describes the general usage to which non-traditional teams can be put, as well as the areas of organizational effort in which these teams are used.

3.1.2 Characteristics of Traditional Project Teams

A traditional project team is one in which custom and usage has been demonstrated in the past, primarily from the construction and defense industries. These teams can be described in the following way:

- A substantial body of knowledge exists which describes why and how such teams can be utilized.
- Typically these project teams involve the design, development, and construction (production) of physical entities for the enterprise.
- A traditional life cycle is found in these projects.
- Substantial financial, human, and other resources have to be marshaled for the conduct of these projects.
- Construction projects are the best example of these traditional projects.
- The use of these teams has evolved from an early beginning throughout history, even though the theory of principles and processes were developed early in the 1950s.
- When people think of a team there has been a tendency to think only of project teams. But that perception is changing.

3.1.3 Characteristics of Non-Traditional Project Teams

A non-traditional team has many of the characteristics of the traditional project teams. There are, however, some singular characteristics of these teams:

- The organizational element with which these teams deal is already in existence, usually in the form of organizational processes rather than physical entities.
- The teams are directed to improve the efficiency and effectiveness of an organizational process. The work of the team begins immediately in dealing with a problem and opportunity.
- Although "hardware" is involved, the teams deal principally with the identification and use of resources in meeting organizational objectives and goals.

- The deliverables of these teams usually are reports which outline recommendations for the improvement of the use of resources.
- The teams are used in many diverse enterprise purposes.
- The teams have direct vital links with the design and execution of operational and strategic initiatives in the enterprise.
- Many times the recommendations of these teams brings about significant changes in the individual and collective roles carried out by the members of the enterprise.
- The teams, and the results of their use, can have a major impact on the culture of the enterprise.
- These non-traditional teams deal with, and cause changes in the way the enterprise uses resources to support mission, objectives, and goals.

3.1.4   The Work of Non-traditional Teams

The work carried out by these teams is varied and important in keeping the enterprise's processes efficient and timely. This work is summarized in Table 3.1 and is described below:

TABLE 3.1   Non-Traditional Teams

Market assessment
Competitive assessment
Organizational strengths and weaknesses
Benchmarking
Establish performance standards
Vision quest
Stakeholder evaluation
Market research
Product-service-process development
Business process reengineering
Crisis management
Self-managed production initiatives
Resolution of organizational issues
Quality improvement
Audit processes
Senior-level decision making
New business development initiatives

Market assessment. Discernment and development of the likely changes in the enterprise's market.

Competitive assessment. Examining the strengths, weaknesses, and probable strategies of the competition to be melded into the organization's competitive strategies.

Organizational strengths and weaknesses. Discovering and evaluating the competencies of the organization vis-à-vis the competition to include recommended strategies coming out of the analysis.

Benchmarking. Review of the performance of the "best in the industry" organizations to include what operational and strategic abilities enable them to perform so well.

Establish performance standards. Identification, development, and dissemination of the performance criteria by which the organization's ability to produce results is improved.

Vision quest. Discernment of the general direction of the future of the organization in terms of what course to follow to reach a desired end.

Stakeholder evaluation. Find out who the stakeholders are and what their likely interest and capacity is to influence the organization's competency.

Market research. Assessment of the possible and probable opportunities for improved or new products and services for the organization.

Product-Service-Process development. The simultaneous development of product, service, and organizational process initiatives for the organization to support its objectives and mission.

Business process reengineering. Used to bring about a fundamental rethinking and radical redesign of organizational processes.

Crises management. Teams that are appointed and trained to deal with real and potential organization crises.

Self-managed production initiatives. Using teams to improve manufacturing (production) operations.

Resolution of organizational issues. Ad hoc teams that are used to solve organizational problems or opportunities.

Quality improvement. Use of teams to improve and integrate quality improvements in products, services, and processes.

Audit processes. Teams that evaluate the competency of organizations, programs, projects, and organizational processes.

Senior-level decision making. Using teams of senior executives to enhance the synergies in the development and execution of organizational strategies.

New business development initiatives. Teams that are used to explore the design and development of new business ventures for the organization.

From the foregoing list, it should be clear that teams are an organizational design strategy that can deal with a wide variety of operational and strategic initiatives, and they have been successful. *Fortune* magazine noted that "The ability to organize employees in innovative and flexible ways and the enthusiasm with which so many American companies have deployed self-managing teams is why U.S. industry is looking so competitive." (Rahuyl Jacob, "Corporate Reputations," *Fortune*, March 6, 1995, pp. 54–64.)

3.1.5   The Personal Impact of Teams

Careers are being impacted by the growing use of both traditional and non-traditional teams. There are enhanced opportunities for more people to try their hand at management and leadership positions. People who serve on these teams will be expected to bring extraordinary knowledge, skills, and attitudes to the team, and to the organization such as:

- Have ability to work with diverse groups of stakeholders
- Have sufficient technical skills to work on the team and gain experience in the application of their skills to dealing with change in the organization
- Gain understanding of what counts for success in the organization, to include a better understanding of what is needed to be "profitable"
- Ability to leverage their knowledge, skills, and attitudes through enhanced opportunities for communication challenges, networking, building alliances, and become a contributing team member
- Recognize that being successful in a career depends less on organizational position, and more on the competencies brought to the organization.

3.1.6   Key User Questions

1. Has the management of the enterprise given consideration to the use of non-traditional teams to deal with product, service, and organizational processes?
2. If non-traditional teams are not being used in the organization, what is the reason for such non-use?

3. Do the senior members of management understand and appreciate what teams can do for the organization?

4. If teams are being used on a regular basis, what results have such teams produced—if any?

5. Have the roles of the supervisors and other management personnel been changed as a result of the use of teams?

3.1.7 Summary

In this section, the use of non-traditional project teams has been explored and examined, particularly examining the alternative uses to which teams can be put. A wide variety of organizational needs were presented along with a recommendation of how teams could help in the management of these needs. The section closed with an explanation of what service on teams can do for the career of individuals.

3.1.8 Annotated Bibliography

1. Cleland, David I., *Strategic Management of Teams* (New York, NY: John Wiley & Sons, 1996). This book provides considerable insight into the enormous potential of using alternative teams in the management of the organization. It provides the reader with the concepts, processes, tools, and techniques needed to integrate teams into the overall strategic management of the organization. A must read for those managers who wish to use teams beyond the traditional project team application.

2. Wellins, Richard S., William C. Byham, and George R. Dixon, *Inside Teams* (San Francisco, CA: Jossey-Bass, 1994). This book offers a behind-the-scenes look at how 20 of the world's best team-based companies realized results through teamwork. Each case history tells why teams were chosen as a competitive strategy, how teams started, the problems and challenges encountered, what was learned in the use of teams, and how the use of teams impacted the bottom line of the enterprise. Written expressly for managers, team members, and human resource specialists, *Inside Teams* provides benchmarks for designing and using teams.

3.2 REENGINEERING THROUGH PROJECT TEAMS

3.2.1 Introduction

This section examines the basics of reengineering defined as the fundamental rethinking and radical redesign of business processes to achieve dramatic improvements in organizational performance such as cost, quality, service, and speed. (Michael Hammer and James Champy, *Reengineering the Corporation: A Manifesto for Business Revolution* (New York, NY: Harper Business, 1994), pp. 31–32.)

3.2.2 Background

Although the popularity of reengineering teams, per se, has diminished since its inception, the idea has become rooted in the theory and practice of management.

> **Reengineering is an in-depth "out of the box" examination and redesign of organizational processes**

The use of a reengineering team usually centers around key organizational processes of the organization:

- Asking basic questions such as "Why do we do what we do?" and "Why do we do it the way we do?"
- Disregarding existing organizational designs, strategies, policies, and protocols and inventing new ways of doing work.
- Concentrating on organizational processes which are a collection of activities that takes input and creates an output that is of value to the organization and its customers. An order entry protocol is an example of an organizational process.
- Achieving dramatic improvements in organizational performance.

Reengineering starts by asking certain basic questions about the organization's mission, objectives, goals, and strategies, such as the following:

1. What business are we in?
2. Why are we in this business?
3. Why are we working as we do?
4. Are there better ways to do our work?
5. What is the basic organizational documentation that guides the way that we work?
6. What can be changed to bring about enhanced performance in the organization?
7. How can our organization be examined to determine how well our organizational processes are carried out?

3.2.3 The Paradox of Reengineering

When project teams are used as the focus for reengineering strategies, enterprise work can be rearranged or even eliminated. Team members and other members of the enterprise participating in a reengineering initiative can have mixed emotions about the results of such initiative. Their jobs and the jobs of their peer groups may be reassigned or eliminated. Guidelines for dealing with this paradox include the following strategies:

- Keep the people informed about the purpose, process, and potential outcomes of the reengineering initiatives.
- Maximize the participation of the people on the reengineering teams to include both planning and execution phases.
- Ensure that people likely to be displaced or eliminated are provided assistance to deal with such uncertainties in their life.
- Communicate frequently with organization members to include careful listening to their suggestions, problems, gripes, and attitudes concerning the reengineering initiative.
- If possible, benchmark some other organizations that are engaged in reengineering and show how these organizations have fared.
- Tell people the truth, maintain open agendas, share information, and what the likely outcome of the reengineering initiatives could be.

3.2.4 Key Messages of Reengineering

Hammer and Champy have put forth some key messages regarding reengineering:

- Managers need to abandon traditional organizational paradigms, operating policies and procedures, and create new ones centered around the integration of organizational processes to do the job.
- The classical division of labor as a way of breaking work up into small units for assignment to specialists needs to be augmented with analysis of the relevant processes required to create value.
- The traditional, old ways of managing the organization do not work anymore. New paradigms are needed.
- The key to success in the modern organization is how the work processes and the people are aligned.
- Reengineering is a new journey, starting with a new road map.
- Reengineering is not continuous incremental improvement—it is major leaps in improvement.
- Organizational processes are the key to reengineering—traditional organizations focus on tasks, jobs, people, and structures.
- Reengineering requires that the following question be asked continually: "Who are our internal and external customers?"
- Reengineering is a new paradigm that goes beyond the traditional delayering, reorganizing, and flattening organizational strategies that worked in earlier days. ("The Promise of Reengineering," *Fortune*, March 5, 1993, pp. 94–97. This is a book review of Michael Hammer and James Champy, *Reengineering the Corporation: A Manifesto for Business Revolution* (New York, NY: Harper Business, 1993).)

3.2.5 Project Team-Driven Reengineering

Process reengineering is a project-team endeavor. Since reengineering work cuts across the functional elements of the enterprise, and even goes out to other organizations that work with the enterprise, a project team provides an appropriate organizational design to focus on the work of reengineering. Other organization teams can contribute to a successful reengineering activity:

- Project teams to design and build capital equipment and facilities.
- Concurrent engineering teams to provide the means to work through organizational processes and functions in conceptualizing, designing, manufacturing, and marketing goods and services to commercialize improved products and services sooner.
- Benchmarking teams to determine how well the organization performs compared to competitors and "best in the industry" producers.

- Self-Managed Production teams which can bring about dramatic improvements in the quality and output of products and services.

3.2.6 Reengineering Basics

There are a few basics about the characteristics of reengineering teams:

- Specific and measurable objectives and goals have to be established.
- The team members must be committed to the reengineering initiatives.
- The team membership should be drawn from the key functions involved in the reengineering effort, and should be easy to convene, interact with, and communicate with as a participating body of experts.
- A working philosophy should be present, which establishes how people are expected to interact, how decisions will be made, how analysis will be done, and the expectations for results.
- Authority and responsibility needs to be clearly established and understood by all concerned.

3.2.7 Reengineering Life Cycle Phases

A reengineering initiative has several phases in its life cycle.

Figure 3.1 suggests a flow of major events in a reengineering project. These events are described more fully below.

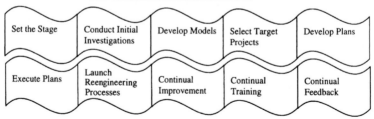

FIGURE 3.1   Engineering life cycle flow.

- **Set the stage** through the development and promulgation of a team plan.
- **Make initial investigations** into the reengineering targets in the enterprise.

- **Develop models** which reflect how and why the reengineering is to be carried out.
- **Select target reengineering projects** such as:
  - Order entry procedures
  - Product and service development
  - Procurement practices
  - Engineering and design of capital facilities
  - Manufacturing efficiency and effectiveness
  - Account receivables
  - Inventory practices
  - Management of projects
  - Materials handling
  - Marketing and sales management
  - Project management practices
- **Develop plans and strategies** for the improvement of identified and desired organizational processes.
- **Execute the plans and strategies** designed to lead to reengineering models in the enterprise.
- **Launch the reengineered process** for the organization.
- **Continually improve** the reengineered process.
- **Pursue continuous training** and indoctrination programs to maintain the effectiveness of the reengineered process.

Reengineering initiatives can have both immediate and subsequent impact on the organization and its stakeholders.

3.2.8 The Trigger Effects of Reengineering

When reengineering is properly carried out there are important factors and forces that are touched off. These include:

- Processes are emphasized and managed.
- Disciplines, functions, and departments become primarily organizations to maintain centers of excellence that provide a focus for operational and strategic processes.
- Single-task jobs disappear and are replaced by multi-skilled jobs.
- Training, retraining, and education become more critical and are success factors in improving productivity.
- People become much less dependent on their supervisors and managers; they become empowered and think and act like managers.

- The ability to produce competitive results becomes the basis for the reward system in the organization.
- Organizations have much less hierarchy, are flatter, and more dependent on team-driven initiatives.
- Values change when people become empowered and work through teams; these values reduce provincialism and territory concerns and place a high value on creativity, innovation, and individual responsibility for results.
- The role of executives changes in part from managers to leaders, who empower, facilitate, coach, teach, and work hard at providing an environment in which people are challenged and see a relationship between their work and the output of the organization.

3.2.9 Key User Questions

1. What is the basis for the launching of reengineering initiatives in the organization?
2. Are properly empowered project teams being used to provide a focal point for the design and execution of reengineering projects?
3. What are the most important organizational processes that should be first examined under the reengineering initiative?
4. Are the key managers in the organization, including the project and functional managers, committed and dedicated to making the reengineering effort a success?
5. What are the alternatives to the design and execution of a reengineering effort in the organization, and do these alternatives hold any promise for improving organizational processes?

3.2.10 Summary

In this section, the use of project teams in a reengineering initiatives was described. A strategy for how such initiatives should be conducted, and guidance for how a project team should manage such initiatives was presented. Finally, the "trigger effects" within the organization undergoing reengineering projects was described.

3.2.11 Annotated Bibliography

1. Hammer, Michael, and James Champy, *Reengineering the Corporation: A Manifesto for Business Revolution* (New York, NY: Harper Business, 1993). This is the book that launched the reengineering initiative. The book contains overall guidance as well as detailed instruction on why and how reengineering projects should be launched.
2. Cleland, David I., *Strategic Management of Teams* (New York, NY: John Wiley & Sons, 1996), chap. 7. This chapter describes how project teams can be used as the focus for reengineering initiatives. Guidance is provided on how teams can be set up, launched, and complete their objectives and goals in reengineering activities.

3.3 CONCURRENT ENGINEERING

3.3.1 Introduction

Concurrent Engineering (CE) is a systematic, simultaneous approach to the integrated design of products/services and associated organizational processes such as manufacturing, procurement, finance, testing and after-sales services. In CE, a project team is organized to represent all disciplines and interests of the product and its processes during the product/service life cycle.

There are clear challenges that contemporary organizations face today in the global marketplace. These challenges are:

- Between 50 and 80 percent of the cost of manufacturing a product is determined during the design phase.
- Well-designed products that were efficiently manufactured may not sell well.
- Overcomplicated designs can cause delays and lead to problems in manufacturing the product and may lead to costly engineering changes.
- By the time a product has left the initial design stage and key decisions about materials and processes have been made, about 70 percent of the cost has been locked in. This means that what happens beyond this point through manufacturing and marketing efficiencies will influence only about 30 percent of the product cost.
- Product life cycles are getting shorter.
- Simplicity in the design of a product is becoming a key factor in how well the product creates value for the customer.

- Efficiency and effectiveness in the manufacture of a product do not make up for poor product design or a marketing strategy that does not consider the customer's needs.

- The importance of product design cannot be underestimated, and the counsel and participation of key stakeholders—such as customers, suppliers, maintenance people, regulatory officials, workers, and others—can be valuable in creating a product design that will please customers, exploit supplier technology, and help after-sales personnel do their job.

- Since customers usually have a sense of their technological needs, their membership on concurrent engineering teams can help in finding innovative and user-oriented ways to design and package the product. In some industries, the majority of technological innovations come from customer insight and needs.

**The benefits of concurrent engineering
are very real**

3.3.2 Benefits of CE

Demonstrated benefits of CE include:

- Reduction of engineering change orders of up to 50 percent
- Reduction of product development time between 40 and 50 percent
- Significant scrap and rework reduction by as much as 75 percent
- Manufacturing cost reduction between 30 and 40 percent
- Higher quality and lower design costs
- Fewer design errors
- Reduction and even elimination of the need for formal design reviews since the product-process development team provides for an ongoing design review
- Enhanced communication between designers, managers, and professionals in the supporting processes
- Simplification of design, which reduces the number of parts to be manufactured, creates simplicity in fixturing requirements, and allows for ease of assembly
- Reduction in the number of surprises during the design process. Figure 3.2 depicts the concurrent engineering process.

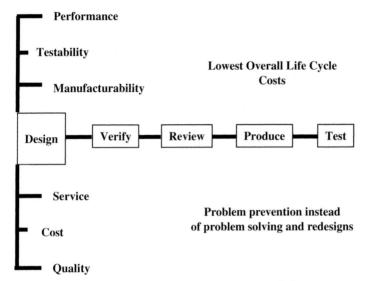

FIGURE 3.2 Concurrent engineering process. (***Source:*** Jon Turino, *Managing Concurrent Engineering* (New York, NY: Van Nostrand Reinhold, 1992).)

3.3.3 Disadvantages of Serial Design

CE is replacing the traditional approach, or serial design (SD) product/ service development process. The disadvantages of SD are:

- Increased costs and schedules
- Untimely product/service commercialization
- Excessive engineering changes to correct problems
- Poor quality
- Limited communication between functional entities
- Lack of common design goals
- Parochialism of organizational disciplines
- Scrap, reworking, recycling, and redoing work increases costs and delays schedules
- Limited perception of the life cycle of the product/service
- A greater number of parts and components is likely
- Customers and suppliers are usually not involved

• No fixation of responsibility within the enterprise for both product/service and organizational process design

Figure 3.3 is a model of the serial design process.

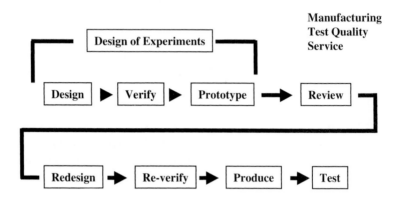

FIGURE 3.3   Serial design process. (***Source:*** Jon Turino, *Managing Concurrent Engineering* (New York, NY: Van Nostrand Reinhold, 1992).)

3.3.4 Advantages of CE

In product/service and supporting organizational process design, there are many advantages:

• It simplifies the design, making the product easier to manufacture.
• It reduces the number of parts, which reduces product cost, improves reliability, and makes after-sales service easier.
• It improves productivity and quality through a greater standardization of components and reduction of inventory.
• It reduces development cycle time, which facilitates a quicker response to changing markets.
• It eliminates redesign work—the design is done right the first time.
• It improves the maintainability and testability of the product.

- It improves competitiveness because of lower costs, earlier commercialization, higher quality, and better satisfied customers (particularly when customers were brought into the design activities).
- It allows for greater introduction of improved technology, particularly from suppliers who became part of the design teams.
- It reduces warranty claims on the product.

3.3.5   Strategies for Initiating CE

- Identification of background literature on CE for distribution to appropriate people
- Conduct of a series of workshops to review the literature and how the strategy in the enterprise will be carried out through the CE process
- Benchmarking of competitors to determine their use and success with CE
- Development and dissemination of objectives, goals, and strategies for the CE initiatives
- Identification of individual and collective roles of team members
- Development of supporting organization documentation to charge and empower the CE teams
- Determination of the decision empowerment of the CE team in their work
- Establishment of how the CE effort will be monitored, evaluated, and controlled by senior managers in the organization
- Determination of how customers and suppliers will be brought into the CE initiative, and the roles they will play on the CE team

3.3.6   Barriers to CE

There are cultural and other obstacles to the successful execution of CE, including:

- The different background of design and manufacturing engineers—two principal players in the CE process
- Lack of a common language among the functional participants in the CE initiative
- A lack of appreciation of the importance of an inter-disciplinary effort in the CE process by the principals involved

- An inherent bureaucratic bias in the organization leading to parochial and provincial perspectives and decisions
- Lack of technical and interpersonal capabilities of the CE team members
- Ineffective communication among the CE team members
- Lack of support and commitment by senior managers
- Premature compromises on designs

It should be noted as well that when CE is launched in an organization, there are likely trigger effects.

3.3.7 Key User Questions

1. Have the shortcomings of serial design been compared to the potential benefits of CE in the organization's competitive environment?
2. Has any benchmarking been carried out on competitors regarding their use of serial design vis-à-vis CE?
3. Have appropriate strategies been developed to guide the CE team in their initiation and completion of CE processes in the organization?
4. Have the benefits and potential problems of CE been discussed with organization customers and suppliers?
5. Have the barriers to CE been discussed with the project team and other principals in the organization?

3.3.8 Summary

The use of CE can provide a distinct advantage in the design and development of products/services and organizational processes. An examination of the benefits of CE vis-à-vis traditional serial design indicates that CE has the potential to favorably impact global competitiveness in that getting products and services to the global marketplace sooner providing a significant competitive advantage.

3.3.9 Annotated Bibliography

1. Cleland, David I., *Strategic Management of Teams* (New York, NY: John Wiley & Sons, 1996), chap. 10, "Concurrent Engineering." This chapter provides an overall view of the CE process to include its advantages, disadvantages, and the project management processes needed to design and execute a successful CE process.

2. C. Wesley Allen, ed., *Simultaneous Engineering* (Dearborn, MI: Society of Manufacturing Engineers, 1990). The chapters in this book represent a wide range of concepts and applications of simultaneous or concurrent engineering in many different situations and industries. The key message is how a user can design and build a closer working relationship among the organizational functions concerned with the design and implementation of concurrent engineering strategies.

3.4 THE MANAGEMENT OF SMALL PROJECTS

3.4.1 Introduction

In any organization that is in motion today, there are many small projects that are used to cope with the changes that are required to make minor adjustments in products, services, and organizational processes. Most of these small projects center around the changes in organizational processes. Project Management Institute's *A Guide to the Project Management Body of Knowledge* (PMBOK® Guide) describes a small project as one whose schedule is less than 30 days. Other characteristics for a small project include a single objective, one principal decision maker, easily defined scope and definition, available funding, and a small team that performs the work on the project. A project is small if it meets the following criteria:

- Three to four months duration
- Dollar value between $5000–$50,000
- Four to five members on the project team
- The team meets daily or weekly
- Not more than 3 to 4 cost centers involved
- Manual methods suffice for the project information
- The project manager is often the primary source.

Some examples of these small projects follow:

- Realignment of production line
- Reengineering of order entity protocol
- Development of information system for marketing function
- Evaluation of procurement practices
- Evaluation of existing practices for customer relationships
- Development of strategy to evaluate vendor protocol

> **Small projects cut across the warp
> and woof of an organization**

In Table 3.2 the basic steps for managing a small project are presented and discussed.

TABLE 3.2 Management of Small Projects—Basic Steps

- Identify the need
- Plan the project
- Collect information
- Analyze data
- Develop and evaluate alternatives
- Present recommendations

3.4.2 Identify the Need

- Identify of the client/sponsor and their perception of the problem.
- Conduct an initial analysis to get an idea of what is involved in the small project.
- Be careful to separate "problems from opportunities."
- Establish tentative objectives and goals for the project.
- Identify the funds that are available for the project.
- Find the initial documentation that describes the problem or opportunity.
- Separate "problems" from "symptoms." A symptom is the circumstances of phenomena regarded as an indication or characteristic of a problem. A problem is a question or situation that presents uncertainty in the work of small projects. For example, a cost overrun is a symptom of an underlying problem, such as inadequate monitoring, evaluation and control systems and processes.

3.4.3 Plan the Project

Every small project needs a plan. The essentials of a small project plan are:

- A summary that can be read in a few minutes
- A list of milestones (goals) identified in such a way that there can be no ambiguity when a goal is achieved
- A work breakdown structure (WBS) that is sufficiently detailed to provide for the identification of all tasks associated with the project
- An activity network that shows the sequences of the work packages and how they are related
- Separate budgets and schedules which are consistent with the work breakdown structure
- A description of the review process
- A list of key project team members and associated stakeholders
- Identify final objectives, goals, and strategies for the project
- Identify what the client or sponsor expects by way of deliverables from the project
- Identify and begin to seek potential answers regarding the key questions surrounding the problem and the project
- Develop a work plan on how and by whom tasks will be performed
- Organize the project team to include identification of individual and collective roles to be carried out by members of the team. The use of an LRC as described in Section 2.2 is useful here
- Become familiar with the organization's work authorization process through which funds are transferred for work on the project to an organizational unit within the organization or to an outside vendor
- Prepare schedules for the work to be carried out
- Come up with a preliminary outline of the expected final report

3.4.4 Collect Information

- Use interviews, surveys, or other data collection mechanisms.
- Develop a bibliography of basic information regarding the problem.
- Study the background information.
- Review miscellaneous data and information regarding the problem and the surrounding circumstances or situations.
- Observe activity to discern what is going on by the people associated with the problem.
- Correlate the data and information that has been gathered.

- Use techniques such as work sampling, work flow, and individual and collective behavior by the people associated with the problem.
- As the strategies for the solution of the problem begin to emerge, conduct a preliminary test of these strategies (policies, procedures, processes, methods, techniques, rules, etc.)

3.4.5   Analyze Data

- Classify the data by some common methodology.
- Question what the data appears to be revealing.
- Count, measure, and evaluate the forces and factors that begin to emerge during the analysis of the data.
- Compare data to the objectives and goals that have been established for the project.
- Look for trends, deviations, and other distinct characteristics of the data.
- Correlate different data that has emerged on the project.
- Conduct quantitative and qualitative assessment of the data. Consider using statistical techniques to assess the data.
- Follow your instincts in terms of what the data is revealing—which elements of data are providing meaningful insight into the problem and its solution.

3.4.6   Develop and Evaluate Alternatives

- Identify a few alternatives that might solve the problem.
- Evaluate these alternatives through the use of informal "cost-benefit" analysis to select the one or two that promise a useful solution to the problem.
- Test the one or two alternatives with the client.
- Select a final alternative.
- Develop implementation strategy.

3.4.7   Present Recommendations

- Prepare report.
- Brief client and/or sponsor.

- Rework as needed.
- Submit final report.
- Send a thank-you note (e-mail) to the project team members and other stakeholders who helped bring the project to a successful conclusion.
- Work with the project team members to prepare a "lessons learned" summary of the project and forwarded to key stakeholders.

3.4.8   Some General Guidance

- A small project can be managed using a scaled-down version of most of the concepts, processes, and techniques used for large projects, except of course for the amount of the resources involved.
- Communicate with the project stakeholders at all times.
- Have regular reviews of how the project is progressing.
- Don't surprise the client or sponsor: keep that person informed of all activities regarding the project, both good and bad.
- Keep in mind the following: If you were the client or sponsor, what would you like to know about the project, its progress, and its final "deliverable"?

3.4.9   Key User Questions

1. Does management recognize that any change, whether in progress or anticipated, can be dealt with effectively by using the ideas put forth in this section on the management of small projects?
2. Have the people who are managing, or are expected to manage small projects, had training in the concept, processes, and techniques of project management?
3. If the management of the organization is not using small project management concepts and process to manage minor change initiatives in the organization, what is being used?
4. Does the organization have a published policy and protocol on how small projects will be managed, such as the development of a work authorization process for the transfer of funds?
5. Do provisions exist on how the "lessons learned" study for each project will be passed on to future managers of small projects?

3.4.10   Summary

In this section, a simple protocol was put forth on how small projects could be managed. A series of work packages of such projects were presented along with the major actions likely to be needed when managing these projects. The point was made that small projects can be managed much like large projects except for the degree of resources that are involved.

3.4.11   Annotated Bibliography

1. Abramson, Bertran N., and Robert D. Kennedy, *Managing Small Projects* (TRW Systems Group, 1969). This publication, although old, is an excellent source as a reference on how to manage small projects.
2. Bass, Lawrence W., *Management by Task Forces* (Mt. Airy, MD: Lomond Books, 1975). Before project management gained the recognition that it has received in the last couple of decades, the use of task forces was championed to deal with interdisciplinary opportunities. These task forces were indeed teams, many of them being small projects as we know projects today. Author Bass provides a manual on the operation of interdisciplinary task forces, in particular how to provide for a coordinated exercise of a diversity of skills under the guidance of a task force or team leader. The book was published over 25 years ago, it provides insight into how task forces—or small projects—can provide for improvement in the management processes in organizations.

3.5   MANAGING MULTIPLE PROJECTS

3.5.1   Introduction

Managing multiple projects is done for economic reasons and the most efficient use of resources. An organization may have many small, unrelated projects that must be completed and the products delivered to customers. These small projects require, in some form, a project plan and resources to complete the work. Assignment of a project leader and project team for each project may not be the most efficient means of obtaining the products.

The reader should note the similarity between this section and the section on Program Management. Both sections go into the key concept of an umbrella focus for multi-projects.

Some planning is needed for all projects and the major framework of planning, such as communications planning, can be the same for all projects. Also, there may be one customer for the products of several of the projects. This commonality of project features allows single point management with fewer resources.

> **Multiple projects deal with multiple changes in the organization and its environs**

3.5.2 Benefits of Managing Multiple Projects

Organizations seeking more efficient and effective ways of managing projects will measure their success by the benefits achieved. Generally speaking, the following benefits can be derived from multiple project management:

- More efficient use of resources when one person can bridge several projects for assignment of people to tasks
- More efficient use of the project leader when small projects under his/her control can be planned, managed, executed, controlled, and closed out without delays between projects
- Faster delivery of products through dedicated effort and prioritization of small projects within the grouping of multiple projects
- More efficient reporting of project progress through briefings on several projects at one time and using a similar reporting format
- Improved project management of projects through continuous learning on a series of small projects
- Improved process through repeatable practices on a series of small projects
- Managing time and resources through a single project schedule that balances resources against project priorities
- Flexibility in adjusting individual project pace to meet delivery requirements

Other benefits may be identified in different organizations that greatly affect such areas as profitability, customer relations, project management effectiveness, and strategic goals. These benefits may be directly or indirectly related to future business opportunities as well as improving current situations. Depending upon the method and degree to which a company chooses to implement multiple project management, the benefits may be more or less.

3.5.3 Grouping Projects for Management

Grouping projects for management under a single project leader has advantages when the grouping follows some basic principles. These principles of managing multiple projects should be followed or there will be increased difficulty in bringing the projects to successful completion. Figure 3.4 shows some of the considerations for grouping projects.

The descriptions of these principles are as follows.

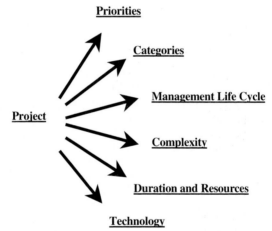

FIGURE 3.4   Principles for grouping projects.

- Project Priorities. Grouped projects should have similar priorities. Priority, the urgency of need for a project, dictates the order in which the project will receive resources and the order in which it should be delivered. Mixed priorities can easily rank a low priority project in such a manner that it receives no resources. The danger is that low priority projects will not be completed.

- Project Categories. Grouped projects should be of similar category. Category, the size of a project measured in duration, dollar value, or resources required, is the organization's method of identifying projects that have a major impact on business. When large and small projects are mixed for managing, there will be an imbalance in the implementation of the projects. Large projects may receive more than their share of resources because they are perceived as being more important. On the other hand, small projects may receive more than their share of the resources because they can be finished sooner and give the allusion of progress.

- Project Management Life Cycle. Grouped projects should have similar life cycles. Although the projects may be in different phases of completion, a similar life cycle provide a consistency for planning and execution. This similarity in life cycle supports the identification of improvements in process for continuous learning.

- Project Complexity. Projects grouped for multiple project management should be of relative simplicity. Complex technical solutions may require more effort and management, which could divert attention away from other projects.

- Project Duration and Resources. Grouped projects should be of relatively short duration, typically less than three months for the complete life cycle, and require few resources. The number of resources required for a project should be less than six persons. A greater resource requirement may divert resources from other important projects.

- Technologies in the Projects. Technologies of projects should be similar and it is best if the projects follow one major discipline. Mixed technologies require different skill sets that are usually not compatible to use across projects. Any mix of technologies will limit the efficiencies gained from managing projects in a group.

3.5.4   Examples of Managing Multiple Projects

The first example was observed in 1994. A Midwestern company had as many as 250 projects each year that were unique in planning, execution, and close-out. The company's business dictated that all projects start and close within the calendar year. The number of active projects at any one time could easily exceed 150 and range in size from $1,000 to more than $15,000,000. All projects were of similar technology and complexity, but followed no single methodology for planning, execution, and close-out.

The situation was recognized as being random and lacked visibility by senior management. Few efficiencies were achieved and some projects failed to meet critical delivery dates. Project managers were qualified engineers with little or no training in project management. Planning was typically a statement of work and some milestone dates.

This company instituted a project management system that required uniform planning and project documentation prior to project execution.

The goal was to achieve project savings of 15 percent or more so additional maintenance projects could be performed. Senior management also wanted greater visibility into project plans prior to approving funding.

Through an external consultant, planning standards and template schedules were prepared. The ten engineers were trained in the fundamentals of project management and given the tools to plan individual projects. The external consultant incorporated the individual plans into a master schedule to determine interfaces and conflicts. Senior management and the project managers (engineers) had access to the master schedule to determine where slippage was happening and where conflicts surfaced.

This uniquely tailored method of multiple project management resulted in nearly 17 percent savings the first year and expected additional savings the following year. Projects were being managed to expectations from an approved plan and senior management exported the concept to other divisions of the company. Continuous improvement was possible in the technical area as well because records of accomplishments identified areas for improvement.

The second example was in 1995. A major international company had lost control over its projects in several countries. The projects were of various sizes and many were dependent upon another project. This situation dictated that something be done to identify the level of planning, the interfaces between projects, and the funding required to complete individual projects.

Managing multiple projects in this environment was defined as one single manager overseeing all the projects and coordinating activities for project interfaces and milestones. The interfaces and milestones were placed in a master schedule for a top level view of the total work. Individual project managers worked on components of the total work within the constraints of the interfaces and milestones.

Because the individual project schedules and plans were in different formats, a standard schedule format was developed. Milestones were assigned and owned by the senior managers at the director level, the same managers as those having responsibility for the budgets. Scheduling conventions were developed and published for all project managers to ensure consistency on the master schedule.

Reporting procedures were standardized for all projects in the more than 15 different countries. These procedures were designed to obtain weekly reports of progress against the master schedule and to provide decision-making information where there was a variance. All reporting was to be accomplished on electronic mail.

This example resulted in the senior management establishing the operating parameters for all the projects and allowed the project managers to manage to milestones. Project managers had responsibility for budgets and meeting technical parameters of the projects as well as maintaining progress within the schedule. This company's concept of managing multiple projects used a single manager at the top with project managers functioning in several countries to meet the cost, schedule, and technical requirements. It was a loose method of bringing all the small projects into alignment with a master plan.

3.5.5 Managing Single Projects vs. Multiple Projects

Managing a single project that could be included in a multiple project environment may be done. There should be rational reasoning for including projects within a multiple grouping as well as identifying a project for intensive management as a stand-alone. When a single project is of such importance that it requires dedicated actions, then manage the project as a stand-alone.

Examples of stand-alone projects could be any of the following:

- Project requires dedicated attention because of urgency of need and criticality for the organization. Failure has major negative impacts.
- Project is required to be completed as the first project because it affects all other projects.
- Project is so technically complex that it requires special attention. Many changes are expected and the scope is tenuous.
- Project is a showcase that will need exclusive attention by the project manager.
- Project is a new type or technology for the organization.

Stand-alone projects may consume more resources and use them less efficiently, but there are compelling reasons for some to be executed in that manner. It is an informed decision when all the facts are weighed and the criteria dictate that a project be excluded from multiple management and be accomplished as a stand-alone.

3.5.6 Key User Questions

1. How can the principles of managing multiple projects benefit your organization with improved productivity?
2. How would you group multiple projects in your organization for management and what criteria would you use?
3. How would managing project schedules from a master schedule affect your organization and what benefits are envisioned?

4. What priority system would you use to first rank projects by their urgency of need and secondly to group for execution?

5. What is the largest-size project within your organization that could be included in a multiple project management arrangement?

3.5.7 Summary

Managing multiple projects does not have a single methodology, but has the flexibility to tailor the project planning, execution, control, and close-out to meet an organization's needs. There are principles of managing multiple projects that should be used or there should be a plan to compensate for the variance. Combining these principles with the business needs allows an organization to create the system needed.

Typically, small projects are grouped for efficiency and the effective use of resources to accomplish project work. This grouping of projects ensures single management of several projects and the ability to move resources across project boundaries without delay and to maintain a continuous flow of work going for the resources. Other benefits are achieved through continuous improvement of project capabilities as well as repeatable planning processes.

3.5.8 Annotated Bibliography

1. Ireland, Lewis R., "Managing Multiple Projects in the Twenty-First Century," *Proceedings of the Project Management Institute,* Upper Darby, PA, October 1997, pp. 471–477. This article describes requirements for managing multiple projects. It gives examples of project priority and category as well as the rationale for grouping projects. Typical management schemes for managing multiple projects are described and concepts amplified.

2. Lewis, James P., *Mastering Project Management* (New York, NY: McGraw-Hill, 1998). This book is a primer on project management. The author suggests several elements that influence success in the management of projects, including systems thinking, scheduling uncertainty, managing quality, improving management processes, and the role of power and politics. He opines that successful project management requires technical, process, and psychology skills. The book is valuable reading for those who manage a single project, and is particularly useful for the management of multiple projects where a system context comes into play.

3.6 SELF-MANAGED PRODUCTION TEAMS

3.6.1 Introduction

A self-managed production team (SMPT) is a team organized and dedicated to managing and creating the goods and services that are provided by an enterprise. The term "production" is used in the generic sense in that any organization produces goods and services whether it be through a manufacturing or other "production" function. SMPTs are found in a wide variety of enterprises, including industrial, marketing, insurance, universities, retail, and construction, to mention a few. These teams, in their self-managing role perform a wide variety of management and administrative duties in their area of work including:

- Design jobs and work methods.
- Plan the work and make job assignments.
- Control material and inventory.
- Procure their own supplies.
- Determine the personnel requirements.
- Schedule team member vacations.
- Provide backup for absentees.
- Set goals and priorities.
- Deal with customers and suppliers.
- Develop budgets.
- Participate in fund planning.
- Keep team records.
- Measure individual and team performance.
- Maintain health and safety requirements.
- Establish and monitor quality standards and measures.
- Improve communications.
- Select, train, evaluate, and release team members. (Paraphrased from David I. Cleland, *Strategic Management of Teams* (New York, NY: John Wiley & Sons, 1996), p. 170.)

Some SMPTs do their own hiring and firing—of course done under the oversight of the Human Resources office. Team members often interview prospective new team members to gain insight into the candidate's technical and social skills.

> **People can manage themselves given the
> opportunity and suitable guidance**

3.6.2 Implementing SMPTs

Preparing people and the organization for the use of SMPTs is usually handled in several phases as portrayed in Figure 3.5 and described below:

Conceptual Phase—in which the idea of how teams are appointed, trained, and operated is considered. The principal work carried out during this phase includes:

- Develop a bibliography on such teams for reading by the people involved in the organization.
- Benchmark other organizations using such teams to learn their experiences to include "successes" and "failures."
- Demonstrated commitment by senior managers on the use of such teams as key organizational design elements.

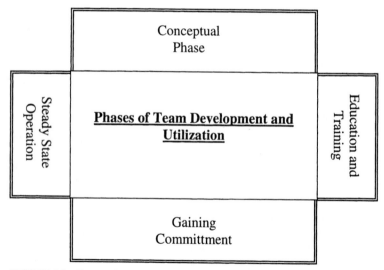

FIGURE 3.5   Phases of team development and utilization.

- Design of the training program to enhance the knowledge, skills, and attitudes of the prospective team members.
- Organization of the team to include a delineation of the overall picture of authority and responsibility that the teams are expected to assume and carry out.

Education and Training Phase—Execution of the training programs, which typically include the following subjects:

- Technical skills
- Social and interpersonal skills
- Management processes skills
- Decision making and execution competency
- Conflict resolution
- Team building
- How the teams can function as integrated, effective organizational design for improving production capabilities

Gaining Commitment Phase—in which the teams are given guidance on:

- How they will operate
- How the work will be done
- How the team is expected to carry out their technical and managerial responsibilities to include:
  - Build and maintain the communications networks and information systems required to do their job.
  - Develop a high level of esprit de corps within the team through better understanding of how individual and collective roles can be carried out in a synergistic manner.
  - Develop the policies and procedures needed to do the team's work.
  - Solidify their roles as technologists and managers in the creation of value for customers.
  - Develop and propagate a cultural ambience in the team that encourages ongoing improvement of production processes through the active participation of all team members.
  - Work with the first-level supervisors in helping those individuals understand how their new role departs from the traditional boss role to one of being a teacher, consultant, or mentor, whose principal purpose is to provide the resources and environment and then get out of the way of the team's work. (David I. Cleland, *Strategic Management of Teams* (New York, NY: John Wiley & Sons, 1996), pp. 165–166.)

Steady-State Operations Phase—in which SMPTs become a way of life in the enterprise. Activities to support this way of life include:

• The management system of the enterprise and the management system for the SMPTs have been fully integrated.
• The teams are fully committed to their work to include continuous improvement in operations.
• Performance evaluation to include merit pay raises have been fully accepted by the team members and the organization.
• Teams are fully recognized as key elements in the operational and strategic management of the organization.

The basic reason for using SMPTs is that the people doing the work know the most about how the work should be done. The use of SMPTs facilitates the development of a cultural ambience that encourages the fullest integration of knowledge, skills, and attitudes to enhance the organization's competitive competence. Table 3.3 provides insight into how the use of teams changes the existing cultural ambience.

TABLE 3.3 Self-Managed Production Teams versus Traditional Production Systems

Self-Managed Production Teams	Traditional Production Systems
Team-driven culture	Individual-driven culture
Multi-skilled members	Single-skill workers
Team purposes	Individual purposes
Team commitment	Manager commitment
Shared resources, results, and rewards	Individual results and rewards
Focus on entire work area	Narrow specialization
Limited levels of hierarchy	Many hierarchical levels
Shared information	Information limited
Rotating leadership	One leader/manager
Team controlled	Manager controlled
Team executed management functions	Managers execute management functions
Teamwork as a way of life	Limited teamwork
Process driven	Function driven
Continued self-appraisal	Limited self-appraisal
Customer driven	Task driven

Paraphrased from David I. Cleland, *Strategic Management of Teams* (New York, NY: John Wiley & Sons, 1996), p. 167.

3.6.3 The Change Factor

The use of SMPTs causes many changes in the manner in which production work is planned, organized, and controlled. A major change impacts the role of the first-level traditional supervisor. Traditionally such a supervisor has been responsible for the following duties:

- Counseling and guiding employees
- Planning, scheduling, budgeting, and rescheduling
- Managing performance and evaluating worker performance, including merit assessments and changes in financial reward for the workers
- Ensuring product quality
- Handling safety management
- Maintaining housekeeping of the production area
- Recruiting, selecting, and training workers
- Ensuring that equipment is operational
- Monitoring and facilitating production processing and product fabrication and assembly (Paraphrased from David I. Cleland, *Strategic Management of Teams* (New York, NY: John Wiley & Sons, 1996), p. 173.)

In contrast, in organizations that use SMPTs, the role of the first-level supervisor becomes interdependent with the operations of the teams. The key responsibilities of the first-level supervisors arising out of the interface with the SMPTs include:

- Determining training requirements and providing for the training to be carried out for those people who need it
- Facilitating team and employee development
- Facilitating team problem solving
- Coordinating communication with team and other stakeholders
- Facilitating change at the production level
- Understanding enough of the technical side of manufacturing to be able to ask the right questions and know if the right answers are being given
- Facilitating meetings
- Resolving conflicts
- Providing an environment that facilitates motivation
- Providing leadership of people in their area of responsibility
- Monitoring
- Teaching

Not all teams work well—some fail. Failure is commonly caused by lack of management commitment to the change process needed to go from traditional production organizational design to the team-driven paradigm. Some of the more common reasons for failure include:

- Inadequate information to do the work
- Cultural resistance to the use of team: people prefer to do things the way they have always been done
- Fear of an unknown outcome of teams in regard to the individual, the culture of the enterprise, and the ability of the enterprise to produce a product or service
- Individuals are unable to change from the traditional style to the team way of doing things.
- Teams are perceived as a threat to existing managers since the teams are likely to undercut and devalue managerial responsibilities—with the threat of blurring authority, responsibility, and accountability.
- People believe that teams depreciate relationships between managers and subordinates.
- Unions may oppose teams since teams are likely to shift their allegiance from the unions to the organization.
- Although the concept of teams may look great, major problems will arise in execution that cannot be solved without a major loss of production performance.
- Empowered SMPTs will not work unless the managers are willing to share control-an unlikely concession on the part of most managers.
- Teams often are launched in a vacuum, with little or no training or support, no changes in the design of the work of the team members, and no new supporting systems, like E-mail, to facilitate communication. (Paraphrased from David I. Cleland, *Strategic Management of Teams* (New York, NY: John Wiley & Sons, 1996), p. 174.)

3.6.4 Key User Questions

1. Will the organization be able to design and manage the changes required to go from a traditional production organizational design to one using SMPTs?

2. Have the proper plans been developed for the introduction of SMPTs to the organization—and can these changes be managed and implemented in an effective manner?

3. Have the traditional first-level supervisors "bought into" the new paradigm using SMPTs? If not, why not?

4. What has been done to change, and respect the cultural changes that come about when SMPTs are used?

5. Has a clear delineation of the management, technical, and administrative duties of the SMPTs been established?

3.6.5 Summary

In today's global competitive marketplace, every advantage must be explored to increase the efficiency and effectiveness with which the production function can be carried out. The use of SMPTs has been described in this section to include how to plan for, organize, and use such teams. Included in this assessment was a brief examination of the new role of the traditional first-level supervisor in working with SMPTs. Such teams provide for an organized focus dedicated to creating and managing the goods or services that are provided by the organization.

3.6.6 Annotated Bibliography

1. Bursic, Karen M., "Self-Managed Production Teams," in David I. Cleland, *Field Guide to Project Management,* 2nd ed., (New York, NY: John Wiley & Sons, 2004). This chapter reflects a major research project into the origins and use of self-managed production teams. Examples are given of how such teams are used, the challenges and dedication required to design and use them, and some of the major benefits likely to come to the organization using such teams.

2. Fisher, Kimball, *Leading Self-directed Work Teams* (New York, NY: McGraw-Hill, 1993). This timely book reviews the challenges, intricacies, and rewards of using self-directed work teams. It provides a step-by-step strategy on how team leadership skills can be developed by people aspiring to use such teams in the design and execution of their strategies.

3.7 BENCHMARKING TEAMS

3.7.1 Introduction

Benchmarking is a process carried out by an interdisciplinary team which compares the organization with competitors and "best in the industry" performers. Benchmarking carried out through an interdisciplinary team is usually of several types:

- *Competitive benchmarking:* In this process, the five or six most formidable competitors are evaluated to gain insight into their strengths, weaknesses, and probable competitive strategies.
- *Best-in-the-industry benchmarking:* The best performers in selected industries are studied and evaluated.
- *Generic benchmarking:* Business strategies and processes are studied that are not necessarily appropriate for just one industry. Information sources for such benchmarking can come from a wide variety of public and private sources to include organization records, site visits, periodical literature, interviews, customers, suppliers, regulatory agencies, seminars, and symposia, to name a few.

A benchmark is used as a reference point

3.7.2 What to Benchmark?

Virtually any area of the organization can be benchmarked. Suggested areas for benchmarking include:

- Product/service and process development strategies
- Organizational design
- Marketing
- Market penetration
- Product/service quality
- Manufacturing (production) capabilities
- Sales
- Organizational process competencies

- Financial practices
- Executive development
- Human resources
- Plant and equipment

3.7.3 Leadership of the Benchmarking Team

The benchmarking team should be organized with appropriate delegation of authority and responsibility. A benchmarking champion should be assigned to provide credence to the effort. Such a champion can provide several key benefits:

- Provide leadership within the enterprise for the planning and execution of the benchmarking initiative
- Ensure that the benchmarking results are integrated into the operational and strategic considerations of the organization.
- Provide the required resources for the benchmarking effort including designating the authority and responsibility of the people doing the work.
- Keep other key managers informed of the progress that is being made on the benchmarking initiative, including the probable outcome and the potential use to which the information from benchmarking can be put.

After the benchmarking work is completed, take the lead in assessing the effectiveness of the work, including the lessons learned, so that future benchmarking strategies can be improved. Membership on the benchmarking should include:

- Team leader
- Functional specialists
- Customer/sponsor
- Champion
- Facilitator
- Support people, such as legal, management information systems and clerical support

3.7.4 Standard Steps in Benchmarking

The steps in benchmarking typically include those indicated in Table 3.4.

TABLE 3.4 Standard Steps in Benchmarking

Determine what areas should be benchmarked.
Decide who are the most relevant competitors.
Decide who are the best performers in the industry.
Develop a benchmarking plan.
Organize the team.
Collect the information.
Analyze the information.
Determine the performance gaps.
Disseminate the findings.
Determine the relevancy of the findings.
Integrate the findings into strategies.
Prepare execution plans.
Execute the plans.
Maintain ongoing benchmarking.
Continuously improve the benchmarking process.

3.7.5 Benchmark Pitfalls

In general, the conduct of a benchmarking process can be carried out with minimum disruptions; however, there are a few pitfalls to be avoided:

- Inadequate charge of the team's authority and responsibility, and neglecting performance standards, objectives, goals, and strategies
- Failure to link the benchmarking team's efforts and the cooperating enterprise's objectives, goals, and strategies
- Having too many people on the team, resulting in duplication of effort, ambiguous authority and responsibility, increased costs, and a general disenchantment with the benchmarking purpose and process
- Not keeping the client informed of progress and results
- Inability to concentrate on the performance metrics and issues to be evaluated
- Selecting the wrong cooperating organizations and/or processes to benchmark
- Collecting too much data compounded by an inability to sort through and select the relevant and important data
- Neglect of an analysis of the meaning of the quantitative data that is collected.

3.7.6 Key User Questions

1. Has the benchmarking team in the organization been organized giving due attention to individual and collective roles, authority and responsibility delegations, and a standard process to manage the team?

2. Have the areas for benchmarking been selected which have a synergy and relevance with the "business" that the organization pursues?

3. Have the benchmarking results been compared to the strengths and weaknesses—and probable strategies of the sponsoring organization?

4. Has thought been given to what data will be collected during the benchmarking process, and how that data will be compared to comparable data of the sponsoring organization?

5. Have benchmarking team leaders been selected who will become champions for the benchmarking process?

3.7.7 Summary

Benchmarking makes sense as a key to gain insight into organizational performance through comparing an organization with competitors and with the best-in-the-industry performers. It is integral to the management process of monitoring, evaluating, and controlling the use of resources directed to organizational purposes. Benchmarking is a process that can be used to complement many team-driven initiatives in the organization, such as total quality management, concurrent engineering, project management, self-directed manufacturing, organizational designs, business process reengineering, and new business development actions.

3.7.8 Annotated Bibliography

1. Cleland, David I., *Strategic Management of Teams* (New York, NY: John Wiley & Sons, 1996), chap. 9, "Benchmarking: Using Teams to Compare." This chapter provides a summary of the major considerations in the design and execution of benchmarking initiatives in an enterprise.

2. Camp, Robert C., *Benchmarking: The Search for Industry Best Practices that Lead to Superior Performance* (Milwaukee, WI: Quality Press, American Society for Quality Control, 1989). This was the groundbreaking book on how to carry out benchmarking initiatives in an organization. Detailed examples are provided which show the reader how to relate benchmarking to one's own circumstances. Case histories provide examples of actual benchmarking investigations from begin-

Figure 3.8 depicts the minimum items that must be addressed in a successful project plan.

Project Plan

1. Problem
2. Need for Change
3. Goals for Change
4. New Organization

5. Schedule for Change
6. Participants
7. Milestones
8. Celebration

FIGURE 3.8 Project planning for change.

3.8.4 Negative Reaction to Change

When organizational change is not managed in such a manner that the affected people understand the need for change and are participants in making the change, adverse emotional reactions will occur. Adverse reactions to change can delay or stop the change. Recognizing the stages that the adverse reaction takes can help in dealing with the issues.

If people are not part of the plan for change, they will at best be neutral and at the worst be obstructionists. When people are informed of the decision for change, or discover the dictated change, they are shocked that someone is changing the status quo. This triggers several stages.

Figure 3.9 shows the stages that individuals will step through if they are not a part of the solution. This individuals will lose many hours of productive effort attempting to block the change or deny the change is happening.

- Disruption of Work. The people are disrupted from their normal duties and struggle to understand why the change is needed. This may bring current work to a halt and most time may be spent discussing the announced change with colleagues.

- Denial of Change. The people may deny the announced change as a rumor or a test. They will typically deny the change is happening and will ignore the information being given them.

- Realization of Change. A realization that there is change will often turn to anger and frustration because they do not understand why this change is affecting them without a reason for changing. They may lash out and say hurtful things to others.

FIGURE 3.9   Stages of resistance.

- Negotiating Change. Negotiating is the next stage that people seek to attempt to avoid the perceived negative consequences of the change. Denial and other reactions are set aside and the people attempt to deal with reality.
- Accepting Change. A feeling of helplessness and depression may follow the negotiating. Some testing of the situation may cause greater understanding of the change and a transition to acceptance of the change. Acceptance of the change means the people work in a productive manner within the new structure and put aside the old structure.

These stages are typical of unmanaged change and the negative impacts that change can have on the organization's work force. Better methods of managing change can offset the lost productivity and disillusionment of the work force while giving better results. A disciplined process for change yields better results.

3.8.5   Using Project Management as a Change Agent

Project management can provide structure to change and can provide visibility into the reasons for change as well as the process for change. The life cycle for organizational change is perhaps different from the typical product life cycle. The life cycle probably does not have the same time line because of the need for human response to situations.

A project life cycle for organizational change has five phases as shown in Figure 3.10.

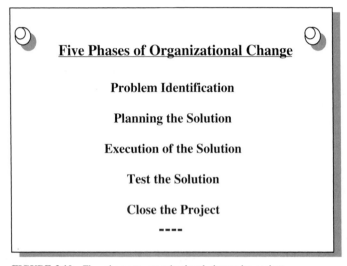

Five Phases of Organizational Change

Problem Identification

Planning the Solution

Execution of the Solution

Test the Solution

Close the Project

FIGURE 3.10   Five phases to organizational change by project.

- Problem Identification—identifying the problem and selling it to stakeholders. This phase could be collecting information from stakeholders to fully develop the problem statement and fleshout the root causes of affects on the organization. Stakeholders, working close to the production side of the business, should be able to provide pertinent information that redefines or confirms management's assessment of the problem. This phase may take a considerable amount of time if the problem is complex or there are many stakeholders. Early involvement of stakeholders improves problem identification and solution buy-in.

- Planning the Solution—develop a plan that solves the problem and involves stakeholders in the planning. The plan must have goals and an anticipated outcome. The details of the plan need to address each area identified by the stakeholders and provide an answer to each legitimate item submitted by stakeholders. The time line for implementation should be flexible and be able to accommodate breakthroughs with the change to people's work and routine.

- Execution of Solution—implement the plan to effect the changes. Commitment and reinforcement is needed throughout the implementation. Stakeholders need to have the purpose and reasons for change restated as the changes occur. Short-term successes need to be celebrated and communicated to the stakeholders to reinforce that the plan is working and there is a consistent advance on the solution.

- Test the Solution—assess the solution and its positive and negative impacts on the organization. Correct any negative aspects of the change, but only adjust the planned change as absolutely required. Maintain the consistency of purpose and demonstrate the plan is working toward the solution.

- Close the Project—celebrate the completion of the change and inform all stakeholders of the successes. Stakeholders must recognize when the project has completed a milestone or the successful completion of the planned change. There should be no doubt that the solution has been achieved.

Organizational change is the start of something new and the completion and close out of a former situation. The change must supplant the former organizational activity and remove it from the "new organization". The beginning of the change is the end of the former state.

The new beginning must be described in the project plan and identify the former activity that is being replaced. Individual stakeholders must recognize the replacement of the old with the new. Four items that can smooth the transition are:

- Be consistent in the direction that the change is taking and frequently reinforce the problem and solution. This should also clearly demonstrate progress to the new organization.

- Plan for and celebrate short-term successes. Early successes in reaching clearly identified progress by the stakeholders are examples of reinforcing that the solution is working.

- Symbolic gestures, such as ribbon cutting or new logos, can signal the transition and reinforce the new image and the new organization. Stakeholder involvement in developing and participating in the symbolic gestures ensures better support for the change.

- Completed organizational change projects should be celebrated and the success communicated to the stakeholders. This celebration signals the completion of this change and successful advances for the organization.

3.8.6   Continuous Change of the Organization

There may be a requirement to institute several changes, either in series or parallel, for an organization. These changes may be required for a product line or for alignment of the organization's structure to better meet business opportunities. Changes of this type may be viewed by stakeholders as continuous change and instability in what management wants the organization to do.

Like a single change, the problems need to be sold to the stakeholders and obtain their acknowledgement that these are problems that dictate change. Agreement on the problems sets the scene for developing solutions. Obtaining stakeholder information on how to solve the problem is critical to the success of implementing the changes.

Change can be managed similar to phased projects or multiple projects. The problems and their respective solutions are planned for implementation. Special emphasis must be placed on communicating the problems and solutions with the anticipated changes. Communication of actions and reinforcing how each part fits into the overall organization's improvements is essential. Always celebrate successes and highlight them as milestones on the road to a new organization.

3.8.7 Using Project Management

Project management techniques lend themselves well to managing change. Establishing firm, measurable goals and documenting the plan, start the resolution of the problem. Schedules and budgets lend themselves to providing visibility into the process and establishing a positive control over the tasks. Risk, for example, is another area that will assist in organizational change and contingency planning when adverse events occur.

Project planning allows the stakeholders to participate and feel they are a part of the solution to an organization's change. Stakeholders provide direction through participation and know the details of change. This promotes support and eases the transition to the new organization. Stakeholders can be a part of building the new while shedding the old.

The visibility gained through project management techniques will show the openness and trust of the organization to those most affected. Gantt charts will permit everyone to track the progress and anticipate future work efforts. Communications will be enhanced through a project communication plan that describes who are the stakeholders and their interests in the change.

Linking change procedures with project management techniques will give a greater capability to successfully transition to a new state. Planning with project management tools and defining the transition eases what can be a difficult task to deal with human emotions. Facts will replace speculation when project management techniques are employed.

3.8.8   Key User Questions

1. How do you get stakeholders involved in changes to an organizational structure?
2. Are organizational changes managed both from the physical and emotional aspects? How will you manage future changes?
3. How do you sell the problem of the need for change first, and then sell the solution to those affected by the change?
4. What types of resistance have you identified in past organizational changes and how will you deal with these in the future?
5. How would you communicate changes to the entire organization when everyone is affected by a planned change?

3.8.9   Summary

Organizational change will be disruptive and the outcome will be unpredictable if there is not a disciplined process that involves the stakeholders. Surprise changes and continuous changes will often lead to loss of productivity and disruptive influences by the stakeholders because they do not understand the problem that triggered the change.

Project management principles, concepts, techniques, and tools will assist in the change process.

Understanding the positive and negative effects of the implementation of change can lead to the use of such project management techniques as "identifying the problem," "developing goals for the change," "communicating the plan to stakeholders," "involving stakeholders in developing the solutions," and "involving stakeholders in effecting the change."

3.8.10   Annotated References

1. Bridges, William, *Managing Transitions* (New York, NY: Addison-Wesley, 1991), chap. 1, 3–7. This book structures the way change is implemented in an effective manner and addresses the resistance to change. Examples of how to manage change and how to ensure change is made are useful tools for all managers.
2. Conner, Daryl R., *Managing at the Speed of Change* (New York, NY: Villard Books, 1995), all chapters. This book identifies the need for managing change in organizations at the rate of change of the operating environment. The process of change and the affects on individuals are major themes. One can easily identify with the examples of change and see how to avoid the adverse reactions.

SECTION 4
THE STRATEGIC CONTEXT OF PROJECTS

4.1   SELLING PROJECT MANAGEMENT TO SENIOR MANAGERS

4.1.1   Introduction

Selling project management to senior leaders requires that they recognize a problem and the need for project management as the best solution. The problem that senior leaders have is managing from a strategic point of view and getting the tactical work to align with those high-level goals. Project management, although mature with more than 50 years of formal recognition and growth as a process and discipline, is still developing.

Bridging the problem from a high-level view to the execution of work may be difficult with the need to have a strategic focus by senior leaders. Implementation of the goals in detail is often not understood by the senior leaders and left to the mid- to lower-level leaders. There is not the ability to cascade the thinking from top to bottom.

> **Senior managers set the tone for the use of project management in the organization**

4.1.2 Background on Strategic Planning

Senior leaders have the non-delegatable responsibility to establish the principles of work and practices for the organization to follow to achieve success for their business. It is important for senior leaders to establish a system to measure the effectiveness of the strategic planning, through the business process, to the operating level. Through measurement, senior leaders know what is working and what is not, where successes are most needed and where lower priorities may exist.

Senior leaders are concerned with and conduct strategic planning that is far removed from the operating level. Strategic planning, however, uses the historical, tactical information from these operating level systems. Information flow is piecemeal and inadequate for setting the strategic direction and future of the organizations.

Currently, senior leaders seek solutions through improved communication of information and tools to measure performance. What is needed is a management strategy that uses operating units to perform the work and measure performance, analyze the effectiveness of the work being performed, and generate information for senior leaders. Project management does all this and is the choice of many senior leaders today for intensively managing critical aspects of the business.

4.1.3 Recognizing the Problem

Before senior leaders accept any solution, they must recognize the problem and be willing to act on that problem. The problem is clearly identified: "more than 90 percent of organizations fail to effectively communicate and execute their strategic plans because the necessary management and communications systems are not in place." Organization success is hinged on three requirements:

- Having a management system that emphasizes accountability and control
- Taking advantage of available resources
- Promoting cultural values dedicated to continual improvement that is supported by an appropriate management system

Current management systems are not delivering the desired results and the communication of strategic goals from senior leaders to operating levels is weak. The layers of management between the senior leaders and operating levels preclude effective information flow down and the required performance flow back to the top.

One perceived solution to the communication breakdown is to buy new information management tools. Better information systems are seen as the need while the efficiency and effectiveness of the operating unit has not changed. More timely and accurate reporting of the performance data is only one part of the challenge to senior leaders.

The management system and operating units retain the same structure and there is little or no improvement in productivity or meeting customer needs.

4.1.4 The Solution

Senior management is responsible for its management systems locally and globally. These management systems should be mutually supportive of the strategic goals as well as taking advantages of improvements across the enterprise. The management system of choice must be supportive of both local and global environments.

Many organizations have adopted project management as the system of choice and made significant gains in productivity and performance. Other organizations have accepted project management as one of its options without giving full support. These organizations have "hired" expertise through bringing in several successful project managers from other organizations. The hired experts bring a variety of methodologies and practices that create conflicts.

Developing project management as a core competency of an organization is a dedicated effort. Typically, organizations buy the tools rather than knowledge and experience. Many organizations do not receive full value because they buy tools. Figure 4.1 compares the most frequent and the most effective sequence of project management implementation.

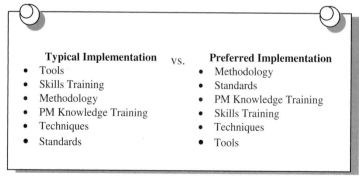

Typical Implementation	vs.	**Preferred Implementation**
• Tools		• Methodology
• Skills Training		• Standards
• Methodology		• PM Knowledge Training
• PM Knowledge Training		• Skills Training
• Techniques		• Techniques
• Standards		• Tools

FIGURE 4.1 Project management implementation sequence.

4.1.5 Benefits of Project Management

Project management, as a process to meet an organization's needs for performing a variety of work, has provided significant benefits. The general list of benefits for all organizations are as follows:

- Balance competing demands and prioritize the work that provides the most advantage to the organization
- Positive control over work progress through a tracking and control function
- Positive communication of needs to project team and frequent feedback to the customer on progress
- Estimates of future resource needs through the life cycle of the project
- Early identification of problems, issues, and risks to project work
- Early understanding of scope of work
- Goals and measures of success for each project
- An integral performance measurement capability to compare planned to actual progress

4.1.6 Comparison of Operations to Project Management

A comparison of operations to project management capabilities, as depicted in Fig. 4.2 is helpful to identify the advantages of project manage-

Operations	Project Management
Uses existing systems	Tailors system to fit requirements
Repetitive work functions	One-time work functions
Relies on stable standard procedures	Uses standards and processes to deliver customer's unique needs
Quota by numbers	Driven by the need for the end product
Focus on maintaining repetitive functions	Focus on taking opportunities for change
Line management with span of control on "owned" resources	Flexible work force with temporary help, only as needed

FIGURE 4.2 Operations vs. project management.

ment under different conditions. Operations typically focuses on maintaining the status quo and producing a given set of products. Project management, however, serves as a process for change and delivering unique products.

4.1.7   Selling Project Management

Selling project management as the management system of choice requires that the seller provide facts and figures to demonstrate the advantages. The advantages must give a better solution than the current system and demonstrate significant benefits to senior leaders. There must be a perceived problem to be solved and all elements of that problem must be solved.

The simple model for selling project management is:

- Identify senior leaders' problem with the organization.
- Low productivity
- Uncontrolled work being done
- Insufficient accurate information on progress
- Work doesn't follow strategic direction
- Work doesn't take advantage of product changes
- Identify causes of senior leaders' complaint
- Work force not keeping pace with technological advances
- Poor definition of work and poor control over work
- Weak or no communication system
- Work driven by perceived requirements based on historical information rather than the strategic direction
- System too rigid to accommodate product changes except on a planned change over

Develop a comparison of current operations, feasible fixes to current operations, and a project management driven approach. See the following illustration and example in Fig. 4.3.

The differences can be documented using a similar model that captures the current problems with the operations, identifies the changes within the current management system to correct weaknesses, and identifies the benefits of transitioning to a project management system. While senior leaders can be influenced by cost figures, there are few data collected that truly represent the situation.

Soft data are often required to justify a transition to project management. Some of these soft data areas are:

Current Operations ⇨	Feasible Changes to Current Operations ⇨	Feasible Project Management
Low productivity.	Train personnel to perform faster and better.	Tailor workforce to meet the productivity needs of the organization. Some training may be required.
Inflexible in meeting changes to product.	Change system to halt work while changes are made.	System anticipates change and has a controlling mechanism to accommodate changes.
Strategic goals not being met.	Change the process for directing and approving work to ensure goals are identified and followed.	Link all projects to strategic goals and ensure linkage during plan approval process.
Performance information accurate and timely for senior leaders.	Establish a communication system that will collect the required information and provide same to senior leaders.	Distill/consolidate project reports for senior leaders. No change from usual management process.

FIGURE 4.3   Comparison of operations to project management.

- Reduced time to market for products
- Improved information reporting
- Better visibility into the product build process
- Reduced human resource stress
- Rapid response to product changes
- Continuous improvement opportunities
- Improved productivity (qualitative)
- Improved return on investment (qualitative)
- Alignment of tactical work with strategic goals

Other areas may be applicable within different industries. Reduced risk through improved risk management procedures, for example, may be an area that is important. Improve product image through improved customer communications could be another area.

4.1.8   Key User Questions

1. When existing systems do not meet the needs of senior leaders, how can project management be sold as the solution?
2. If strategic goals are not being met by existing systems, how can project management fill the gap especially in the area of communication of information upward?
3. How can projects be linked to strategic goals and what is that process?
4. What are some of the inherent benefits of project management when an organization has rapidly changing products, both in features and in different products?
5. How can senior management take immediate advantage of project management to fill the need for continuous improvement?

4.1.9   Summary

A project management system will consistently outperform standard operations functions and provide better results. The dynamic business environment where "more than 90 percent of the organizations fail to effectively communicate and execute their strategic plans" provides many opportunities to sell project management as the management system of choice. Organizations need the discipline of project management while receiving the benefits of a flexible, tailored work force to complete work.

Selling project management to senior leaders requires a comparison of the benefits and risks to be developed for each organization. These benefits over current operations will provide the reason for change. Project management must be sold as the solution to existing weaknesses and it must be demonstrated as to how these will be overcome.

4.1.10   Annotated Bibliography

1. Caldwell, Chip, "The Roles of Senior Leaders in Driving Rapid Change," *Frontiers of Health Services Management,* Ann Arbor, MI, Fall, 1998, pp. 35–39. This article discusses the role of senior management in establishing and cascading down the principles of work for the organization. Measurement tracking is proposed as bottom-up, top-

down method to ensure focus on what is working and what is not working.

2. Gaiss, Michael, "Enterprise Performance Management," *Management Accounting,* New York, Dec. 1998, pp. 44–46. This article emphasizes the shortfall in communicating requirements from the top to the project level. It touches on three success criteria: (1) a project management process, (2) using available assets, and (3) promoting a continuous improvement culture.

4.2 PROJECT PARTNERING

4.2.1 Introduction

Partnering in projects has emerged in recent years as a means of sharing equally in managing large projects by two or more organizations. One organization may initiate partnering to gain additional capability for a project as well as share in the risks of large, complex projects. Partnering is also a means of combining information, such as in research and development projects, to improve the chance of success.

Partnering may be accomplished between public, private, and for-profit and not-for-profit organizations. A government agency may, for example, partner with a private organization to conduct research. Another example is for a private, for-profit professional association to partner with a not-for-profit company to develop a commercial product from intellectual property.

There is no limit to project partnering by organizations. A common goal and complementing capabilities are the basis for a project partnership. Finding the talent needed to perform specific project work and developing cooperative arrangements for the partnering are needed for a working partnership.

> **A partnership is an affiliation of parties who seek a common purpose**

4.2.2 Types of Project Partnering Arrangements

Project partnerships may take many forms of cooperative agreements. The formality and contractual relationship is determined by the needs of all partners. Some examples of these relationships are depicted in Table 4.1.

TABLE 4.1 Types of Project Partnering Arrangements

Relationship & binding documentation	Description of possible working arrangements	Remarks
Formal and Contract	Two companies obligate themselves to perform parts of a project. The allocation of work is based on expertise in the type of work to be performed.	One or both companies must commit to the customer for delivery. One or both may sign the contract with the customer.
Formal and Consortium	Two or more companies obligate themselves to perform project work through a single contract that joins them in a separate legal entity.	The consortium represents itself to the customer as the contracting entity. The individual identities of the companies may not be visible to the customer.
Formal and Contract	One company may bid on a project and use another company's resources. The resources are excess to the "loaning" company, but become a part of the project under the direction of the host company.	The arrangement is invisible to the customer. The "loaning" company has an obligation to provide qualified resources to the project and may or may not provide supervision.
Informal and Agreement	One company bids and wins work on a project. One or more other companies agree that they will contribute to the success of the project through selected work.	Second tier companies are invisible to the customer. Work performed by second tier companies is accomplished as needed, but the customer does not know other companies are involved.

The number of arrangements is left to the organizations desiring to partner and work together. One important aspect is the visibility that the customer has of the accomplishment of different work packages. In some partnering, the customer wants to be aware of who is doing the work and in other situations, the customer is only concerned with quality of the work.

4.2.3 Examples of Project Partnering

There are many examples of project partnering that demonstrate the concept and the future of this type of business relationship. Figure 4.4 shows several examples of project partnering.

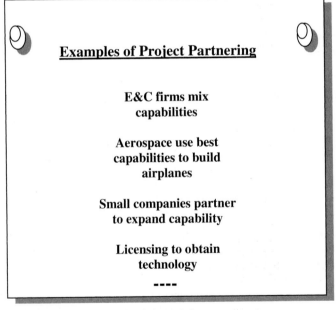

FIGURE 4.4 Examples of project partnering.

- Engineering and construction firms partner to obtain the best mix of talent and capability for projects. Partnering to obtain excellence in project control for major projects is often found in major projects such as the Super Collider Project in Dallas, Texas.

- Aerospace companies partnered to build a stealth bomber on a $4 billion project. Several companies worked together to develop the best in stealth technology. Technology was the driving force for partnering. The combination of companies could not solve the technical problems and the project was cancelled when the cost was estimated to exceed nearly $7 billion.

- Three small companies combined their capabilities to bid and win a project requiring expertise in computer technology, computer network operations, and procurement knowledge. The project, although relatively small, combined the talents from all three companies to perform the required tasks.

- Licensing of intellectual property of a professional association to a company to build software products is a current project. The association's standards were licensed to a company for the specific purpose of expanding the distribution of knowledge in the standards and to generate a modest profit for the association. This is partnering in that the professional association retains review over the products developed by the company.

4.2.4   Managing Partnered Projects

Customers are concerned with the management of the project work and who will be responsible for such items as reports, corrective actions, changes to the project, and overall project direction. Strong management capability builds confidence with the customer, while a weak or vague project management structure erodes confidence.

Figure 4.5 shows typical types of arrangements for managing partnered projects. A detailed discussion follows.

Some management structures for partnered projects are:

- Steering Committee—senior managers from all partnering organizations. This committee reviews progress and sets direction. A project manager or co-project managers are designated for all partnered work and report to the Steering Committee.

- Project Manager and Deputy Project Manager (or Co-Project Manager)—two individuals are designated from the partnering companies to lead the project. These individuals may report to their respective company executives or to a steering committee.

- Project Manager—a single individual is appointed as the project manager for all the project work. Project team members report to the project manager for performance issues.

FIGURE 4.5 Managing partnered projects.

These three management structures can be modified to meet any situation. Large projects need senior guidance from a group such as a steering committee while small projects would only require a single project manager. The management structure must meet the needs of both the project and the customer.

4.2.5   Technical Aspects of Partnered Projects

One of the primary reasons for partnering is to gain additional technical capability. A customer wants assurances that the project will be technically successful and that the performing organization has the capability to accomplish the work. Partnering gives that assurance by bringing the best talent from more than one company.

Principles of partnering will guide companies to better work solutions and better products for the customers. The following is a list of principles:

- Work allocation—divide work so the work packages are performed with an integral team where possible. Attempting to develop teams on a short-term basis may not be the most efficient use of talent and expertise. So, a team from one of the partnering companies should be assigned work packages commensurate with its skill.

- Project control—a team should be assigned responsibility for controlling the project work. This team may be individuals from different companies, but should have discrete work responsibilities. The project control team must report to the project manager.

- Project management—there needs to be a single person responsible for managing the work. Some work may be performed under the supervision of a manager, but that manager is responsible to the project manager for delivering a product component. All elements of the project must be managed from a single person with the authority to direct and accept work.

- Company participation—company managers must not, as individuals, become involved in the direction of the project. Direction must be through a consolidate body, such as a steering committee or senior management representative group. Individuals may bring the wrong solutions to the partnered project.

- Customer interface—like any project, partnered projects must have a single interface with the customer. This may be the project manager, the chair of the steering committee, or the elected representative from the management group. In some situations, there are two levels of customer interface. The strategic direction and liaison is from the senior steering committee and the daily interface is between the project manager and the customer's representative.

4.2.6   Benefits in Partnered Projects

Partnering has benefits that exceed those that a single organization may derive from a project. The benefits can be short-term or long-term, with a major impact on future business. Some of the benefits for the partnering organizations are:

- Technology—partnering with a high-technology exposes personnel to state of the art technology and processes. These technologies may be used in future projects or they may serve as information when seeking future partnering relationships.

- Small to large companies—small companies partnering with large companies learn improved methods of managing and bidding on projects. The transfer of knowledge is valuable to small companies in future work.

- Managing projects—all partnering organizations gain skills and knowledge on how to better manage projects. The cross-pollination of skill, knowledge and abilities helps organizations improve the project management capabilities.

- Efficiency—availability of a large resource pool should give the right skills for the work and lead to a more efficient level of work. This efficiency should yield benefits in that rework and other waste could be significantly reduced.

- Monetary rewards—improvements in management and work processes from a wide base of several companies should result in improved profits. The efficiency and effectiveness of the work will cost less to perform and yield a better margin.
- Corporate image—if companies in partnered projects are visible, the efficiency and effectiveness may improve the image as quality performing organizations. This image and reputation can be used in future bidding and project work, whether partnering or alone on a project.

4.2.7 Key User Questions

1. What are some of the reasons that an organization would want to participate in a partnered project?
2. Why would an organization loan people to another organization for a cooperative effort in partnering a project?
3. What are the major management structures for guiding a partnered project at the top level?
4. When would it be appropriate to have someone other than the project manager be the customer liaison in a partnered project?
5. In a major project, who would the project manager report to and why?

4.2.8 Summary

Project partnering takes on many forms based on the desires and creativity of the partners. The arrangements will vary according to the project and the visibility of organizations may be more or less, depending upon the needs for the partnered project. Organizations, public and private, for-profit and not-for-profit, can create partnerships to meet the needs of a small, medium, or large project.

Managing a partnered project is typically driven by the desires of the customer. Customer confidence and visibility into the project dictate the management structure as well as the size of the project. It is not uncommon to have a two tier management structure, a project manager and a steering committee to which the project manager reports.

4.2.9 Annotated Bibliography

1. Department of Energy, Super Collider Project, Dallas, TX (c. 1988–1993). This project is an excellent example of many different compa-

nies being involved in cooperative work. It is not an example of part-nering per se, but how many companies share the responsibility for different elements.

2. Department of Defense, Stealth Bomber Project, Washington, DC (c. 1986–1994). This project used four major companies in partnership and hundreds in subcontracting roles. This example shows a complex ar-rangement that failed to achieve the desired results and had to be ter-minated.

4.3 PROJECT STRATEGIC ISSUE MANAGEMENT

4.3.1 Introduction

In a project, a strategic issue is a condition or pressure, either internal or external, that has the potential for a significant influence on one or more factors of a project during its life cycle, such as its cost, schedule, and technical performance objectives as well as financing, design, engineering, construction, and operation. Strategic issues can arise from many different stakeholder groups: customers, suppliers, the public, government, inter-venors, and so forth. Some examples of strategic issues include:

- The need to reduce time required to design, develop, and produce an automobile—resolved through the practice of Concurrent Engineering
- The failure to recognize the environmental-related and political issues in product development, which, in the case of the US Supersonic Trans-port Program led to the cancellation of that program by the US Congress
- The need to develop strategies for a "turn-around" project underway at Eastman Kodak—remedial action was required to reduce and eliminate resistance to change from the employees
- The emergence of a key environmental-related factor which held poten-tial to delay or cancel a project, as in the case of the discovery of an earthquake fault offshore from the construction site of a nuclear gener-ating plant

> **A strategic issue is a point of contention involving a project**

4.3.2   Managing Project Strategic Issues

There are four key phases involved in the management of strategic issues for a project. These phases are portrayed in Fig. 4.6, and discussed below.

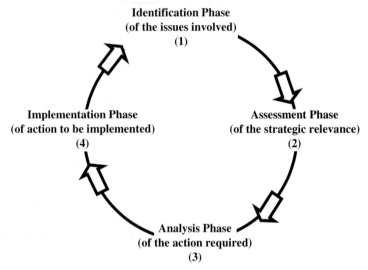

FIGURE 4.6   Key phases for managing strategic issues for a project. (Adapted from David I. Cleland and Lewis R. Ireland, Project Management: *Strategic Design and Implementation*, 4th ed. (New York, NY: McGraw-Hill, 2002), p. 204.)

4.3.3   Identification Phase

Chances for the identification of potential issues facing the project can be enhanced if in the project planning a "work package" is established for the identification and management of the strategic issues. At each review of project progress, consideration should be given to whether or not any strategic issues have arisen that could impact the project. By reviewing the project stakeholder interests in the project, the likelihood of strategic issues coming forth is enhanced.

4.3.4   Assessment Phase

In this phase, judging an issue's importance is essential. Several criteria can be used to assess the importance of an issue, such as:

- Strategic relevance, or magnitude and duration of the issue
- Actionability—considers whether or not the project team has the knowledge and resources to deal with the issue. A project may face an issue over which it has little control. In such issues, keeping track of the issue and considering its potential impact may be the only realistic strategy that can be pursued. For example, in government defense work, contracts can be cancelled at the convenience of the government
- Criticality of an issue relates to the importance of the potential impact of the issue on the project. If an issue appears to be non-critical, then continued monitoring of the issue may be all that is needed
- Urgency of the issue regarding the time period in which something needs to be done

4.3.5   Analysis of Action

In the analysis of strategies and actions required to deal with the project, seeking answers to the following types of questions can be useful:

- What will be the probable effect of the issue in terms of impacting the project's cost, schedule, and technical performance objectives?
- What might be the impact on the project of the key stakeholders involved in the issue?
- What specific strategies have to be developed by the project team to deal with the issue(s)?
- What resources will need to be committed to deal with the issue?

4.3.6   Implementation of Action

In this phase, a plan of action is implemented. The action plan and its implementation can be dealt with as a sub-project and involves all of the elements in the management of a project, such as planning, organizing, direction, control, and so forth. An important part of the implementation phase is a review of the strategic issues during the regular review of the project's progress.

4.3.7   Key User Questions

1. Do the members of the project team and senior managers understand the concept of project strategic issues, and the need for the active awareness and management of such issues?

2. Are the potential strategic issues involving the project(s) in the organization identified and brought into the management processes of the project and the organization?

3. Has a work package been established on the project(s) that is dedicated to the management of strategic issues as outlined in this section?

4. Are post-project appraisals used to identify historical strategic issues and help foster an understanding of the role of strategic issues in the management of the organization's future project?

5. Have any strategic issues in the portfolio of projects in the organization been identified and reported to senior management for their awareness, and suggestions on how the issues might be resolved?

4.3.8 Summary

A strategic issue in a project is a condition or pressure that has the potential to have a significant impact on the project. Strategic issues can arise from within the organization, and from outside, such as major concerns or forces that the project stakeholders consider of importance to the project. In this section a few ideas were presented on how a project team could better manage the strategic issues that their project faces. Strategic issues can arise at any stage in the life cycle of the project. The project manager should be aware of the concept of project strategic issues, and provide proactive leadership in dealing with the strategic issues that are likely to impact the project.

4.3.9 Annotated Bibliography

1. Cleland, David I., and Lewis R. Ireland, *Project Management: Strategic Design and Implementation,* 4th ed. (New York, NY: McGraw-Hill, 2002), chap. 7, "Strategic Issues in Project Management". This chapter provides a description of the concept of strategic issues, and suggests a management process on how such issues can be identified and managed.

2. King, William R., and David I. Cleland, *Strategic Planning and Management Handbook* (New York, NY: Van Nostrand Reinhold, 1987), chap. 15, "Strategic Issue Management." This chapter provides relevancy and a rigor to the management of strategic issues. The identification and the assessment of strategic issues is described in a proactive manner that is more likely to result in strategic change, as compared to a more passive approach to the management of these issues.

4.4   PROJECT STAKEHOLDER MANAGEMENT

4.4.1   Introduction

Project stakeholders are those individuals, organizations, institutions, agencies, and other organizations that have, or believe that they have a claim or stake in the project and its outcome. Stakeholders in the political, economic, social, legal, technological, and competitive environments have the potential to impact a project. Project managers need to identify and manage those stakeholders likely to have an influence on the project's outcome.

A stakeholder is a party who has a claim regarding a project

4.4.2   Examples of Stakeholder Influence

- Actions by environmental groups that delayed progress on the design and construction of nuclear power-generating plants in the US.
- On the James Bay Project in Canada, special effort was made to stay sensitive to social, economic, and ecological issues coming from vested stakeholder interests on that project.
- Proactive action by environmentalists, tourists, and government agencies motivated the project team to take special care to protect and preserve the scenic well-being of an area during the construction of a highway through Glenwood Canyon in Colorado.
- Stakeholder action caused the rerouting of a major highway in Pittsburgh, PA to preclude the razing of an old church that had historical value to the community.
- Failure on the part of the project team to manage the political interests of key US congressmen and environmental groups who developed such organized opposition to the US Supersonic Transport Program that it was cancelled by lawmakers, even though the technological feasibility of the aircraft had been demonstrated.

4.4.3   Evaluating Potential Stakeholder Influence

Developing a strategy to consider the potential impact of project stakeholders can start with seeking the answers to a few key questions such as:

- Who are the project stakeholders, both primary and secondary?
- What stake, right, or claim do they have in the project?
- What opportunities and challenges do the stakeholders pose for the project team?
- What obligations or responsibilities does the project team have toward its stakeholders?
- What are the strengths, weaknesses, and probable strategies that the stakeholders might use to realize their objectives?
- What resources are available to the stakeholders to implement their strategies?
- Do any of these factors give the stakeholders a distinctly favorable position in affecting the project outcome?
- What strategies should the project team develop and use to deal with the opportunities and challenges presented by the stakeholders?
- How will the project team know if it is successfully "managing" the project stakeholders?

4.4.4 A Model of the Project Stakeholder Management Process

The process of dealing with stakeholders focuses around the application of the management functions (planning, organizing, motivating, directing, and controlling) to potential stakeholder issues. Figure 4.7 shows a model of such process.

4.4.4.1 Identification of Stakeholders. A project usually faces two kinds of stakeholders: (1) Primary stakeholders who have a contractual or legal obligation to the project team; and (2) Secondary stakeholders who usually have no formal contractual relationship to the project team but have, or believe that they have, a stake in the project or its outcome. Figure 4.8 is a model of these stakeholders.

The key authority and responsibility of the primary stakeholders include:

- Providing leadership to the project team
- Allocating resources for use in the design, development and construction (production) of the project results
- Building and maintaining relationships with all stakeholders
- Managing the decision context in the design and execution of strategies to commit project resources

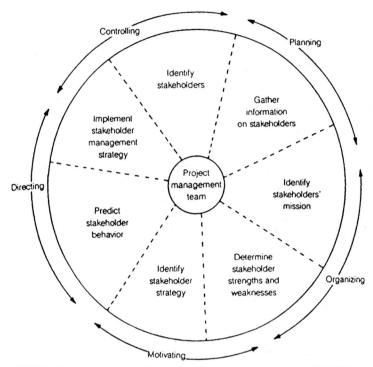

FIGURE 4.7   Project stakeholder management process. (***Source:***   David I. Cleland and Lewis R. Ireland, *Project Management: Strategic Design and Implementation,* 4th ed., (New York, NY: McGraw Hill, 2002), p. 174.)

- By example, set the project's cultural ambience, which emphasizes the best of people in providing high-quality professional resources to the good of the project
- Maintain ongoing and effective oversight of the project's progress in meeting the schedule, cost and technical performance objectives, and where necessary reallocating and reprogramming resources as needed to keep the project on track
- Periodically checking the efficiency and effectiveness of the project team in doing the job for which they were hired

Secondary stakeholders can be difficult to manage. Some of the more obvious characteristics of these stakeholders include:

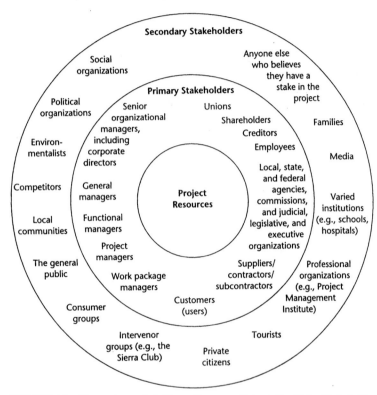

FIGURE 4.8 The project stakeholders. (*Source:* Cleland, David I., "Stakeholder Management," chap. 4 in Jeffrey K. Pinto, ed., *Project Management Handbook* (Newtown Square, PA: Project Management Institute, and San Francisco, CA: Jossey-Bass, 1998), p. 61.)

- There are no limits to where they can go and with whom they can talk to influence the project;
- Their interests may be real, or perceived to be real as the project and its results may infringe on their "territory";
- Their "membership" on the project team is ad hoc—they stay so long as it makes sense to them in gaining some advantage or objectives involving the project;
- They may team up with other stakeholders temporarily to pursue their common interest for or against the project's purposes;

- The power they exercise takes many forms such as political influence, legal actions, such as court injunctions, emotional appeal, media support, social pressure, local community action, and even scare tactics;
- They have a choice of whether or not to accept responsibility for their strategies and actions.

4.4.4.2   Gathering Information of Stakeholders.   Information about project stakeholders can come from a wide variety of sources to include:

- Project team members
- Key managers
- Business periodicals including *The Wall Street Journal, Fortune, Business Week, and Forbes*
- Business reference services, including *Moody's Industrial Manual, Value Line Investment and Security*
- Professional associations
- Customers and users
- Suppliers
- Trade associations
- Local and trade press
- Annual corporate reports
- Articles and papers presented at professional meetings
- Public meetings
- Government sources
- Internet

4.4.4.3   Identifying Stakeholder Mission.   This involves a determination of the mission of the stakeholders that support their vested interests in the project. For example, the mission of a stakeholder group on a highway project in Pittsburgh, PA was to "Change the route of the freeway highway to avoid razing of the historical church."

Stakeholders can be supportive or adversarial—with degrees of influence in between. Both types require managing—to strengthen the project by gaining continued support—or to reduce as much as possible the likely adversarial impact of the stakeholders on the project.

4.4.4.4   Determining Stakeholder Strengths and Weaknesses.   The determination of the stakeholders' strengths and weaknesses is essential to ascertaining how successful they might be in influencing either the project or its outcome. A stakeholder's strengths might include those listed in Table 4.2.

TABLE 4.2 Stakeholders Strengths

- Availability and use of resources
- Political and public support
- Quality of their strategies
- Dedication of the stakeholder members

Conversely, the weaknesses of a particular stakeholder group might include those listed in Table 4.3.

4.4.4.5   Identification of Stakeholder Strategy.   A stakeholder strategy is a prescription of the collective means by which resources are dedicated for the purpose of accomplishing the stakeholder's mission, objectives, and goals. These prescriptions stipulate what resources are available, and how such resources will be used in an organized fashion to accomplish the stakeholder's purposes. The use of legal means to stop construction on a project because of claims of adverse impact on the ecology is one principal strategy used by stakeholders. Another means is to seek support from political bodies to influence the project. Picketing construction sites, letter writing campaigns, paid radio or TV advertisements are other means that stakeholders can use. Once an understanding of the stakeholder's strategy is gained, then the chances are increased to predict what the stakeholders' behavior will be.

4.4.4.6   Predicting Stakeholder Behavior.   How will the stakeholders use their resources to affect the project? This is the key question to be answered. There are many alternatives available to the stakeholders to attain their end purposes. These can include:

- Influencing the economic outcome of the project—as a construction union might do when a new manufacturing plant is being built

TABLE 4.3 Stakeholders' Weaknesses

- Lack of public and political support
- Lack of organizational effectiveness
- Inadequate strategies
- Uncommitted, scattered members
- Poor use of resources
- General incompetence

- Attaining specific objectives of the stakeholder—such as the protection of the environment, which is part of the mission of the Sierra Club
- Legal actions to influence the outcome of the project—such as work stoppage orders until stakeholder claims of safety are resolved
- Political actions through gaining the support of legislators—such as having governmental projects funded for construction in their political districts
- Health and Safety—often brought up when new drugs or medical protocols are being developed

4.4.4.7 Implementing Stakeholder Management Strategy. The final step depicted in Fig. 4.7 in managing the project stakeholders is to develop protocol for the implementation of the resources dedicated to managing the stakeholders. Once the implementation phase has been embarked on, the project team needs to:

- Be sure that the key managers and professionals appreciate the impact that both supportive and non-supportive stakeholders may have on the project outcome.
- Manage the project review meetings so that the stakeholder assessment is an essential part of deciding the project status.
- Keep in touch with key external stakeholders to improve the chances of determining stakeholders' understanding of the project and its strategies.
- Ensure a clear evaluation of possible stakeholder response to major project decisions.
- Provide ongoing, current reports on stakeholder status to key managers and professionals for use in developing and implementing project strategy.
- Provide a proper security system to protect sensitive project information that might be used by adverse stakeholders to impact the project's success.

4.4.5   Key User Questions

1. Project stakeholders must be managed—taking the form of a specific "work package" in the management of the project. Is this being done on your projects?
2. Both primary and secondary project stakeholders exist. The project team must know who these stakeholders are, and what their likely strategies will be in order to manage them. Is this being done on your projects?

3. Who are the stakeholders with which this project must deal, and what is their perception of the nature of their stake in the project?

4. There are specific sources for the collection of information on the project stakeholders, which can be used in the development of countervailing strategies for dealing with the stakeholders. Have the sources been developed on your projects?

5. Every review of the project should include an assessment of who the project stakeholders are, and how well they are being managed. Is this being done on your projects?

4.4.6 Summary

The management of project stakeholders is a vital "work package" in planning for and implementing the use of resources on the project. An individual should be assigned a specific "work package" responsibility on the project team to identify, track, and manage the project during its life cycle.

4.4.7 Annotated Bibliography

1. Cleland, David I., "Stakeholder Management," chap. 4 in Jeffrey Pinto ed., *Project Management Handbook* (Newtown Square, PA: Project Management Institute, and San Francisco, CA: and Jossey-Bass, 1998). This chapter deals with the concept of project stakeholders to include identification and management of such stakeholders. The chapter provides basic information concerning how project stakeholders should be considered by the project team during the life cycle of the project.

2. Padgham, Henry F., "The Milwaukee Water Pollution Abatement Program: Its Stakeholder Management," *PM Network*, April 1991, pp. 6–18. This article describes how project stakeholders were managed on a major construction project.

4.5   THE STRATEGIC MANAGEMENT OF TEAMS

4.5.1 Introduction

The essence of managing an organization from a strategic perspective is to maintain a balance in using resources to accomplish organizational mission, objectives, and goals. The key is to balance operational competence, strategic effectiveness, and functional excellence. Figure 4.9 shows the linkage between strategic effectiveness, operational competence, and functional excellence for teams. These terms are defined below:

Operational Competence is the ability to use resources to provide customers with high quality products and services so that sufficient profit is realized to advance the state-of-the-art in such products and services, and have sufficient funds left over to develop new strategic initiatives for the organization.

Strategic Effectiveness is the ability to assess the need for, and develop future products, services, and organizational processes. Alternative project teams, as described in Section 3 can be used to assess the possibilities and probabilities in the use of resources to ensure the organization's future.

Functional Competence is the ability to maintain state-of-the-art ability in the use of resources to support the organization's current and future needs.

A balance must be maintained among the organization's Operational Competence, its strategic effectiveness, and its functional excellence, as

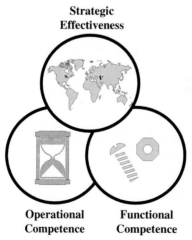

Strategic Effectiveness

Operational Competence **Functional Competence**

FIGURE 4.9 Balance between strategic management challenges. (***Source:*** David I. Cleland, *Strategic Management of Teams* (New York, NY: John Wiley & Sons, 1996), p. 5.)

depicted in Fig. 4.9. This balance can be facilitated by providing appropriate project team linkages within the organization.

Project teams can make a difference in the strategic management of an organization

---

4.5.2   The Team Linkages

There are many team-driven organizational initiatives that are endemic to the strategic management of the organization. Some of these initiatives are discussed below.

- Reengineering teams used to bring about a fundamental rethinking and radical redesign of business processes. The output of these teams is improved business and processed changes in the organization

- Crisis management teams that serve as a focus for any crises that might arise in the organization's activities.

- Product/process development teams which provide for the concurrent design and development of products, services, and organizational processes. Effective use of such teams result in higher quality products, services, and organizational processes, developed at a lower cost and leading to earlier commercialization and greater profitability

- Self-directed production teams which manage themselves and when team members are adequately prepared and used can result in greater efficiency and effectiveness in the manufacturing or production activity of the organization

- Task forces which are *ad hoc* groups to solve short-term organizational problems or exploit opportunities to support the operational and strategic well-being of the organization

- Benchmarking teams used to measure the organization against the most formidable competitors and industry leaders. When properly used the results of the work of these teams can be improved operational and strategic performance

- Facilities construction teams which design, develop, and construct capital improvements for the organization. This is the "traditional" use of project teams which have reached considerable maturity in industries such as construction, defense, engineering, research, and in government, educational, health systems, and economic development agencies

The use of alternative project teams has been necessary in part because of the way that jobs and strategies are changing in contemporary times.

4.5.3 Jobs Are Changing

Jobs are changing from fixed specific areas of responsibility to a team effort. The use of part-time and temporary workers is increasing; the organizational design today is undergoing major changes. Some of the more important other job changes include:

- An organizational design is used that is more flexible than the traditional structure based on specialization of work.
- There is increased reliance on alternative teams to deal with the change affecting the organization, particularly in the development of new and improved products and organizational processes.
- There is increasing acceptance of the belief that traditionally structured organizations are inherently designed to maintain the status quo rather than respond to the changing demands of the market and competition.
- There is less reliance on job descriptions or supervisor directions; rather the team members take their direction from the changing demands of the team objectives.
- Workers develop the knowledge, skills, and attitudes to deal with individual effort and the collective responsibility of the team.
- Team members report to each other, look for facilitation and coaching from the team leader, and focus their effort on the resources needed to meet team objectives.
- A single job is important only insofar as it contributes to the bundle of capabilities needed to get the team's work done.

Along with job changes, the traditional roles of managers and supervisors are changing. In fact, such traditional roles are becoming an endangered species. Team leader members are performing many of the traditional duties of these previous "in-charge" people. Contemporary managers are becoming mentors, facilitators, teachers, coaches, and other roles which are different from the "in-charge" nature of earlier periods. Team leaders and members are performing many of these traditional manager roles such as:

- Planning the work and assignment of the tasks by members of the team
- Evaluating individual and team performance in doing the work

- Moving toward both individual and team rewards
- Counseling poor team performers
- Participating in key decisions involving the work the team is doing
- Organizing the team members regarding their individual and collective roles
- Taking responsibility for the quality of the work, team productivity, and efficiency in the use of resources
- Developing initiatives to improve the quality and quantity of the output
- Seeking better ways of doing the work and, in so doing, discovering creative and innovative means of preparing for the team and the organization's future
- A form of team building is required to prepare the organization for the use of teams.

4.5.4   Preparing the Enterprise for the Use of Teams

There are some important fundamentals that managers need to recognize when preparing the enterprise for the use of teams in the organization. These include:

- An ability of the different teams to work with each other and to respect each other's territory
- Recognition of the high degree of interdependence among the members of the team, and how that interdependence can be used as a strength for integrated team results
- A willingness of the team members to share their information and work with each other, leading to a high degree of collaboration among team members
- Recognition of the deliberate conflict usually found in team work—a conflict, often stemming from different backgrounds, which exists because of substantive issues and not interpersonal strife
- A more efficient way for team members to proactively seek ways to work out problems and opportunities through team members to reach a result in which each member sees part of his or her work in the solution
- A better understanding of the culture of the team, its thought and work processes, and greater willingness to give unselfish support to the team's objectives and goals
- Enhanced communication among the team members about their individual and collective work on the team

- Better understanding of the purpose of the team, its relationship with the work of other teams, and how everything comes together to support the mission, objectives, and goals of the enterprise
- Greater opportunity for a higher degree of esprit de corps, a sense of belonging on the part of the members, and pride in working together to accomplish desired ends

The responsibilities with which the members of the design and implementation team are charged for the design and implementation of a team-driven strategy in the organization include the following:

- Providing for the transfer of the organization's vision and values to the teams
- Developing documentation for the teams, including a supporting charter, authority-accountability-responsibility relations, work redesign, and general cultural support
- Providing both conceptual and work linkages between the teams and the other organizational elements
- Evaluating existing reward systems for compatibility with the team organizational design
- Becoming the champion for the transformation of the organization from its traditional configuration to one characterized by team modes, values, and processes
- Providing general support, including allocation of resources
- Evaluating how the teams will interface with the functional expertise, strategic management, and cultural characteristics of the organization
- Recommending training initiatives to support the transition to a team-driven organization
- Assessing the existing supporting technologies, such as computer and information systems, to support the team architecture

4.5.5 Key User Questions

1. What are the likely advantages to the organization from the use of alternative teams described in this section?
2. Do the members of the organization understand and accept what alternative teams can do for the organization?
3. Has a specific strategy been developed to prepare the members of the organization for the use of teams and their likely individual and collective roles while serving on such teams?

4. Do the members of the organization understand the things in the organization that need to be changed to use teams effectively?

5. Do the members in the organization understand the meaning and probable use of teams to contribute to the organization's operational efficiency, strategic effectiveness, and functional excellence.

4.5.6 Summary

Teams working in cross-functional and cross-organizational environments are making major contributions to the operational and strategic well-being of contemporary organizations. The decision to use teams in the organization's strategy should be carefully assessed, to include the potential value of such teams, how the enterprise has to be prepared for using teams, and how the oversight of teams will be carried out. Changes in the culture of the organization will be profound when teams are introduced and used as elements of organization strategy.

4.5.7 Annotated Bibliography

1. Cleland, David I., *Strategic Management of Teams* (New York, NY: John Wiley & Sons, 1996), chap. 1, "The Concept and Process of Strategic Management," and chap. 2, "Getting the Organization Ready." In these chapters, the concept and process of strategic management is described along with guidance on how to prepare the organization to use teams. A substantial portion of the material in this section has been paraphrased from this reference.

2. Parker, Glenn M., *Cross-Functional Teams* (San Francisco, CA: Jossey-Bass, 2003). This book describes the challenges to managing teams considering the recent explosion of global and virtual teams. The book provides practical guides to include team rewards and recognition, communications technology, and multicultural and virtual-team issues, among other potential issues of managing teams in a wide variety of different applications. Many timely and relevant examples are given which provide a pragmatic relevancy to the book's messages.

4.6 SENIOR MANAGEMENT AND PROJECTS

4.6.1 Introduction

Senior managers, including general managers, have the residual authority and responsibility for the management of the organization, and are responsible to the Board of Directors. Since projects are building blocks in the design and execution of strategic management initiatives, these managers should pay particular attention to why projects are used in the organization and how well they are managed.

> **Senior managers have the residual responsibility and accountability for the management of projects in the organization**

4.6.2   Senior Management Responsibility for Projects

- Ensuring the adequate organizational design has been established
- Ensuring that the cultural ambience of the organization is evaluated to determine how well such culture will support the use of projects
- Providing resources to update the knowledge, skills, and attitudes of people associated directly with the projects
- Maintaining oversight over the development and acceptance of a plan of action for each project
- Seeing that an adequate information system exists that contributes to the planning and execution of projects
- Ensuring a system for monitoring, evaluating, and controlling the use of resources on projects to include an assessment of how well projects are meeting their cost, schedule, and technical performance goals and objectives
- Maintaining an ongoing consideration of where projects have, and will continue to have, a strategic fit in the organization's mission

The authority and responsibility of the senior managers of the organization and its directors are similar. If the directors of the enterprise are going to do their job in maintaining oversight over the planning and execution of projects, then the senior managers must ensure that the adequate management systems are in place in the organization, and that the directors are kept informed on the status of major projects underway.

4.6.3   Which Projects?

In an ongoing organization there can be hundreds of projects, of all sizes, and for different purposes in the product, service, and processes of the enterprise. General guidelines for which projects senior managers should be directly concerned will include:

- New product, service, and organizational process projects which have the promise of providing a competitive advantage in the marketplace
- Product/service projects which contain the potential for breakthroughs in technology, significant reduction of development costs, or the promise of bringing the product to earlier commercialization
- Projects that require the commitment and obligation of substantial enterprise resources, which if inappropriately applied can reduce the financial or competitive well-being of the organization
- Projects that are linked to strategic alliances with other organizations such as those designed to share development risks, or penetrate local markets in the global marketplace
- Projects which are closely linked to such strategies as downsizing, restructuring, cost reduction, productivity improvement, investment opportunities, or acquisitions, mergers, and alliances with other organizations

Senior management must develop a means for surveying the major projects underway, and determine which are sufficiently important that their personal involvement in planning for and execution of project strategies is required.

If adequate oversight of the firm's projects is not carried out, there is a risk of failure of one or more projects. Such failures could impact the competitive advantage that the organization holds in the market place. Some of the more common failures on the part of the oversight of senior managers is shown in what follows.

4.6.4 An Inventory of the Causes of Failures

- A lack of appreciation of how projects are inexorably tied to the operational and strategic trajectories of the organization
- Unwillingness to become involved in assessing how well the projects are being planned and carried out
- Ignorance of how projects should be managed in the spirit of a "systems approach" as outlined elsewhere in this Handbook
- Not providing resources that are needed to support the projects, as well as failing to understand and articulate how the organization's projects are linked to functional and other organizational units in the organization

- Not requiring that all projects in the organization should be reviewed on a regular basis following the model for monitoring, evaluation, and control suggested in Section 7 of this Handbook
- Not providing suitable recognition of project efforts
- Not understanding, and thus likely not appreciating, the impact that project stakeholders can have on in-house projects
- Neglecting the commitment of resources to train project teams in the management of projects
- Not being a role model for how prudent and effective management can be carried out in the organization

4.6.5 Senior Management Review by Life Cycle Phases

Conceptual Phase is the period when a conceptual framework for the project is being determined to include probable cost, schedule, technical performance objective, and potential strategic fit. Major responsibilities of senior managers include:

- Evaluating potential "deliverables" of the project to include probable strategic fit in the organization's businesses
- Determining the ability of the organization to provide the resources to support the project during its life cycle
- Determining the capability of the organization to provide trained people to serve on the project team as well as provide human resource support from the functional entities of the organization
- Selecting of the "right" project manager who is clothed with the authority and responsibility to manage the project
- Ascertaining if adequate planning has been established for the project
- Finally, seeing that the project and its deliverables are appropriately linked to the operational and strategic purposes of the organization. Require contingency planning for the early termination of the project.

Execution Phase is the phase when resources are committed to the project and its design, development, and execution is underway. Major responsibilities include:

- Provisioning of resources to support the project
- Giving the project manager and the project team the freedom to manage the project without interference from other managers

- Providing oversight of the project's cost, schedule, and technical performance progress followed by appropriate feedback to the project manager and project team
- Maintaining contact with key stakeholders such as project customers, suppliers, and regulatory agencies
- Ensuring that other managers in the organization are committed to supporting the project through the allocation of needed resources
- Allowing the project manager and team members enough freedom to follow their muse in finding innovative and creative solutions to problems and opportunities
- Providing a buffer to guard the project against the inevitable politics that come up in any organization and its stakeholders
- Providing suitable rewards and inducements to bring out the best performance in the people supporting the project
- Determining if a project audit review would make sense

Post-Project Phase is the phase when the project's results are being integrated into the operational and strategic business of the organization. Major responsibilities during this phase include:

- Ensuring the successful strategic fit of the project into the affairs of the organization
- Ensuring that appropriate measures have been taken for the after-sales support of the project results
- Doing a post-project review to gather information on how well the project was managed to include a collection of "lessons learned" that can be passed on to other project management initiatives
- Examining the rationale for the project to support organizational purposes

4.6.6   Key User Questions

1. Do the senior managers understand the key role that they have in the management of projects in the organization?
2. Do the senior managers recognize that projects are building blocks in dealing with product, service, and organizational process change.
3. Given such recognition, have these managers developed a philosophy and protocol for the ongoing regular review of projects?
4. Senior managers are vital links between the enterprise projects and the organization's board of directors. Have these managers provided infor-

mation and a protocol on how the board should carry out its responsibility in maintaining oversight of the planning and execution of projects.

5. Have provisions been provided for the post-project reviews to help the organization do a better job of managing projects in its future?

4.6.7   Summary

In this section, the role of senior managers in maintaining oversight of the planning for and execution of projects was presented. Such managers have key responsibilities in the planning for projects to include an assessment of what the likely strategic fit of the project will be. The role of these managers was described in the three key phases of a project, with appropriate oversight responsibilities that senior managers should carry out during these phases.

4.6.8   Annotated Bibliography

1. Hartley, Kenneth O., "The Role of Senior Management," in David I. Cleland, ed., *Field Guide to Project Management,* 2nd ed. (New York, NY: John Wiley & Sons, 2004). This chapter provides basic philosophies and protocol on how senior managers can best execute their fiduciary responsibilities in the management of projects in the organization.

2. Snyder, James R., "How to Monitor and Evaluate Projects," in David I. Cleland, ed., *Field Guide to Project Management,* 2nd ed. (New York, NY: John Wiley & Sons, 2004). See This chapter provides insightful guidance into the monitoring and evaluation of projects by senior executives. The chapter is an excellent primer on the responsibility of senior executives to keep abreast of key projects in their organization.

4.7   *THE BOARD OF DIRECTORS (BOD) AND MAJOR PROJECTS*

4.7.1   Introduction

Once a project is funded and organization resources are being used to design, develop, and construct (manufacture) the project, the BOD has an important responsibility to maintain surveillance over the progress that is being made on the project. By maintaining such surveillance, the BOD gains valuable insight into how well the organization is being prepared for its future. The projects over which the BOD and senior management of the oganization should maintain approval and oversight include:

- New product, service, and organizational process projects, which hold the promise of giving the organization significant competitive advantage in the marketplace
- Projects which require the commitment of substantial resources such as new facilities, restructuring, reengineering, or downsizing initiatives
- Projects that lead to strategic alliances, research consortia, partnering, and major cooperative efforts with other organizations
- Major initiatives that promise to provide initial or expanded presence in the global marketplace
- Projects, such as concurrent engineering initiatives, which promise to provide earlier commercialization of products and services
- Projects that lead to potential major changes in the organization's mission, objectives, goals, or strategies

By maintaining oversight over these kinds of projects the BOD members gain valuable insight into how well the organization is preparing itself for the future.

The BOD has principal fiduciary oversight responsibility for major projects in the organization

---

4.7.2   Some BOD Inadequacies Regarding Projects

- On the Trans-Alaska Pipeline System (TAPS) an "Owner's Committee," similar to a BOD, did not maintain oversight of the TAPS project, particularly with respect to strategic decisions on the project to include:
  - The development of a master plan for the project
  - Early integrated life-cycle project planning
  - Design and implementation of a project management information system
  - Development of an effective control system for the project
  - Design of a suitable organization

- In the design and construction of nuclear power generating plants all too many utilities had BODs that neglected to exercise "reasonable and prudent" oversight over the planning and execution of nuclear power plant projects to support corporate purposes.
- The Washington Public Power Supply System (WPPSS) defaulted on interest payments due on $2.5 billion in outstanding bonds in part because of the failure of its directors. Communication at the senior levels of the company, to include the BOD, tended to be "informal, disorganized, and infrequent."

However, on some nuclear power plant projects, BOD served in an exemplary fashion, such as in the case of the Pennsylvania Power and Light Company where ongoing surveillance was conducted over the planning and construction of the Susquehanna nuclear plant project.

4.7.3   Key Responsibilities of BOD Regarding Projects

- Set an example for the ongoing review of projects that support organizational purposes.
- Provide senior managers guidance in the strategic management of the organization as if its future mattered.
- Assure the strategic fit of ongoing projects with the strategic direction of the organization.
- Ensure that projects are recognized as building blocks in the design and execution of strategies, and that the linkages of projects with other initiatives in the organization are carried to help ensure the organization's future.
- Maintain ongoing and regular review of major projects, and through doing this, help to motivate the general managers and project managers to do the same with respect to their projects.
- Be available to meet with key project stakeholders, such as project customers, should the need arise during the life of the projects.
- Review the key elements of the plan for major projects.
- Require formal BOD briefings or the status of the project at key points in the project's life cycle.
- Visit the construction site on major projects. By so doing, an important message will be sent to the project team, and to other stakeholders on the project.
- Direct that the necessary independent audits are carried out for the projects that need such audits to determine their status and progress.

4.7.4 Project Information for the BOD

To carry out their responsibilities, the BOD members need certain information that is provided in an orderly and regular fashion. Examples of this information follows:

- Presenting information and issues on the project progress and status prior to BOD meetings so that the members have time to study the material prior to the meeting
- Providing for clear reporting on project information to the BOD members, avoiding burying the information in other corporate records to be reviewed at the BOD meeting
- Allowing time at the BOD meeting for a full discussion of the relevant project-related issues and decision matters
- Ensuring that the BOD members take the time and care to determine the "strategic fit" of major projects underway in the organization

4.7.5 Key User Questions

1. On what basis are the major projects in the organization reviewed by the BOD members?
2. Have the review of the major projects in the organization by the BOD resulted in any cancellation, redirection, or acceleration of the use of resources on the project?
3. Do the members of the BOD receive timely and relevant information about the status of the projects in the organization, which enables them to make informed judgment about the progress and status of such projects?
4. Are any of the major projects in the organization likely candidates for a BOD directed audit?
5. Have any of the major projects in the organization been the target, or might become the target, of litigation proceedings?

4.7.6 Summary

In this section, the role of the BOD with respect to major projects in the organization was presented. A case was made that an ongoing surveillance over the planning and execution of projects would provide the BOD members insight into the effectiveness with which the organization is preparing for its future. Suggestions were made concerning how the role of the BOD regarding projects could be strengthened.

4.7.7 Annotated Bibliography

1. Cleland, David I., and Lewis R. Ireland, *Project Management: Strategic Design and Implementation,* 4th ed. (New York, NY: McGraw-Hill, 2002), chap. 5, "The Board of Directors and Major Projects." This chapter provides a prescription on how major projects should be reviewed by the BOD. Information, strategies, and policies are suggested to improve the review of projects by the BOD.

2. Shultz, Susan F., *The Board Book* (New York, NY: AMACOM, American Management Association, 2001). This book recognizes some major failings of corporate directors in the execution of their fiduciary responsibilities for the enterprise. The author believes that "For nepotism, lethargy, and plain old misguidance, few corporate entities can top the board of directors." Starting from that belief, Ms. Shultz provides a practical, reader-friendly guide for developing and managing on-track, top-notch corporate boards that can serve as models for recognizing and discharging board responsibilities.

4.8 PROGRAM MANAGEMENT

4.8.1 Introduction

Program management is an organizational strategy that groups projects together at a higher level when they have a common purpose in supporting the goals and objectives of the organization, and eventually contribute to the operational or strategic purpose of the organization. Sometimes these groupings take on different names, such as "megaproject," "multiproject," or, increasingly "program."

These groupings are all directed to a bundle of integrated projects and initiatives that are building blocks in the design and execution of organizational strategies. The program provides oversight of the interrelated ad hoc projects that, taken together, provide for new products, services, or organizational processes that have a common purpose. Having a program structure in the organizational design of the enterprise means that projects can be conceived, designed, and built in an interdependent manner. To support a larger purpose of the organization, an explicit philosophy of program management guides the development of the individual projects as well as a program structure.

Programs are umbrellas!

4.8.2   Programs vis-à-vis Projects

There are differences and similarities between *Programs* and *Projects*.

- Programs are composed of a mixture of interrelated projects and other initiatives to create something that does not exist in an organization's products, services, or organizational processes.
- Programs are key "choice elements" in the strategic management of the enterprise.
- Programs become integrated into the ongoing business of the organization.
- A matrix organizational design may be used for the management of a program.
- Programs serve as a cover and protector for projects that are used in the operational and the strategic management of the enterprise.
- Programs can be used as an effective management strategy to prepare the organization for its future.
- A program manager has line jurisdiction of the assigned projects—but at the same time may have a matrix organizational design with other programs in the enterprise.

4.8.3   Projects

- Projects are ad hoc initiatives to create something for the enterprise that did not previously exist.
- Projects have a defined life cycle.
- When completed, projects become integrated into the ongoing business of the organization.
- A matrix organizational design is usually used for the management of a project.
- Projects are building blocks in the design and execution of enterprise strategies.
- A project when completed contributes to a goal or objective of the enterprise.

Programs and Projects	
Programs	Projects
Composite of projects	Ad hoc initiatives
Key "choice" elements	Defined life cycle
Ongoing strategy	Building blocks
Cover/protector of organizational initiatives	Matrix design
Line jurisdiction over projects	Change management

4.8.4 Evolution of Programs

Sometimes programs are conceived and developed on an ad hoc basis with only a general sense of how they might fit together as an entity. Then, as the projects gain maturity in the enterprise, managers realize that something bigger is emerging. To capitalize on this larger opportunity, an informal Program Office is set up to coordinate the family of projects. Another alternative way of a program structure to emerge is as a starting point to design and develop a high-level strategic objective from which a program mission statement can be developed. From this mission statement a family of initiatives, most of which become projects, come forth to provide a foundation for the development of future strategies for the enterprise.

Alternatively, a program structure can emerge as a starting point to design and develop a high-level strategic objective from which a program mission statement can be developed. From this mission statement through the planning process a family of initiatives, most of which become projects, comes forth to provide a foundation for the development of future strategies for the enterprise.

Another alternative is when ad hoc projects developed as building blocks in the design and implementation of organizational initiatives seem to have a common purpose. A program theme or mission statement becomes the logical output to support the organization's future.

As projects are reviewed within the context of the program, a higher-level review of the program as an element of organizational strategy is required. When programs are established as an element of organizational strategy, a designated program manager is required, who may be designated as a profit center manager when the program is of sufficient size and importance to the organization.

4.8.5 Example of Program Evolution

A manufacturing company realized that competitive pressures dictated that it review the systems nature of how manufacturing was carried out in the organization. A competitive analysis and an assessment of the firm's strengths and weaknesses revealed that the following weakness existed:

- A serious obsolescence in the manufacturing equipment caused by the failure of the company to have an ongoing equipment program to update its ability to apply competitive state-of-the-art technology to improve the quality and reduce the cost of producing its product lines.

- The competency of the manufacturing people was not comparable to that of competitors. The obsolescence was aggravated because the people working with manufacturing equipment were not aware of the contemporary state-of-the-art technology.

- The manufacturing plants were outmoded facilities, which would require updating as the company acquired modern equipment and upgraded the knowledge and skills of its manufacturing workers and specialists. As new equipment and the skills of the manufacturing personnel were updated, the marketing, financial, and R&D strategies of the company would have to be evaluated.

- As the manufacturing competency of the company began to improve, an analysis of the manner in which manufacturing was executed needed to be conducted.

- Finally, during the preliminary evaluation of the company's manufacturing capability, the question arose as to whether the company should do in-house improvement of its competency or, alternatively, acquire small companies that were already manufacturing at the leading edge of the advancing manufacturing systems technology.

Given these conditions, a manufacturing improvement program was established to provide central management in the analysis and execution of improved manufacturing capabilities. A senior executive of the company was appointed as a Manufacturing Improvement Manager. This manager designed and developed a project-driven strategy, which consisted of the following "building block" project approaches:

- Project teams to design, develop, and oversee the updating of the company's manufacturing facilities.

- A project team to investigate and coordinate the development and use of self-managed production teams to replace the traditional organizational structure of the company.

- A project team to develop the requirements for the updating the knowledge, skills, and attitudes of managerial, supervisory, and technical professionals in the organization.

- A team to develop and execute a strategy for the replacement of the traditional equipment with modern manufacturing equipment, to include supporting systems such as process planning and control, inventory management, equipment layout, and production control systems.

- Each of the functional areas was evaluated by reengineering teams to determine what changes were needed by way of financial strategies, product marketing, order entry protocol, quality improvement, and related activities.

As the program and its supporting projects developed, regular reviews of the individual projects as well as the overall program status were carried out by the program manager and senior managers of the enterprise.

Over a period of three years, the company was able to improve its manufacturing capability to the point where its products, services, and organizational processes were competitive in the marketplace.

One of the key measures used in the program/project structure was the ongoing involvement and oversight carried out by the senior managers of the enterprise. Without such involvement it is doubtful that the company's high degree of success could have been achieved. Of course, in order to carry out an active ongoing review of the program/project developments, an information system had to be designed to provide managers at all levels of the status of their development work.

4.8.6   Another Example

Projects are the principal building blocks in a program. An aerospace corporation, the Lockheed Martin company, considers programs to be the vehicle for achieving customer satisfaction and shareholder value. The components of its major programs range from $1 million commercial information technology support programs to multibillion-dollar government contracts. In this corporation, one program, the F-35 Joint Strike Fighter (JSF), involves an incredible complexity of sub-programs and projects. Valued at an approximately $200 billion contract, the program spans decades, involves several major partners, and support organizations in more than 27 states and the United Kingdom. A multitude of support projects by way of key equipment and logistic support systems are involved. The overall JFS is managed through a Program Management Council. This top-level organization is composed of key business leaders from each of the company's business areas and corporate functional units and meets every two months for a program review. In such a large program the development and deployment of standard management processes, tools, and techniques is a major management challenge.

4.8.7   Continued Explanation

Programs provide a focus for various projects and activities having a common goal to support organizational objectives. For example, in the case of a product cost improvement program, the associated projects and activities leading to a single defined goal for the enterprise might be as follows:

- Quality improvement project
- Production planning and control improvement project
- Plant layout redesign project
- Employee relations project

The idea of a program is not precisely defined; it is closely related to the objectives of the organization. One clear characteristic of a program is that it is *output-oriented.* A program is first defined in terms of what the organization is trying to achieve.

The US Office of Management and Budget (OMB) used the following principles for the initial development of appropriate output:

- Program categories are groupings of agency programs which serve the same broad objective or generally similar objectives, such as improvement of higher education.
- Program subcategories are subdivisions which should be within each program category, such as the improvement of science and technology education.
- Program elements are subdivisions of program subcategories, such as research funding support in science and technology.

The interaction of the programs, operating units, and staff functions of an organization in program budgeting is illustrated in Fig. 4.10, which gives a hypothetical description of the US Department of Defense. Each of the Armed Forces and each staff function conceivably cuts across each major program. Thus, budgeting on a program basis allows duplications can be eliminated and valid requirements for the accomplishment of objectives to be determined. For example, both the Navy and the Air Force contribute toward the strategic retaliatory mission via submarine-based

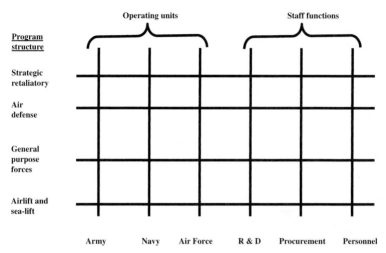

FIGURE 4.10   DOD program structure.

missiles and long-range bombers. Also, all staff functions must exert efforts toward accomplishing these objectives.

In the civilian sector, we could construct a hypothetical corporation to illustrate a similar interaction through the program structure. Figure 4.11 depicts the interaction of operating units, staff functions, and programs for a corporation in the same fashion that the previous model does for the US Department of Defense. One major program of the fictitious organization is plant nutrition. The corporation's chemical products division is involved, as are the agricultural marketing division (which may sell bulk fertilizers to farmers) and the consumer products division (which may sell both lawn care chemicals and equipment to individual consumers). Similarly, in the animal nutrition program, the chemical and consumer products division might be concerned with pet food products.

Although PPBS (Planning, Programming, and Budgeting) has been formally dropped by the US federal government, the basic ideas of PPBS have stayed around. The experiences of the US federal government in PPBS have served as a model for non-governmental organizations to use. The overall system of PPBS was designed to enable each agency of the government to:

• Make available to top management more concrete and specific data relevant to broad decisions

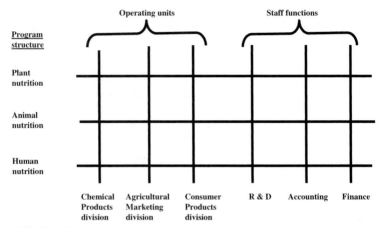

FIGURE 4.11 Corporate program structure.

- Spell out more concretely the objectives of government programs
- Analyze systematically and present for agency head and presidential review and decision possible alternative programs to meet those objectives
- Evaluate thoroughly and compare the benefits and cost of programs
- Produce total rather than partial cost estimates of programs
- Present on a multi-year basis the prospective costs and accomplishments of programs
- Review objectives and conduct program analyses on a continuing, year-round basis, instead of on a crowded schedule to meet budget deadlines

4.8.8 User Questions

1. Have the managers of the organization ever considered the use of a program management approach to bring a focus to the management of projects and other initiatives in preparing the enterprise for the future?

2. If a program structure is not being used in the organization to bring a focus to new initiatives in the organization, what philosophy and organizational design are being used to accomplish the needed convergence?

3. If the organization uses the program approach in its strategies, how well do the appropriate managers maintain oversight of the program approach and the results likely to be accomplished from using this approach?

4. Considering the program/project organizational design approach in the organization, has adequate documentation by way of policies, processes, protocols, and procedures been published and distributed to the principal stakeholders in the enterprise?

5. Do the principals using the program/project organizational design approach understand the resulting "matrix management system" that will likely come forth?

4.8.9 Summary

In this section the program management structure has been summarized. A standard definition and approach to this structure was described, to include the role that projects play in such an approach. Several examples of the use of the program management structure were given to show how programs provide a focus for the convergence of projects and other initiatives in preparing the enterprise for its future. One of the examples given was that initiated by the US government during the early 1960s.

4.8.10 Annotated Bibliography

1. Davies, C., A. Demb, and R. Espejo, *Organization for Program Management* (New York, NY: John Wiley & Sons, 1979). This book, although published over two decades ago, provides one of the earlier descriptions of program management in the non-governmental context. The theme of the book is that a program approach provides a more comprehensive concentration of resources and attention to key organizational initiatives than can be achieved through normal traditional management processes. The authors provide insight into how and why program management emerged as a new way of managing resources within the context of project management.

2. Gray, Roderick J., "Alternative Approaches to Programme Management," *International Journal of Project Management,* vol. 15, no. 1, 1997, pp. 5–9. This short and relevant article describes the different approaches to the grouping of projects. It suggests that there may be circumstances in which traditional hierarchical management structures may not act to improve performance, and suggests some of the alter-

native choices to be made within the context of program and project management.

4.9   PORTFOLIO MANAGEMENT FOR PROJECTS*

4.9.1   Introduction

Portfolio management for projects uses the characteristics and attributes of projects to categorize and assess their value to the enterprise. This identifies specific characteristics of projects to determine the positive and negative factors that determine objective selection criteria. In turn, the selection of projects gives a balanced portfolio containing different characteristics that aligns with the enterprise's strategic goals and operational needs.

> **Project portfolio is more than size—small, medium, and large**

Projects are typically categorized as small, medium, or large. These categories take on different meanings based on the organization's definition of project size. Size is relative and is only one parameter for describing projects. There are other important characteristics that must be defined for the organization and used to build a portfolio of projects that support enterprise productivity and growth.

Portfolio management for projects is a strategic decision that determines favorable characteristics before a project is selected. Selected projects must have a strategic fit with the enterprises strategic objectives and goals and an organizational fit with the enterprise's competencies and be building blocks to the organization's business line. Within the selected projects are other parameters that have favorable indications of success that create a balance between such characteristics as high, medium, and low risk.

---

*Adapted from David I. Cleland and Lewis R. Ireland, "Portfolio Management for Projects," in *Project Management: Strategic Design and Implementation*, 4th ed., (New York, NY: McGraw-Hill, 2002), pp. 210–217.

4.9.2 Background on Portfolio Management

The financial community learned some time ago that the best position for investments is a diversification among stocks, bonds, and cash. Within the investments are different degrees of risk of loss and different degrees of opportunity for gain. A balance between the type of investment and the opportunity for gain or loss, consistent with an individual's goals, comprises a portfolio. For example, an investment portfolio of $100,000 could be diversified as $75,000 in stocks, $23,000 in bonds, and $2,000 in cash. An investor might be willing to accept the risk of gain or loss in the stocks and take advantage of the higher degree of return than one could earn on bonds or cash.

Project portfolio is the future of enterprise strategy

Project portfolios are gaining acceptance in the organizations to balance the types of projects with their relative benefits. Each organization needs a tailored portfolio that aligns with its strategic goals, operational needs, and organizational competency. Basic solutions of balancing size and risk to fit the organization's capability could be a start. More sophisticated models may be needed to emphasize and capitalize on the organization's direction and the rate at which it may need to grow.

4.9.3 Developing a Project Portfolio Model

The project portfolio model needs to be specific for an organization. The table below provides an example of a relatively fundamental model that could be a start point for an organization. The items listed are examples of areas that could comprise an organization's portfolio model.

Organizational Project Portfolio Model (Example)

No.	Item	Target
1	Small-sized projects	32
2	Medium-sized projects	12
3	Large-sized projects	4
4	High-risk projects	2
5	Medium-risk projects	6

No.	Item	Target
6	Low-risk projects	40
7	Projects related to competencies	48
8	Number of project customers	>12
9	High-technology projects	2
10	R&D projects	5

In this table, there are three sizes listed with targeted numbers of project defined for each size. There is a tendency to have the majority of projects in the small size, perhaps because they are quick to complete, the organization has a lot of experience in small projects, there is a need for a steady revenue stream, or this may be the majority of the market for this organization's products and services.

Project risk is a major consideration in any endeavor. The three listed levels of risk are not further defined, but the trend is to target low-risk projects. Whereas low-risk projects may be the preferred type, typically low-risk projects are accompanied by low profit margins and called "bread and butter" work. High-risk projects, on the other hand, should have greater profit margins. Why is the project high-risk? It could be that this is a new technology for the organization or that a lot of uncertainty is associated with the project.

Matching projects to competencies is important for performance. Penalties for late delivery or not meeting functionality requirements can materially impact an organization. Furthermore, an organization should be continually building on its core competencies and expanding with growth into new product areas. Performance on projects is directly related to the organization's competency maturity.

Diversity in project customers is a plus. Having only one or two customers places an organization in an untenable position if a customer no longer desires the marketed products or the customer should close his/her business. Several customers increase the work of managing stakeholders, but take away the risk of the sudden loss of all or a major share of the business by loss of one or two customers.

High-technology projects typically give an organization new products and byproducts, but often at an investment cost if the product is not one of the existing product lines. Growth and better positioning through products that are pushing state-of-the-art can be beneficial both in terms of profits and in corporate image. High-technology products may, however, require a large investment to make advanced features and functions.

Research and development projects may be required to explore methods of manufacturing new products or defining new service features. This is an investment in the organization's future unless a contract can be won that permits charging to develop new products and services. Typically, advanced products and services would be defined in the marketing strategy for an organization or some goals set in the strategic plan.

Project selection and initiation would typically follow the model outlined in the table above. A greater amount of detail would be required, such as detailed definitions of project size, risk criteria, and organizational competencies.

4.9.4 Project Selection Criteria

Criteria for project selection need to be developed and tailored consistent with the organization's size, business line, products and service offerings, customer base, and other factors. In the following table, some examples of project characteristics and associated criteria are suggested. These represent examples of project characteristics that an organization may use to select projects. These are not all the criteria that might be used, however, and other criteria should be considered.

Organizational Project Portfolio Selection (Example)

Characteristic	Criteria
Project size	Projects mix will consist of small (less than $10K), medium (between $10K and $100K), and large (more than $100K) projects. • Small projects will represent 70 percent of the total projects. • Medium projects will represent 27 percent of the total projects. • Large projects will represent 3 percent of the total projects.
Core competencies	Projects will be selected based on fitting into one of the organization's four competencies. Exception to this will require approval of the Board of Directors.
Business risk	Projects (products) will have a high degree of success (90 percent or more) before being selected).
Project risk	Project goals will have a 70 percent or greater chance of success.

Characteristic	Criteria
Technology	Projects requiring new technology will be compatible with existing core competency growth plans.
Profitability	Projects will have an expected profitability of greater than 15 percent.

These criteria offer a very conservative approach to project selection and a project with sufficient return on investment might be selected. For example, business risk is 10 percent or less where in actual situations there may be enough uncertainty about the project to reduce the probability of success. This may result in an organization selecting a project with perhaps only 60 percent chance of success.

The project selection process may have many variables, some used alone to eliminate a project and others in combination to select or eliminate. Again, this is an organization's strategy as to which projects to select and which to avoid. Some variables and their explanations are described in the table below for consideration by organizations constructing a project selection model.

Elements of a Project Selection Process

Variable*	Comments
Profit margin	Organizations need to earn a profit on their project work when the resultant outcome is a product or service being sold. Organizations typically set a profit goal for projects based on factors such as risk, degree of difficulty to complete the project, the type of work, and whether there is a usable byproduct of the project.
Project risk	Organizations must assess the risk of the project. The risk may be whether the product will meet market expectations or whether the project can be completed within established goals for cost, schedule, and technical performance.
Process change	Organizations may use projects to optimize their organizational processes. Using projects to upgrade or establish new processes may be the most cost- and time-effective method.

Variable*	Comments
Resources	Human and non-human resources need to be considered. Human resources may or may not have the requisite qualifications to perform on a project. Special material resources or tools may or may not be available to complete the project. In some instances, current human resources may be insufficient in quantity to complete the project within the required time frame.
Financial considerations	Cash flow may be negatively impacted by large initial expenditures on a project. The cost of labor or outsourcing of work can also have major financial impacts for an organization.
Building block	Question whether the project is a building block for the organization by further development of core competencies and contributing to the organization's success. Or is the project something that neither contributes to building the organization nor is within the organization's overall purpose?
By-product	Question whether there are byproducts that can be used in subsequent projects or whether byproducts may be used to enhance the organization's future capabilities.
Technology	Question whether the organization understands technology and is building a business based on it. Also, the degree of maturity of a technology or whether the technology is to be developed is key to the project selection.
Project duration	Question whether the project duration fits into the normal work arrangements and whether there are only long- or short-duration projects in the organization.
Size (relative to organization)	Question whether the size (dollar, resource, duration) is right for the organization. Organizational structure may find the project to be too large for the capability or too small for the type of management used.

Variable*	Comments
Corporate image	Question the image that the corporation will get when taking on a project.
High competition	Question the degree of competition for the project or the product and whether this project is a declining market area.
Client	Question whether the new project is for an existing client or a new one. Determine whether the organization's business is centered around one or two clients where any loss of a client would have a major impact on the organization.
Life cycle phases	Question whether the project's life cycle phase provides for continuity of work or whether there is interrupted flow work. Interrupted flow work typically requires more resources and costs more.
Core competency	Question whether the project is within the organization's established core competencies or whether it is initiating a new core competency. Performing existing competencies is easier than starting new ones.
Urgency of need	Question the urgency of need to determine whether delivery is possible within the time frame desired. Also, determine whether there will be resources available to complete the project.
R&D	Question R& D projects to see if there are too many or too few and whether their focus is on the right or wrong areas. Determine whether new projects duplicate efforts or lead to enhanced or new products.

* *Note:* These variables are purposely in a random order. The order of items in the table does not place any weight on a variable or suggest that a variable may apply to an organization's strategy. Further, organizations must develop their specific model and may use any or all of the above variables.

Once project selection criteria are established for the type of projects determined by an organization's portfolio of needs, a review of all projects can be made. Using a model of what is ideal for the organization, one can then compare the differences to establish a direction for change. Seldom will the model replicate the actual situation.

For purposes of illustration of the concept of a portfolio of projects, the table below gives an example of what an enterprise could have for project parameters.

Example of a Portfolio of projects

No.	Item	Target	Actual	Difference
1	Large-sized projects • Low risk • Medium risk • High risk	 3 1 0	 2 0 1	 −1 −1 +1
2	Medium-sized projects • Low risk • Medium risk • High risk	 10 4 1	 8 6 2	 −2 +2 +1
3	Small-sized projects • Low risk • Medium risk • High risk	 22 14 6	 19 15 	 −3 +1 −6
4	Internal process change • Organizational change • Continuous improvement	 4 2	 1 1	 −3 −1
5	Internal product development • New product • Improved product	 4 7	 0 2	 −4 −5
6	Internal ad hoc • Benchmarking • Competitive analysis • Marketing effectiveness	 1 1 1	 0 0 0	 −1 −1 −1

This table shows some differences that may have significant impacts. There is a slight trend toward selecting high-risk projects in all three sizes. This may send a signal that project risk is a possible future problem for the organization. The second area of concern is that there is a shortfall in the internal projects that build on the organization. This may suggest that there is no budget being allocated to enhance the organization's capability to do business and operational decisions are focused on projects that generate revenue.

This example of a summary of projects gives information for an organization and should provide direction for future project selection. It is also possible that this information could be compared to the strategic goals to determine whether they are realistic for the product, service, and market.

Using a portfolio of projects gives senior management a better understanding of the character of the organization's work. The portfolio focuses attention on projects that are meeting selection criteria as well as those that exceed the established boundaries. Portfolio management provides that framework to guide the organizations to better decisions about work and the nature of the work. It also provides the opportunity for improved productivity and growth.

4.9.5   Key User Questions

1. How is project portfolio management linked to the strategic plan of an organization?
2. How is project size determined for selecting as a candidate to be included in a project portfolio?
3. Who in the organization is responsible for managing the project portfolio and why is this position responsible?
4. Why would it be imprudent to select only projects with a high payoff to include in the portfolio?
5. What is the role of the project manager in managing the project portfolio?

4.9.6   Summary

Project portfolio management is the development of a balanced set of projects for an enterprise through a tailored model that aligns project selection with strategic goals. Projects are selected based on other considerations, such as the organizational fit of a project and the organization's competency to complete the selected project.

Project portfolio strategy requires that a specific model for the balance of projects be defined and used to select projects. This unique model meets the needs of the organization as expressed in the strategic goals and organizational business plans. Alignment of project selection to meet these goals is critical to survival and growth in a highly competitive world.

Project portfolio management adds value to any organization by providing for understanding the types of projects selected and maintaining a balance of projects consistent with the strategic goals and objectives. Project portfolio management also highlights adverse trends for an organization when such items as too much risk is accepted through individual projects selected outside of the strategic and organizational fit.

4.9.7   Annotated Bibliography

1. Senn, Ann, "Portfolio Management," *Chain Store Age,* vol. 78, no. 10, Sec. 2, October 2002. Senn provides a comparison between Internet Stock and Information Technology projects. Senn addresses the need for a diversified portfolio and the process of selecting new projects within a field of many. Using the Information Technology discipline as an example, the author makes the point that projects are selected without regard for objective selection criteria.

2. Hoffman, Thomas, "Balancing the IT Portfolio," *Computerworld,* vol. 37, no. 6, February 10, 2003, pp. 25–26. Hoffman provides an excellent review of tools that are used in project portfolio management to track progress of projects, manage resource assignments, and conduct scenario variations. Hoffman discusses the pros and cons of the various software tools offered by major tool vendors to include the features and functions used to track and control multiple projects.

SECTION 5
PROJECT LEADERSHIP

5.1 OVERVIEW OF PROJECT LEADERSHIP

5.1.1 Introduction

A project leader is that individual who leads a project during its life cycle and accomplishes the project's technical objectives on time and within budget. To lead any organizational effort, both a *presence* and a *process* are required.

5.1.2 Key Characteristics of Proven Project Leaders

- They have their act together.
- They are visible to the team members they lead, and are on top of everything.
- They are available to their team members to listen, debate, and gather information for decision making and execution.
- They are able to say, "Let's do it," when the time is right.
- They are decisive, and have a track record of making and executing the right decisions.
- They see the best in the competencies of the team members.
- They work at making things simple, and avoid making things complex.

- They are fair and patient.
- They work hard in their leadership role.

5.1.3 Project Leadership vis-à-vis Managership

Warren Bennis in "Good Managers and Good Leaders," *Across the Board*, October 1984, pp. 7–11, proposes a distinction between these two roles as paraphrased thusly: " A leader does the right things (effectiveness) and a manager does things right (efficiency)." Taking Bennis's distinction and fleshing it out provides the following characteristics of leadership and managership:

Leadership is the capacity to lead

5.1.3.1 Leadership

- Develops and sells a vision for the project
- Copes with operational and strategic change on the project
- Builds reciprocal networks with relevant stakeholders
- Develops a cultural ambience for the project team that facilitates commitment and motivation
- Sets the general direction of the project through collaboration with project stakeholders
- Perceives broad issues that are likely to impact the project, and then works with the team members in accommodating these broad issues
- Becomes a symbol of the project and its purposes
- Becomes the principal project advocate in working with stakeholders
- Does the right things

5.1.3.2 Managership

- Copes with the complexity of developing and implementing a management system for the project
- Maintains oversight of the efficient and effective use of project resources

- Designs and develops the management functions of planning, organizing, motivating, directing, and controlling within the context of a project management system (PMS), for the project
- Reprograms resources as needed to maintain a balance for supporting the project
- Monitors the competence of project team members to include guidance to these individuals for the improvement of their knowledge, skills, and attitudes
- Ensures that the communication processes involving the project work effectively
- Maintains oversight to ensure that project monitoring, evaluation, and control are carried out
- Does things right

Project managers must both *lead* and *manage*. In carrying out these two roles, competency in the following is required:

- Having a general understanding of the technology that is involved in the project
- Having those interpersonal skills that facilitate building a cultural ambience for the project team and its stakeholders that reflect trust, loyalty, commitment, and respect
- Understanding the management processes and its application to the project
- Being able to see the "systems" context of the project
- Being able to make and implement decisions involving the project
- Being able to produce the desired results on the project

The competency to serve as both a project leader and a project manager is dependent on the individual's knowledge, skills, and attitudes.

5.1.4 An Experienced Viewpoint

At a meeting of experienced senior project managers, the participants were asked to write down a phrase, word, or sentence that described a "good project leader" and a "poor project leader." The results are shown in Table 5.1. The contrast between good and poor project leaders is evident. Members of the project team should ask themselves how they would describe their leadership style, and whether they would fall under the good or poor leadership column.

TABLE 5.1 Good and Poor Project Leaders

Good project leader	Poor project leader
Positive attitude; recognition; knowledgeable supervisor	Uses authority position title to direct people—does not understand or solve. Does not listen effectively, ignores or rejects input not politically acceptable. Changes scope or direction at will while blaming others for doing the wrong things.
Interested in personal aspects of employees (family situations, etc.), anticipates concerns (problems) before they become evident. Excellent role model; decisive.	Does not ask for help; does not set an example for the followers; does not know the technical aspects of the process.
Clearly communicates a vision of what is to be accomplished, challenges and motivates. Key is that manager gives measurable parameters by which to chart programs. A "results-oriented" manager.	Lets the managers run the business in an undisciplined manner. Does not stay on top of the problems when they arises. Cares only about the bottom line. Does not commend, only criticizes
Can see a vision of the future for the business, communicates it to the people involved in the business, and allows the personnel involved to make the contributions toward the goals. Exhibits trust, support, and willingness to take blame and suffer disappointment, yet still trust and support.	Does not translate the vision (if there is one)—does not explain why. Pays little attention to implementation—"That's just the month's buzzword."
Helps subordinates set a direction for work and allows them to grow toward that goal. Is a mentor, not a master. Knows where everyone is going and why everyone wants to get there—and is able to follow.	Does not listen to other's ideas. Does not know how to constructively criticize. Expects perfection. Does not recognize or compliment a job well done. Discourages creative thought and new ideas.
	Totally focuses on self promotion. Unenthusiastic. Cannot communicate vision or ideas.
Sensitive to the effects of decisions on everyone involved. Emphasizes teamwork. Recognizes individuals and groups for contributions. Tries extra hard to relate to subordinates.	Is not people-oriented; shows lack of interest; is not forceful enough; has no vision and/or way to implement a vision.
	Communicates through the combination of yelling, waving and pointing of hands, and a dissatisfied look on his or her face. When something does go right, says, "That's not bad, but just make sure you don't do this." In essence, speaks in negative terms only. A true believer in Theory X, but does not even know it.

TABLE 5.1   Good and Poor Project Leaders (*Continued*)

Good project leader	Poor project leader
Listens to thoughts and ideas of subordinates. Does not antagonize, but offers criticism. Brings harmony to historically battling departments. Does not just sit in office, but goes out into the field.	
Recognizes a good job and how to do a better job. Notes problems or flaws created.	
Treats coworkers as human beings as people rather than just another cog. Asks for input and thoughts on problems; allows subordinates to spend time on projects of their choice. Does not point a finger to place blame, but says, "The problem cannot be undone—what can be done to prevent it in the future?"	

5.1.5 Key User Questions

1. Do the team members understand the differences, and similarities of project managership and project leadership?

2. Are there any strategies in the organization underway to develop in the project team members the competencies of both leaders and managers?

3. Where do the project managers in the organization fall in the table above?

4. How would the project leaders/managers perceive their roles based on the insight gained from a careful perusal of the table above?

5. Is the interpersonal style of the project team members such that competencies can be developed and carried out which have both leadership and managership characteristics?

5.1.6 Summary

In this section, a brief examination of the characteristics and differences of project leaders and project managers was given. The individual who has the authority and responsibility to maintain oversight of the making and execution of decisions on a project is expected to develop the characteristics and competencies of both a leader and a manager and know when to use each competency appropriately.

5.1.7 Annotated Bibliography

1. Cleland, David I., and Lewis R. Ireland, *Project Management: Strategic Design And Implementation,* 4th ed. (New York, NY: McGraw-Hill, 2002), chap. 16, "Project Leadership." This chapter describes some of the background on the evolution of leadership and presents some of the fundamentals on leadership that have been developed by theorists and practitioners to explain the nature and process of leadership. The chapter also contains some references that could be used to develop a fuller understanding of project leadership and the difference between leadership and managership.

2. McGregor, Douglas, *The Human Side Of Enterprise* (New York, NY: McGraw-Hill, 1960), pp. 179–189. This book describes some of the important issues involved in leadership including powerful insight into the importance of attitudes in understanding and practicing leadership.

5.2 LEADERSHIP IN PROJECTS—A FURTHER PERSPECTIVE

5.2.1 Introduction

Leadership is such an important topic that the authors have elected to provide a second perspective on the subject.

It has been said that managers can be trained, but leaders must learn from example. Learning from example requires that existing leaders act as role models for future leaders. Future leaders can then develop while following others.

Leadership also requires that individuals set and adhere to the highest levels of honesty and integrity. Without these two critical characteristics or traits, one can never achieve the full measure of leadership. Others will not follow a person who is neither honest in his or her actions or conduct themselves without integrity. The question is asked "How many times can I lie to you and still have your trust?" The answer is always zero times.

5.2.2 Leadership Fundamentals

Leadership is defined by the military as "the process of influencing others to accomplish the mission by providing purpose, direction, and motivation." This same definition can easily apply to project leadership with some minor modification. Project leadership can be defined as "the art of influencing others to perform project work by providing purpose, direction, motivation, and coaching to individuals and the project team." Project leadership, like military leadership, incorporates the highest standards of honesty and integrity in dealing with people to build trust and confidence.

- **Purpose** provides the team members the general scope of the project and the tasks to be performed. This is the "story" of the project's life and its highlights. Using purpose as the guide, team members then can be a part of the process.

- **Direction** provides team members information and describes the tasks, task assignments, and priorities to be accomplished. This also includes the standards to be applied to the work and the expectations for work that is completed. Direction guides the team members in performing the work to the required levels of workmanship.

- **Motivation** provides the team members information on the importance of the project work and instills the will to complete that work. Motivation provides the focus on accomplishing the project's work under adverse conditions.

> **Good followers usually become good leaders**

- **Coaching** provides the development of team members to continuously improve their knowledge, skills, and abilities in the project management profession. Coaching is accomplished through setting the example, demonstrating methods of performing work, and counseling individuals on acceptable standards of work.

Team members have fundamental expectations from their leaders. These expectations include as a minimum the following, as depicted in Fig. 5.1.

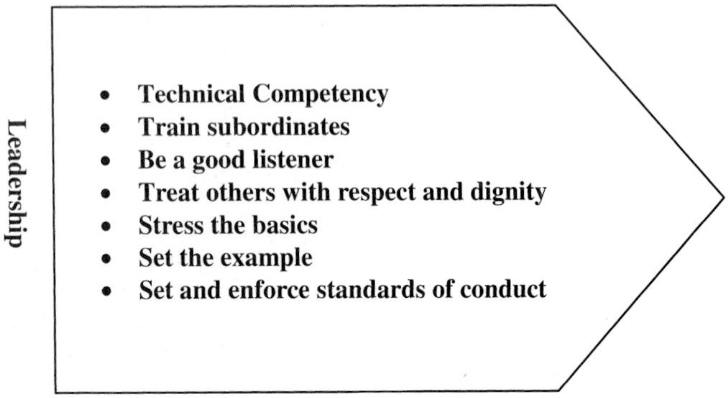

- Technical Competency
- Train subordinates
- Be a good listener
- Treat others with respect and dignity
- Stress the basics
- Set the example
- Set and enforce standards of conduct

Leadership

FIGURE 5.1 Team member expectations of a leader.

- **Demonstrated technical competence**
  Leaders must demonstrate competency in the project management profession through knowledge of the discipline. Individuals expect and demand that leaders be confident of their ability to plan and execute projects.

- **Training subordinates**
  Leaders must move beyond formal training programs to reinforce the principles and practices of the profession. Leaders must take time to share experiences and the benefits of their knowledge with individuals. Leaders build on the people's capabilities.

- **Be a good listener**
  Leaders must take time to listen to team members with equal attention as one would give a senior manager. Listening identifies the important issues of the team members and permits resolution through problem solving. Leaders know their team members through listening.

- **Treat others with respect and dignity**
  Leaders must always show concern and compassion for team members. Each individual must be treated better than the leader expects to be treated. The leader is in a position to demonstrate the respect and dignity to team members more than the team member is to the leader.

- **Stress the basics**
  Leaders must demonstrate mastery of the fundamental skills of project management such as organizing, planning, assigning work, tracking work, and decision making. These and other skills must be taught to team members through actions and by repetitively demonstrating them.

- **Set the example**
  Leaders must set consistently high values and abide by these standards.

By example, the leaders encourage a commitment to these same values and individuals will emulate the leaders.

- **Set and enforce standards of conduct**
  Leaders must know and enforce the standards of conduct established by the organization. Fundamental rules for safety, conduct toward co-workers, and work standards must be reinforced and any deviations promptly corrected.

Leaders do not compromise on the fundamental concepts of leadership and do not settle for second best performance. Through example and values, leaders succeed with project teams and by successfully completing projects. Leaders accept nothing less than the best from their team members.

5.2.3   Project Leadership

The nature of complex projects, both the technical aspects and the organizational aspects, place significant demands on the project leader. Project leaders are also faced with vague or ill-defined situations, which must not be transmitted to the team. Organizations need project leaders who:

- Understand the dimensions of a project within its organizational environment
- Provide purpose, direction, and motivation to their teams
- Show initiative to take advantage of opportunities
- Are technically competent organizers of teams
- Are willing to take calculated risks to advance the project while taking advantage of opportunities.
- Have the will to win and not let small obstacles delay projects
- Build a cohesive team
- Communicate effectively, both orally and in writing
- Are committed to the project and its completion

5.2.4   What is Leadership?

The most essential element of project management is competent and confident leadership. Leadership provides purpose, direction, motivation, and coaching to the project team for successful project completion, as depicted in Fig. 5.2. All four areas must be demonstrated for the project to be considered.

Leadership = Purpose + Direction + Motivation + Coaching

FIGURE 5.2   Leadership components.

The mandate for competent project leadership is simple and compelling. Failed projects are not as much technically challenged as they are leadership challenged. This leadership is both within the project and senior management decision making.

Leadership may be characterized by four factors: those being led, those leading, the situation, and communication.

- **Those being led** are typically the project team. However, others can be influenced by the leader and should also be considered in this category. How the individuals are led depends upon the level of training and skill levels that they possess. The leader must differentiate between those who have the abilities to do the job and those who don't. There is also a need to differentiate between those who can perform and those who won't perform.

  Individuals with lower skill levels will need more coaching and often more detailed instructions on their work than those who have mature skill levels. The leader must encourage and reward those individuals with the lower skill levels as they demonstrate increased capability. Those with mature skills must be rewarded for their accomplishments.

  Individuals who have the ability to perform at the proper level, but fail to perform must be assessed as to why there is a shortfall in accomplishment. This shortfall may result from a lack of understanding or other legitimate factors. An individual who has the ability, but refuses to perform the work must be counseled and, if necessary, disciplined. It is often a delicate balance between identifying those who do not perform for lack of information and those who are unwilling to perform.

- **Those leading.** Leaders must not only lead, but they must be seen to lead. Leaders must be visible to those being led and there must be demonstrated acts that make clear who is leading. The term "leader" implies that one is in front of the project team and that person is responsible for the project.

  Leaders must have an honest understanding of who they are, what they can do, and what they know. Own strengths, weaknesses, capabilities, and limitations must be recognized by the leader so that self-control and discipline can be exercised. Leaders must capitalize on strengths and capabilities while avoiding the weaknesses and limitations.

  Being a leader means continually improving one's self and growing in all aspects of interpersonal skills. A leader must demonstrate character and trustworthiness to those being led to ensure there is confidence in his ability to lead. No orders, directives, words, or other statement of position will make one a leader. Only those being led can designate a person a leader. The following, extracted from a speech made to newly commissioned Army Officers, clearly points to leadership.

> I feel a tinge of regret that I am not young enough to be sitting out there as one of you. You have so many years of challenge and adventure to look forward to. So many of these years are behind me. Soon you will meet . . . your troops. What do we expect from you as officers, commanders, leaders? We expect of you unassailable personal integrity and the highest of morals. We expect you to be fair-to be consistent . . . to treat each soldier as an individual, with individual problems. And we expect you to have courage . . . the courage to stand up and be counted-to defend your men when they have followed your orders . . . to assume the blame when you are wrong. Your orders appointing you as officers in the United States Army appointed you to command. No orders, no letters, no insignia of rank can appoint you as leaders. Leadership is an intangible thing; leaders are made, they are not born. Leadership is developed within yourselves. (SGM John Stepanek, 1967, quoted from "The Gates of Hell")

- **The situation.** Leaders find the situation differs in encounters with individuals. The first act is to assess the situation prior to taking actions. There is an old maxim: "Praise in public, criticize in private." A leader may want to compliment an individual on some accomplishment in front of peers. On the other hand, a leader may wait until there is privacy before counseling an individual for poor behavior.

  Leaders look for desired outcomes in all situations. It is seldom proper behavior for a leader to shout or raise the volume of his/her voice. Leaders must stay calm and clearly communicate the desired outcome. Shouting is only appropriate in emergency situations where this sudden shift of behavior will result in immediate response for safety of human life.

  A leader selects the proper timing for corrective counseling. Some behavior requires immediate response by the leader while other behavior can wait. This time of counseling is often dependent upon the pace of other activities. Regardless of the timing, a leader should always address both excellent behavior and poor behavior. Counseling and praise may be deferred, but must always close-out issues in the work environment.

- **Communication.** Leaders must be effective communicators, whether the communication is oral or written. Communication is an interchange of information between individuals. Leaders must use all means to effectively communicate with senior management, peers, team members, customers, and other stakeholders. Three means of communicating are listed below.

- Words or semantic notations—written or oral words or other understood sounds.
- Illustrations—graphics, pictures, charts or other forms that convey some meaning.
- Mathematics—two types.
  - Logic and structure deals with general approaches to problem solving
  - Content deals with specific problem solving.

All forms of communicating can be effectively placed within these three categories. The selection of the forms of communication to be used will typically be a mixture rather than just one. The mixture will depend upon the audience, the purpose of the communication, and the distance between the parties.

Communication is affected by an individual's beliefs and values. Beliefs are what a person knows, expects or suspects. Values are what a person wants, desires, and prefers. Project leaders must consider values and beliefs to be effective and avoid attempts to change them.

Communication between leaders and team members will be affected by the values and beliefs of the listener. The confidence and trust placed in the leader by the team member will also enhance or detract from the leader's communication abilities. It can be seen that the leader's reputation for honesty, integrity, and openness materially affect communication with everyone.

5.2.5 Principles of Leadership

Eleven principles of leadership identified for project leaders are shown in Fig. 5.3.

5.2.6 Leadership Traits

Leaders possess and demonstrate traits that are of significant value in gaining the willing obedience, confidence, respect and loyal cooperation of the project team members. Figure 5.4 identifies the traits of a good leader.

Principles of Leadership for Project Leaders

- Know yourself and seek self-improvement
- Be technically competent
- Seek and take responsibility for your actions
- Make sound and timely decisions
- Set the example
- Know your team members and look out for their well-being
- Keep your team members informed
- Develop a sense of responsibility in your team members
- Ensure tasks are understood, supervised, and accomplished
- Build the teams capabilities
- Accept tasks only within your team's capabilities

FIGURE 5.3   Principles of leadership for project leaders.

Leadership

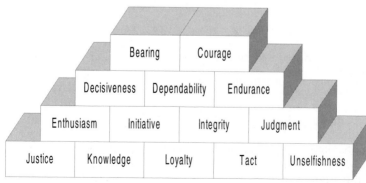

FIGURE 5.4   Leadership traits.

5.2.7   Key User Questions

1. What characteristics do you look for in a project leader?

2. How many people have you known who qualify as true project leaders, and why do you consider them to be leaders?

3. How many principles of leadership do you see being demonstrated within your organization?

4. If you are an appointed leader, do you meet the test of being considered a leader by those who you lead?

5. What is your greatest strength and your greatest weakness as a leader (whether you are currently a leader or not)?

5.2.8   Summary

Leadership is obviously a greater combination of skills, knowledge, and abilities than the more common management concepts. Leadership includes the demonstration of characteristics and traits, which instill confidence in those being led.

Leaders are not born and leaders are not trained. Leaders learn from the examples of others and emulating those positive characteristics that influence others to willfully perform their duties. Leaders must demonstrate their willingness to deal openly and honestly with the project team members.

Leaders are self-made people. Neither training, nor education, nor appointment to position will make a leader. A person earns the right to be called a leader by those being led. Earning the right to be called a leader is accomplished through extraordinary effort to demonstrate values and characteristics that build confidence.

5.2.9   Annotated Bibliography

1. Headquarters, Department of the Army, Washington, DC, Field Manual 22-100 *Military Leadership,* chap. 2, 4–6, App. A and B (July 31, 1990). This document shows the principles of leadership, factors affecting leadership, and what it means to be a leader under difficult situations. It gives guiding principles for what a leader must know and do. Appendices A and B are the result of studies of leadership throughout the spectrum of leadership responsibilities.

2. Oh, Christopher K., "The Gates of Hell" (May 22, 1996), unpublished. (Internet access). This is a study of what it takes to be a leader. It uses

the background of a military environment, where life-and-death decisions are made daily. The factors of leadership, both visible and invisible, are examined for consistent successes that a person may follow.

5.3 COACHING PROJECT TEAM MEMBERS

5.3.1 Introduction

Coaching is an important aspect of leadership and team building. It is both praising good behavior and correcting disruptive behavior. Coaching is for individuals and for the team as a whole, and is a skill that is developed and improved over time by leaders.

Coaching is an essential leadership competency. Knowledge of the techniques and tools of coaching are essential for a person to assume the role of project leader. Coaching competency gives the project leader an improved opportunity to successfully complete the project as well as build stronger individuals and teams.

5.3.2 What is Coaching?

Coaching is defined as "those leader actions taken to develop the professional capability of project team members through personal intervention to change or reinforce behaviors." Counseling is often used synonymously with coaching. Coaching, however, entails actions that reinforce the positive behaviors and attempts to change the negative behaviors. Counseling may often refer an individual to the activities of healthcare professionals, spiritual leaders, and other specifically trained professionals.

Coaching in the project environment specifically focuses on behavior that affects work performance. This is generally encouraging individuals and the team to perform in an expected manner. It may require that coaching extend into the personal life of the individual when a personal situation affects performance. For example, a person with a marriage problem may not see his/her performance on the job as being primary. The stress of a marital problem can easily diminish productivity to low levels.

A coach is a teacher and counselor

5.3.3 What is the Project Leader's Role as a Coach?

The project leader is charged with delivering the benefits of the project to stakeholder and more specifically to the customer or client. This charge entails organizing and maintaining a competent project team to perform the project work. Guiding this team to be a productive work unit requires leadership.

Leadership must first set the expectations for the project team and communicate the standards that the team is to follow. Setting expectations may be accomplished at the project kickoff meeting and entails addressing the scope of the project, specific requirements, target delivery dates and describing how the project will proceed. It is also important to set expectations as to the project team's standards of conduct.

Once the expectations are set, the project leader must enforce them through corrective actions for variances and praise when the behavior matches the expectations. Expectations must be enforced to be seen as being real or needed for the project. When a project leader fails to enforce deviations from the requirements, this person penalizes those who do conform. The project leader who does not enforce the "rules" because of favoritism or other personal consideration will fail to achieve the desired results in developing the team.

5.3.4 Coaching Techniques and Tools

The project leader must be competent in his/her job to be an effective coach to team members. If the coach has personal problems that overwhelm his/her capability to perform, then that person cannot effectively coach others. Competency in leadership, free of external personal problems, is a prerequisite for being a leader-coach. Figure 5.5 shows some of the key coaching techniques and tools.

One of the most effective methods of coaching for performance is to set an example. If the leader sets the workday start time as 8:00 am each morning and arrives consistently late for work, this sends a strong negative message to the team. On the other hand, if the team has to work late and the project leader stays with them this also sends a strong positive message.

Praise is one of the most effective techniques that can be used by the project leader. Praise must be sincere and honest. Praise should be related to the expectations set at the start of the project (or when the team member joins the project). Praise may be for accomplishing a task in a superior manner or it may be used to recognize improved skills. Praise is feedback to the individual or team member that he/she is doing something well.

Praise should be specific and related to work. Some examples of praise could be:

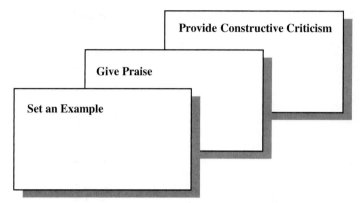

FIGURE 5.5 Coaching techniques and tools.

- "That was a good job, Pete. You finished it a day ahead of schedule. I appreciate your work."
- "Joe, your ability to prepare the weekly reports has improved significantly. This helps us keep all our stakeholders happy with timely reporting."
- "Mary, your participation in the team meetings has improved and the team is benefiting from your knowledge. Keep up the good work!"

Criticism should be used sparingly and in private. Any criticism of individual performance can most effectively be handled in a neutral area. Criticism should avoid the use of personal pronouns so that the situation is addressed as a professional shortcoming and not a personal attack. Criticism must always look for improvements.

Criticism in a constructive manner must be delivered in a calm manner that sticks to the issue at hand. Some examples of criticism could be:

- "Fred, when the team had its kickoff meeting, timely reports were identified as being critical to the success of our project. Your reports have not been turned in so that they could be integrated with the other team members' reports. This affects the entire team and makes it look bad. What can we do about submitting timely reports?"
- "Oscar, during my orientation of new project team members, we agreed that absences from the project would be approved by me or my deputy. Yesterday, no one could find you when the team had an important meeting. Your contribution was needed and placed the team in a bind. What happened and what can we do about the future?"

- "Ellen, there has been a noticeable lack of interest during our team meetings. This morning you were sleeping during the daily update briefing. This sends the wrong message to the entire team and any guests who attend the briefings. What can we do about improving this situation?"

Both praise and criticism are needed in the project environment. Project leaders must anticipate their responses to the situations that develop. Some small items that are overlooked or not corrected may generate larger issues. Critical performance and conduct issues must always be corrected. Anticipating what to praise and what to correct will arm the project leader before hand. For the new project leader, it may be appropriate to also anticipate how a variety of situations can be resolved through coaching.

Project leaders' efforts to develop individuals should be guided by the following four objectives:

- Cause the individual to recognize strengths or shortcomings and define any problems. Patience, sincere interest, clear thinking, and calm demeanor are required by the project leader to accomplish this.

- Have the individual determine possible courses of action, based on facts, to resolve shortcomings and for the individual to select one course of action. Project leaders must use skill, knowledge, and restraint to allow the individual to find the correct course of action.

- Cause the individual to take the appropriate corrective action. This depends upon the individual's commitment to his/her decision and the will to complete the corrective action.

- Have the individual assume full responsibility for his/her decisions and actions. The above three objectives must be met to ensure this object can be implemented. Project leaders must ensure the first three objectives are met before obtaining the individual's commitment to action.

Coaching the team is essential to building collaborative efforts. Praise can be used when the team accomplishes tasks or achieves milestones. It is most effective when the team's efforts are recognized by someone other than the team leader. This can be done by having a senior manager or the customer talk to the team on their accomplishments. When the team knows its accomplishments are recognized and important to others, this builds pride in work.

Criticism of the team is difficult and can be awkward. Shortcomings must be addressed and the team leader must obtain a commitment by the team to change. One method is to describe the shortfall in general terms of the lack of results. Then, ask the team what should have been done to prevent the poor results. Let the team identify how the results occurred. Project leaders, as part of the team, should be prepared to identify their weaknesses as contributing to the poor results. The project leader must obtain a commitment from the team as to their future performance.

5.3.5   Characteristics for Effective Coaching

To develop the attitude and behavior for proper coaching, the project leader must be familiar with the characteristics shown in Fig. 5.6.

- Flexibility—fitting the coaching style to the unique character of each individual and to the desired future relationship.
- Respect—respecting individuals as unique, complex team members with unique sets of values, beliefs, and attitudes.
- Communication—establishing open, two-way communication with individuals being coached, using oral and non-oral actions, gestures, and body language. Effective coaches encourage individuals to speak more than the coaches speak.
- Support—supporting and encouraging individuals through actions and interest in their concerns is essential for the project leader to coach

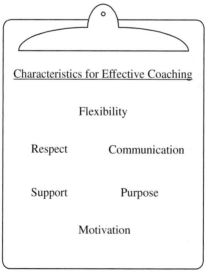

FIGURE   5.6   Characteristics   for   effective coaching.

individuals, all the while guiding them to seek solutions to their problems.

- Motivation—some individuals will seek to be coached while others may be passive. Those who seek coaching are motivated to improve themselves. The passive individuals are more likely to need coaching and can benefit more. Project leaders must seek out those who need coaching, but do not want it.

- Purpose—to develop responsible and self-reliant team members who solve their individual problems.

Project leaders must maintain confidentiality of the conversations conducted during coaching. At the onset, the project leader should make clear to the individual that this is a confidential conversation that will not be repeated to others. The only exception to this, and the individual should be so advised, is unlawful acts. The coach cannot become a part of or an accessory to a crime.

There are limits to a project leader's ability to coach individuals. As most project leaders are not healthcare professionals, it is well to recognize when the coaching should stop and the individual referred to a medical doctor, a minister, or a counselor specializing in the specific problem. The goal is to help the individual become the most effective team member that he/she can be. Referral to the professional who can help the individual is the best that a coach can do for that individual.

5.3.6 Key User Questions:

1. On the projects that you have worked, how effective was coaching in aligning attitude and behavior with project requirements?

2. What type of training do project leaders in your organization receive to qualify them to coach members of the project teams?

3. Could better coaching by the project leader help projects in your organization?

4. What coaching techniques do you use during projects and how effective are they in maintaining a stronger project team?

5. What coaching capabilities would improve your organization's project management competency?

5.3.7 Summary

Coaching is a critical part of a project leader's responsibilities because he/she will spend many hours dealing with team members to develop these individuals into more effective, productive individuals. This development benefits both the individual as an asset that is more highly valued by his/her organization and the project team in functioning as a unified work group. Coaching, however, is often not given the emphasis that it needs in projects and the benefits of coaching are not realized.

Techniques and tools of coaching can be learned by project leaders and should be practiced in a role playing environment to fine tune the skills. Project leaders must recognize the limits of coaching and refer individuals to specific counseling specialists when appropriate. Confidentiality of coaching is a must and builds on trust when it is maintained. Communication where the individual does most of the talking and the project leader listens is the most effective.

5.3.8   Annotated Bibliography

1. *Leadership Counseling,* Headquarters Department of the Army, Washington, DC, Field Manual 22-101, chap. 1–4 (June 3, 1985). This document provides insight as to the importance of counseling and the positive benefits that can be gained from human resources with the proper attitude. It also covers the fundamentals of counseling, to include strengths and weaknesses.

2. Wilkinson, Lloyd (1998), "Assertive Management Course," US Postal Service, Great Falls, MT. This course demonstrates the need to communicate concepts and ideas as well as enforce the organization's policies on human behavior. It is especially important that safety areas be adhered to in all work practices.

5.4   MANAGING CONFLICT IN PROJECTS

5.4.1   Introduction

Conflict will always occur in projects because of the temporary nature of a project team and the number of external interfaces the project leader must manage. Conflict is when there is a disagreement between two or more parties as to an important element of the project's work. This disagreement may be a difference of opinion in the technical solution, the cost of an item, how something will be accomplished, or when something will be delivered.

Human conflict may also take on personal areas that inhibit getting the work done on time and to the customer's satisfaction. Personal conflicts are typically more difficult to manage than professional disagreements. Personal conflicts, however, do negatively impact project goals because they destroy working relationships, which are essential to effective project completion.

> **Conflict is to be expected—and can be managed in the project environment**

5.4.2   Sources of Conflict

Conflict in the project environment is always present because of the nature of projects. The drive to complete a project on time, within budget, and meeting the customer's requirements are competing goals. These goals and their potential for conflict are inherent within the project team as well as with external sources.

People have different agendas and see things differently. Actual and perceived differences can increase the potential for conflict. The project team needs to work toward a single agenda to ensure success for the project. It is the function of the project leader to bring this team together and establish the single agenda.

External sources of conflict are those people who have a competing requirement. Conflict arises when there is a sharing of resources between projects, for example, and the competing manager also has a need for the resources. Competing requirements for the same, limited resources is perhaps the biggest source of conflict in the matrix environment.

Figure 5.7 summarizes some of the areas of conflict within projects and the details are discussed below the figure.

Items that generate conflicts in a project are identified as areas to manage more effectively. These are:

- Project priorities—priorities typically conflict between project leaders and functional managers. This is often as to which project or work has the most urgent need for resources or for the use of some facilities. Organizations do not do well in assigning project priorities and create this situation for the project leader. Organizations will take the position that "everything is priority one." When everything is priority one, there is no priority system.

- Project management methodology—methodology for managing projects typically "modified" during project execution. Some parts are omitted or changed to accommodate the project work. This change creates conflict among the project team as to which is the best method for planning

Some Sources of Conflict

- Project priorities
- Project management methodology
- Schedules and resources
- Sexual harassment
- Jokes

FIGURE 5.7   Some sources of conflict in projects.

and executing the project. "Modified" project management methodology typically omits an important part of the process. Testing is often reduced to "save time."

- Schedules and resources—project dynamics will change the time frame in which resources are needed for the project. Functional managers want a fixed time that resources can be allocated to the project; no more time, no less time, and within the baseline schedule planned for the assignment.

Personalities often conflict and there is not the degree of cooperation that is needed. The most common areas of conflict in recent time reflect upon the misconduct of one or more persons. Some of the more important items that are disruptive and create conflict are identified here:

- Sexual harassment—this is currently defined as the unwanted sexual advances of another, male or female, that creates a hostile environment. The advances may be physical acts or oral statements.

- Jokes—inappropriate humor that degrades another or an ethnic group is inappropriate, whether that person or a representative of the group is present or not. Humor that is off-color or suggestive of sexual acts is also inappropriate.

- Other—inappropriate behavior that is demeaning, draws attention to anyone as a means of degrading, outlandish dress or immodest dress, and other degrading acts should not be tolerated. Gestures and skits that are demeaning should not be tolerated.

5.4.3 Classification of Types of Conflict

Classifying the types of conflict help with the identification and managing each item. There are two classifications of conflict: open and closed.

- Open conflict—when one or more parties challenge another. This is healthy to put the issues forward and find resolution. It is the first step in finding a solution. Open conflict can be recognized and managed.
- Hidden conflict—when one or more parties conceal a difference from another, but will work to actively sabotage or will fail to give full support. Conflict resolution is difficult in this situation because only the sabotage or lack of support are identified, not the reason for the conflict. Until the reason for the conflict is identified, it is difficult to remove it.

5.4.4 Sources of Conflicts

The dynamic nature of projects is a source for conflict. Commitments to perform certain tasks typically have a time and place for the activity. Slips in schedules and changes to the tasks may be sources of frustration when the performing party has little flexibility. Functional managers typically have difficulty with the dynamic schedule changes when they are committing resources to the task.

Getting consensus during meetings is a source of conflict. Consensus in its true meaning is that everyone agrees. The nature of organizational culture is that people do not raise issues when the issue does not affect them. Therefore, consensus is practiced as "if it doesn't affect me, I will not disagree." Thus, consensus does not address issues, but only lets them lie dormant until a task is attempted.

Another approach to consensus is that "no one in the room is harmed by any decision." Anyone who is not in the room may be affected. This is a potential source of conflict and can be extremely disruptive for the project. This approach should be prohibited and everyone should be looking for the common ground for all parties.

Organizational culture is a source of conflict. Organizational culture is the sum of organizational values, some good and some not so good. Individuals use "organizational culture" as an excuse for not performing certain tasks. For example, it is permissible in some organizations to not accept tasks from the project leader unless the person assigned the task agrees with the work. Otherwise, the project leader has a conflict.

Organizational culture also affects decision-making at all levels. The culture will not allow one person to make a decision that affects another without the other person's consent or acceptance. This delays decisions and places work in queue. Conflicts arise when another person cannot perform his/her work because of the delayed tasks.

5.4.5   Organizational Values

While individuals talk about organizational culture, they are actually talking about the accepted or adopted organizational values. Organizational culture is the sum of all organizational values, whether real or perceived.

It is difficult to address "organizational culture" because it is a composite of all real and perceived organizational values. To address conflict, one must identify the organizational value directly creating the conflict or supporting the conflict. The decision is then to emphasize the organizational value or discard it.

5.4.6   Conflict Resolution Modes

Project leaders may embrace one or more of the conflict resolution modes listed below. The modes are useful to understand for improving one's ability to quickly and effectively resolve conflicts. These modes are:

- Withdrawal—the disengagement from an actual or perceived conflict. It is a delaying action that neither solves the conflict nor clarifies it. This is a weak approach to managing conflict.

- Smoothing—the attempt to convince the parties that conflict does not truly exist by de-emphasizing differences and emphasizing commonalties. The perceived conflict is often described as a difference in how people view a situation. This approach to conflict resolution is weak and only serves to reduce the stature of the person using smoothing.

- Compromise—the attempt to resolve the conflict through each party giving up something. This approach assumes that both parties are equally right (or wrong) and both need to give away something as a peace offering. It is usually less than satisfactory to one or both parties.

- Forcing—asserting one's viewpoint at the expense of one or both parties. This may be necessary when both parties refuse to cooperate in the conflict resolution or when there is no time to work through another mode of conflict resolution.

- Problem solving—the attempt to resolve the conflict by defining the problem, collecting facts, analyzing the situation, and selecting the most appropriate course of action. This mode is time consuming, but the most effective if both parties are cooperative and open to resolving the problem.

5.4.7   Preventing Conflict

Conflicts arise when two or more people are operating from different instructions or with different information. Conflicting instructions or information must be obtained from some source. The source of instructions or information may be delivered directly from someone, for example, the project leader, or they may be surmised because of the lack of guidance from the project leader.

Preventing conflicts is perhaps the most effective method of handling situations that can generate conflict. The project leader must ensure that all team members understand what is expected of them and that they are familiar with the project plan. The project leader should also ensure project team members understand the top level of the project's objectives and the concept for executing the project.

Team building that emphasizes "trust" and "confidence" in fellow team members can also reduce conflict. A trusting environment will foster cooperation and reduce the tendency to compete among the team members. Trust and confidence in the project leadership makes for a more cooperative environment.

5.4.8 Key User Questions

1. When two individuals have a personality conflict, does the project leader view this as a problem to be solved or is it personal that does not require professional intervention?

2. There are five identified modes of conflict resolution. What mode, do you use to resolve conflict?

3. Why would the project leader want to interfere when a person is telling off-color jokes and it makes most of the project team laugh to relieve stress?

4. What are some techniques for preventing conflict and who should initiate them?

5. What is the difference between organizational culture and organizational values? Why differentiate between the two?

5.4.9 Summary

Managing people in an effective manner dictates avoidance of conflict—an energy-consuming, destructive situation that stops effective progress. Avoiding conflict is perhaps the best method of "conflict resolution" because it promotes project team trust and cooperation. Prevention of conflict is the best approach.

Understanding the sources of conflict is helpful in planning to either avoid the problem or deal with the problem if it occurs. The five modes of conflict resolution will label actions required under each one to determine which one to use in a given situation. No mode is undesirable if used in the proper situation.

5.4.10 Annotated Bibliography

1. Adams, John R., and Nicki S. Kirchof, *Conflict Management for Project Managers* (Drexel Hill, PA: Project Management Institute Press, 1982), chap. II–V. This monograph summarizes conflict management in projects. It gives causes of conflict and suggested means of resolution.
2. Cleland, David I., Lewis R. Ireland, *Project Management: Strategic Design and Implementation,* 4th ed. (New York, NY: McGraw-Hill, 2002) This book presents information on Conflict Management in many places, as indicated on pp. 81, 111, 166, 190, 209, 234, 237, 241, 242, 245, 265, 273, 275, 280, 293, 365, 395, 491, 509, 513, 520–525, 557, 563, 577, 579, 583, 611, 634.

5.5 TEAM LEADERSHIP

5.5.1 Introduction

Leadership was studied and written about extensively in the twentieth century, and continues to be. There have been thousands of articles and papers about this subject. Several hundred books have appeared that examine the function of leadership in different forms of organizations—government, military, industrial, and political to mention a few. In recent times the difference, if any, between leadership and management has been examined and will continue to be studied in the future. The role of project leadership will be examined here.

5.5.2 What is Leadership?

There have been many definitions of leadership. One study found over 130 definitions of the subject. Another source notes over 5,000 research studies and monographs on the term. There are many traits and processes cited that identify the qualities of effective leadership. A few generalizations about leaders are shown below.

5.5.2.1 General Characteristics of Leaders

- They have their act together—their personal ambition and motivation drive them to succeed.
- They maintain high visibility to the followers they wish to lead. There is no doubt in any follower's mind that the leader is in charge and on top of everything.
- Leaders are available to the followers to listen, debate, and get the needed facts together to make and execute decisions. Once they make a timely decision there is little hesitation in them saying, "Let's do it."
- They see the best in the people and institutions that they lead. Leaders are winners who praise and motivate their followers to action.
- Good leaders try to make things simple—to sort through the complexity of a problem or opportunity, make and implement decisions efficiently and effectively.
- They are fair and patient with their followers.
- They work hard at being a leader, and avoid being perceived as an "absentee landlord."

Characteristics of Leaders
• Have their act together
• Are winners
• Have high visibility
• Simplify things
• Available to followers
• Patient with followers
• Work hard to be a leader

5.5.3 The Difference between Leadership and Management

Leadership is part of management, yet at the same time distinct from it. Warren Bennis, a noted researcher on leadership, offers the following differentiation between leadership and management: "A leader does the right thing (effectiveness); a manager does things right (efficiency)." An effective "project leader" develops a vision for the project, assembles the resources, and provides the inspiration for the project stakeholders. An effective "project manager" ensures the appropriate management systems are designed and used to provide technical and resource support to the project. Indicated below is an expanded description of the difference between project leaders and project managers.

5.5.3.1 Leaders

- Develop and communicate a vision to the stakeholders about the project.
- Build supportive networks with the project team members and other stakeholders.
- Watch out for major patterns and relationships in the project that have a potential for a major impact on the project.
- Develop and communicate a suitable management system for the project.
- Become a symbol and advocate for the project within the organization.
- Facilitate the linkage of the project with the operational or strategic direction of the organization.

5.5.3.2 Managers

- Oversee the design, development, and operation of the management systems to support the use of resources on the project.
- Maintain oversight of the efficiency and effectiveness of the use of resources on the project.
- Facilitate the planning, organizing, motivating, and control systems for the project.
- Monitor the competency of the project team members as well as the competency of the entire team, to include other stakeholders.
- Reassign resources as required to ensure maximum benefit to the project needs.
- Facilitate the professional development of the team members.

Leaders vis-à-vis Managers	
Leaders	Managers
• Develop a vision	• Management system
• Networking	• Efficiency
• Relationships	• Management functions
• Symbol	• Monitors
• Linkages	• Resources

Project managers have both leadership and managership responsibilities. They deal with both operational and strategic issues on the project.

Both leadership and managership require sensitivity to the human consid- erations of the project. A project manager's formal grant of authority comes from organizational policy documentation. The role of the project leader comes from the individual's competency.

The role of project leadership is most challenging when dealing with project stakeholder's who are beyond the organizational project team, such as suppliers, government agencies, unions, professional associations, local community officials, and, when required environmental protection offi- cials. A few guidelines for face-to-face working with these people are offered:

- Be alert to the opportunities for helping these stakeholders, thereby building opportunities for reciprocity, which can help the project.
- Seek opportunities to describe and stress the project manager's role and responsibilities.
- Encourage the stakeholder to identify with the project and its reciprocal needs with that project.
- Nurture the professional relationship that is desired between the stake- holders and the project team.
- Seek opportunities to encourage the reciprocal relationship that is needed between the project team and other stakeholders.

Project leaders will find many situations in which they must ask, en- courage, cajole, praise, reward, demand, manipulate, and in general use ethical processes and interpersonal skills to gain and hold support for the project.

5.5.4 Decisions

An important leadership opportunity that the prudent project leader will assume has to deal with the decision context of managing the project. This leader has to know how to make and implement decisions within the systems context of the project. In such decision-making, certain key fun- damentals are required, such as:

- Define the decision issue or opportunity regarding the project.
- Facilitate the development of the relevant databases required to evaluate fully the nature and timing of the decision.
- Consider alternative ways of using the project resources to bring about a timely and relevant decision.
- Ensure that an explicit assessment of the risk and cost factors surround- ing the decision is considered.

- Evaluate and select the appropriate alternative in the assessment of the decision.
- Develop a plan for how the decision will be implemented.
- Dedicate the required resources to implement the decision.

Of course, the actual making and implementation of a decision are much more complex than is implied above. These fundamentals should be considered as a "way of thinking" in dealing with the inevitable decisions facing the project manager and the team members.

The fundamentals of decision-making are as follows.

- Define the issue or problem
- Assess risk and cost implications
- Build databases
- Develop implementation strategies
- Evaluation alternatives
- Dedicate resources to solution

5.5.5   Key *Modus Operandi*

What are some of the key *modus operandi* and characteristics of the successful project leader? In the material that follows, insight into the answer to this question is given:

- An ability to conceptualize the likely deliverables of the project, followed by a project plan that provides guidelines for making the project objectives and goals a reality.
- A positive attitude by the project manager in spite of bad news and disappointments. Many successful projects have traveled a path of bad news and failure, but survived because of the positive leadership of the project manager.
- Any project can be criticized. The project leader must have a tough skin to deal with the inevitable faultfinding or blame that can come in the management of a project.
- Use policies, processes, procedures, protocols, and documentation along with empowerment of the project team to guide team members in making and implementing decisions in their area of responsibility.
- Be able to assume risk and deal with the uncertainties of the project. To help in dealing with risk, find people in the stakeholder groups who

can help in the analysis of risk and provide recommendations on how the risk can be reduced or eliminated.

- Although the "buck" in the management of the project stops with the project manager, every effort should be made to decentralize the authority and responsibility to make and execute decisions to those stakeholders most qualified to make such decisions.
- Have the tenacity or persistence to seek out problems, opportunities, and decisions involving the project.
- Finally, work hard at mentoring, teaching, coaching, and guiding the people who are in charge of defining and using resources on the project.

5.5.5.1   *Successful Project Manager Key Modus Operandi*

- Conceptual deliverables
- Assume risk
- Positive attitude
- Decentralized management
- Tough skin
- Look for problems and opportunities
- Policy guidance
- Mentor

5.5.6   User Questions

1. Has the organization defined the requirements for effective leadership in the management of projects?
2. What general characteristics does the organization expect of its leaders?
3. Do the principal managers of the organization recognize the difference between leadership and managership? Why or why not?
4. Have the characteristics and *modus operandi* of successful leaders in the organization been identified? Have these attributes been documented and passed on to all of the project managers/leaders in the organization?
5. What expectations do the key user managers have regarding the leadership requirements for dealing with the project stakeholders?

5.5.7   Summary

A brief summary was given of project leadership. A comparison was made between what managers do and what leaders do. Some of the characteristics of leaders were presented. The roles that project managers are expected to carry out with regard to the leadership of stakeholders were also examined. The leadership responsibility of the project manager in the making and execution of decisions was also summarized. Finally, the section closed with an examination of the key *modus operandi* and characteristics of the successful project leader.

5.5.8 Annotated Bibliography

1. Briner, Wendy, Michael Geddes, and Colin Hastings, *Project Leadership* (Aldershot, Hampshire, UK: Gower, 1990). This book begins by explaining why the concept of project leadership is being adopted so widely. The first part of the book examines the project leader's task. It outlines the six key elements of the role and looks at the processes used by effective project leaders. The second part of the book provides advice on how to handle the issues that emerge at each stage of a project's life cycle.

2. Gemmill, Gary, and David Wilemon, "The Hidden Side of Leadership in Technical Team Management," *Research Technology Management,* November-December 1994. The authors recognize that project managers must deal with the functions of management on the projects to which they are assigned, but also cope with a multitude of interpersonal issues always present in team-oriented work environments. This article, based on a field study of 100 project leaders, examines several interpersonal issues found in most technical teams and suggests some options for managing them. If the project leaders deal only with the technical dimensions of the project and are not competent in dealing with the hidden interpersonal issues, the management of the project can be undermined.

SECTION 6
PROJECT INITIATION AND EXECUTION

6.1 PROJECT SELECTION CONSIDERATIONS

6.1.1 Introduction

Although a project manager is usually not directly involved in the selection process of projects to support enterprise purposes, he or she should have a general understanding of some of the approaches that are used to determine which project to initiate. Senior managers have the responsibility to make the selection of such projects regarding:

- New or modified products
- New or modified services
- New or modified organizational processes to support product and service strategies
- Projects that are used to do basic and applied research in a field of potential interest to the organization

A major concern of the senior managers should be to gain insight into the probable promise that projects hold for future competition. Senior

TABLE 6.1 Project Selection Questions

- Will there be a "customer" for the product or service coming out of the project?
- Will the project results survive in a contest with the competition?
- Will the project results support a recognized need in the design and execution of organizational strategies?
- Can the organization handle the risk and uncertainty likely to be associated with the project?
- What is the probability of the project being completed on time, within budget, and at the same time satisfying its technical performance objectives?
- Will the project results provide value to a customer?
- Will the project ultimately provide a satisfactory return on investment to the organization?
- Finally, the bottom-line question: Will the project results have an operational or strategic fit in the design and execution of future products and services?

---

managers, in their evaluation of projects, need to find answers to the questions outlined in Table 6.1.

As the senior managers consider the alternative projects that are already underway in the organization, as well as new emerging projects, the above questions can help the review process and facilitate the making of decisions for which senior managers have the responsibility.

In addition, as senior managers review and seek answers to these questions, an important message will be sent throughout the organization: Projects are important in the design and execution of our organizational strategies!

6.1.2 Strategic and Operational Fit

Senior managers of an enterprise are expected to act as a team in selecting those projects whose probable outcome will enhance the competitiveness of the organization. Managers need to be aware of the general nature of project selection models and processes.

There are two basic types of project selection models—numerical and qualitative.

The numerical model uses numbers to indicate a value that the project could have for the organization, whereas the non-numerical uses subjective perceptions of the value likely to be created by the project.

Project selection models do not make decisions, people do. Such models can provide useful insight into the forces and factors likely to impact the value that the project can provide to the organization. However sophisticated the model, it is only a partial representation of the factors likely to impact the selection of a project. A selection model should be easy to calculate and easy to understand.

6.1.3   Other Factors

The factors to consider in the selection of a project will differ according to the organization's mission, objectives, and goals. In addition to the above questions that need to be asked about the project, other factors can serve as a starting point:

- Anticipated pay-back period
- Return on investment
- Potential contribution to organizational strategies
- Support of key organizational managers
- Likely impact on project stakeholders
- Stage of the technical development
- Existing project management competency of the organization
- Compatibility of existing support by way of equipment, facility, and materials
- Potential market for the output of the project

Managers should use some techniques to facilitate the development of data bases to facilitate the decision process, such as:

Brainstorming—or the process of getting new ideas out by a group of people in the organization.

Focus groups—where groups of "experts" get together to evaluate and discuss a set of criteria about potential projects, and make recommendations to the decision makers.

Use of consultants—to provide expert opinions concerning the potential of the project, such as the availability of adequate technology to support the project technical objectives.

6.1.4   Project Selection Models

The use of appropriate numerical and qualitative models is dependent on the information available, the competency of the decision managers to understand that information, and their ability to understand what the project-selection models can do. There are a few selection models that can be used to guide the decision about project selection made by the managers.

Qualitative methods—when there is general information that can be used in the model.

Q-Sort—technique to rank-order projects based on a pre-selected set of criteria.

Decision-tree model—where a series of branches on the decision tree are used to determine which project best fits the needs of the enterprise.

Scoring models—when sufficient information is available about the potential value of the projects to the organization.

Payback period—used to determine the amount of time required for a project to return to cash flow equal to the amount of the original investment.

Return on investment—where an evaluation of a potential project indicates a given level or rate of return.

Other approaches are available which occur as the result of an emergency need such as:

- A senior manager's "pet" project
- An operating requirement which can be resolved only through the introduction of new facilities, equipment, or materials
- A competitive need to meet or exceed the competitor's performance in the market place
- Extension of a successful product or service capability

6.1.5 Key User Questions

1. Have criteria been developed to use in guiding the senior managers in their selection of projects to support organizational performance?
2. Do people understand that any project selection model is only a guide that provides insight into which projects are likely to best support an organization's mission, objectives, and goals?
3. In addition to the questions that are suggested to use in the process of selecting a project, have other criteria been developed and understood by the decision makers in the organization?
4. Has the project selection process been diluted in the organization because the senior managers are prone to make the decision based on their background and current interests?
5. How successful has the selection of projects been in the past—particularly with respect to the value of the outcome of the portfolio of projects selected to support organizational purposes?

6.1.6 Summary

In this section, a brief insight is given to acquaint the reader with some of the strategies and models that may be used to select projects. The point was made that the selection process was highly personal. The best and most sophisticated models are only a means of influencing the decision of the executive who has to select a portfolio of projects to support the organization. In the end judgment, is supreme—and the manager should not allow any selection models to make the decision. Project selection is revisited in Section 6.5.

6.1.7 Annotated Bibliography

1. Meredith, Jack R., and Samuel J. Mantel, Jr., Project Management: *A Managerial Approach* (New York, NY: John Wiley and Sons, 1985), chap. 2, "Project Evaluation and Selection." This chapter provides an excellent overview of the process of project selection. Both numerical and non-numerical models and techniques are presented.
2. Cleland, David I., and Lewis R. Ireland, *Project Management: Strategic Design and Implementation,* 4th ed. (New York, NY: McGraw-Hill, 2002), chap. 4, "The Strategic Context of Projects." This chapter provides insight into some key questions and criteria that senior managers should be concerned about when they design and execute the decision process in the selection of an organization's portfolio of projects. In addition, the chapter describes a user friendly project selection framework whereby certain criteria are selected and used through a ranking process to select projects.

6.2 *LEGAL CONSIDERATIONS IN PROJECT MANAGEMENT*

6.2.1 Introduction

Project management is like any other management challenge. There are many laws, regulations, protocols, and conventions that have emerged that conditions the project manager's authority in managing the project. Disputes concerning the project can delay, increase schedules or costs, or even result in the cancellation of the project. The project manager has access to a legal system in the US, which usually provides a comprehensive framework that can guide the planning, and execution of a project.

6.2.2   The Legal Framework

The new project manager should recognize the potential for legal relationships that could impact the project. There are some specific agreements relative to the project that require particular attention and consideration such as:

- Contractual agreements with customer(s) and vendors
- Internal work authorization initiatives that provide for the delegation of authority and the transfer of funds to perform work on project work packages. Although these agreements are not "contracts" per se, if care is taken to negotiate agreements with internal organizations of the organization, the chances are enhanced that they will more fully support the project
- Relationships with project partners, such as a joint venture
- Funding agreements in the case of a project that is funded externally. Because the lender or giver of funds has assumed considerable risk, control over the management may be desired, such as an increased role over the use of funds on the project.
- Regulators, whether local, state, or federal, are key stakeholders of the project. The roles and expectations of these stakeholders should be determined as soon as possible, to include definition and agreements for the legal relationship that is expected.
- Project insurers can provide for the reduction of the risks expected on the project. In seeking protection from an insurer, the project manager should carefully negotiate the coverage, limits and liabilities, responsibilities and authorities of the participants, and the means to be used for determining damages, if incurred.
- Licensing agreements involves the use of proprietary technology or other property rights.
- Other stakeholders, such as those defined in Section 4.4, who have or believe that they have some right or share in the manner in which resources are used on the project. These rights or shares are discussed in that section, and will be mentioned here only as a brief reminder:
  - Governments
  - End users of the project results
  - Competitors
  - Investors
  - Intervenors or interest groups
  - Employees
  - Unions

Some of these stakeholders hold a contract with the organization that is sponsoring the project. For example, the employment contract held by employees is conditioned by non-discrimination statutes, and the collective bargaining contract with the union. All of these contracts may restrict the project manager's ability to assign employees to work on the project because of the potential of a lawsuit.

There are many complex, encompassing legal issues in which the project manager can become embroiled.

To guard against putting the project or some of its stakeholders at risk of legal action, the project manager and his/her team should seek an early and continuing relationship with the organization's legal office. In our personal lives, the best time and manner to use lawyers is before we get into trouble! It is no different in project management.

6.2.3   The Contract Structure

Since the success or failure of the project usually centers around cost, schedule, quality, and technical considerations, the performance standards in these areas must be established early in project planning. There are four basic contracts, which the project manager can use:

- Fixed-Price contracts—in which the contractors agree to execute a particular scope of work in a defined time period for a specified price.
- Cost-Reimbursement contract—used when the scope of work is not well-defined, the buyer assumes most of the risk in cost, schedules, and technical performance objectives. Normally these contracts will be reimbursed for all of their costs plus a percentage fee.
- Unit-Price contracts—in which the buyer assumes all risk of changes in the scope, costs, and schedules of the project, and the contractor assumes the risk that the cost of performing a unit of work may be greater than estimated.
- Target-Price contracts—in which the contracting parties establish cost and schedule standards with accompanying rewards and penalties for the seller of products or services.

There are variations for these types of contracts, as well as other specialized contracts that might be used. When the question of which contract to use, or the conditions in which it can be used, see the legal advisor! The potential for disputes in project management is always present. When disputes arise, the counsel of the local legal office should be sought as soon as possible.

6.2.4 Resolution of Disputes

- Mediation—in which a non-adjudicative process in which the disputing parties work with a neutral third party and attempt in good faith to resolve the disputes. The disputing parties are normally not bound to accept any proposed resolution by the third party.

- Arbitration—in which an impartial, binding adjudication of a dispute without resort to formal court proceedings is carried out. The arbitrator's decision is final for all practical purposes. The award the arbitrator decides on will normally be final without resort to court proceedings.

- Litigation—a form that should be used as a last resort. When managed efficiently and effectively, litigation may be more cost effective than arbitration, since the parties do not pay the courts for the adjudicative work.

- Standing Dispute-Resolution board—in which a standing board is appointed at the beginning of a project to evaluate and decide disputes on a real-time basis. This form enables the resolution of disputes in a more timely manner. This type of board can be expensive, and is normally used on very large projects whose time frame extends over many years.

6.2.5 Documenting the Record

On the chance that one or more of the above forms of resolution of disputes will be needed, in addition to simply making good sense, records should be kept backing up decisions on the project. Minutes of planning and review meetings, organizational design selection, monitoring, evaluation, and control meetings should be maintained. In other words, use common sense and document the management of the project—in particular those management actions that have to do with the manner in which decisions are made and executed on the project. Certain information of a project is proprietary and should accordingly be adequately safeguarded.

If a dispute may arise, or has arisen, seek guidance from the counsel of the local legal office on how the documentation relative to the dispute should be researched and provided.

6.2.6 Project Changes

Changes to project cost, schedule, or technical performance parameters should be carefully documented and maintained for use in future disputes. The project manager should build a project change mechanism around the basic factors shown in Table 6.2.

TABLE 6.2 Project Change Mechanisms

- Evaluate how and when the contract changes may impact the project.
- Authorize the change—after coordination with the relevant stakeholders.
- Communicate the change to all concerned parties.
- Modify the existing contracts or work agreements.

6.2.7 Handling Potential Claims

- Be sure that all real and likely changes have been resolved before assuming that the project is a success.
- Near the end of the project, go through an explicit analysis, working with the project team and stakeholders, to ascertain if any claims should be instituted, or if there are any outstanding claims that need to be resolved.
- In all circumstances, see the legal counsel and obtain assistance. Keep in mind that the legal office is an office of functional experts, and like other functional entities, supports the project.

6.2.8 Key User Questions

1. Do the members of the project team understand the legal obligations and liabilities that will buffet the project?
2. Have any thoughts been given to the design and development of measures and protocol to reduce the potential disputes in the execution of the contracts supporting the project?
3. Has an appropriate type of contract(s) been selected for the project? Do all members of the project team understand the meaning of the contract(s)?
4. Has a suitable working relationship been established with people in the local legal office?
5. Do members of the project team understand, and accept, that some stakeholders may have a legal right to intercede in the management of the project?

6.2.9 Summary

The project manager must understand the legal liabilities associated with the project stakeholders. All members of the project team should understand the basics of contracts, disputes, dispute resolution, and how the legal components of the project should be managed. The section closes with the admonition that the help of the local legal office should be sought when becoming involved in any legal matters involving the project.

6.2.10 Annotated Bibliography

1. Speck, Randall L., "Legal Considerations for Project Managers," in David I. Cleland, ed., *Field Guide to Project Management,* 2nd ed. (New York, NY: John Wiley & Sons, 2004). This chapter is a primer on some of the major legal considerations likely to face a project manager. When properly used, it can acquaint the user with some of the forces and factors that play a role in the legal component of projects.

2. Fleming, Quentin W., *Project Procurement Management* (Tustin, CA: FMC Press, 2003). This book is an excellent summary primer on project procurement management. Fleming provides the basics of procurement management to include the subfunctions of Procurement Categories, Planning for Procurement in the management of projects, Selection of types of contracts, Source Selection, and Contract Administration to name a few. The book's size (276 pages) makes it portable, something that could be used in the project manager's work-a-day activities.

6.3 PROJECT STARTUP

6.3.1 Introduction

Getting a project started right is much easier than trying to correct erroneous expectations or redirecting the effort of the project team. For the best solution, it is always better to start with all the project team working in the same direction and committed to making the project a success. Once the project execution begins, the dynamics will keep it moving either in the right direction or the wrong direction.

Project startup is the first opportunity to bring the entire project team into one location and have a mutual understanding as to the project, its goals, the expectations of the customer and senior management, and the relationships within the project team. This is the time for the project leader to set expectations and obtain the commitment of the individuals and the project team to the goals.

**Starting a project on the right course is the first
step to success**

6.3.2 Getting Started

The first task for a project team is to build the project plan. With limited planning experience, the team must take the goals of the project and convert those to a coherent guide from start of execution through project close-out. The planning exercise is critical to obtaining commitment to the project as well as ensuring the team understands the work to be accomplished.

Project planning will require collaboration among the team members to design the solution to the problem and elaborate on the goals for the project. A typical set of tasks for the project team could be:

- Review and analyze the project's goals and other amplifying documentation
- Validate the project's goals and feasibility of meeting the goals
- Identify issues and seek resolution to them
- Identify risks and seek mitigation options
- Develop a product description or specification
- Develop a work breakdown structure
- Develop a project schedule
- Develop a project budget
- Develop supporting plans, including:
  - Change Control
  - Scope Control
  - Risk
  - Procurement
  - Communication
  - Quality
  - Staffing
  - (Other as needed)
- Obtain senior management's approval of the plan.

Project planning by the team has become a joint effort, with some members performing parts that support the overall approach. For example, one member may write the staffing plan and another may write the risk plan. The entire team, however, reviews the final project plan for accuracy and completeness before seeking approval from senior management.

Project plans may also receive a peer review and comment. Often those who have worked in projects of a similar nature will be able to provide a critique of the plan and identify points that need clarification or special consideration.

6.3.3 Project Diary

A project diary is started on the first day of the project. The project diary is the place to record actions and activities that are informal, but have a bearing on the project. It may be viewed as a log of activities that affect the planning, execution and control of the project.

The project diary is the history of many actions that may need to be reviewed and to record information that affects the project. Figure 6.1 is an example of project diary entries.

It is important to maintain the project diary for future reference. Although an informal document, it is a reference to continue follow-up on actions. Also, some issues and problems may be recurring and it is best to have a record of the number of times they occur.

Project Diary			
No. Date Place		Activity/Item	Remarks
1	10/1/03 Conference Room	➢ Held kickoff meeting. Attended by all the team and company president	➢ President committed to supporting project with additional people, if needed
2	10/1/03 Project Site	➢ Vendor advised PL that equipment will be 3 days late in delivery	➢ No impact on project schedule
3	10/1/03 Project Site	➢ Issue identified: Product functionality is inadequate to meet customer's needs. New functionality not specified	➢ Unable to revise product specification until new functionality determined. Advised project sponsor of issue and requested assistance
4	10/2/03 Project Site	➢ Item #3 resolved by project sponsor	➢ Will continue with stated functionality and redesign upon release of version 1.1
5	10/5/03 Project Site	➢ Lead technical person being replaced because of illness	➢ HR Dept. notified and requested replacement NLT 10/10/03

FIGURE 6.1 Example of a project diary.

6.3.4 Project Kickoff Goals

Planning a kickoff meeting should be driven by the goals that one will achieve. A kickoff meeting is not a social event where there is an exchange of pleasantries and backslapping, but it is an important launch point for the project.

Goals for the kickoff meeting should be:

- To establish the project leader as the single point of contact for the project and as the head of the project team. This includes announcing the project leader's authority and responsibilities for the project planning, execution, control, and close-out.
- To establish the project team's role and responsibilities for the project and obtaining commitment, individually and collectively, for the project's success.
- To provide project background information and planning guidance. This includes all information needed by the project team to initiate the next phase of work.

6.3.5 Project Kickoff Meeting

When the project team is first assembled, it is essential that the project structure and purpose be communicated to all team members. The first meeting may be prior to the start of the planning or it may be at the start of execution. If there is a small team at start of planning and a larger team at start of execution, it may be necessary to hold two meetings.

Planning the kickoff meeting to ensure coverage of important items requires an agenda. This agenda could include all or a majority of the following.

- Set the tone and general expectations for the project team—project leader should describe the project in narrative with some graphics and describe the project's importance to the organization. Senior management may also address the importance of the project and its contribution to the organization's business.
- Introduce team members to each other—team members introduce themselves and state their expertise that will contribute to the project's success.
- Set expectations on working relationships—project leader discusses his or her expectations of what the team and team members can expect from him or her. Team members state what others may expect from them.

- Review project goals—project leader reviews the project goals with the team. The goals are expanded upon in terms of whether they are fixed or subject to change for any reason.

- Review senior management's expectations for the project—project leader reviews senior management's requirements for the project and what the team can expect from senior management in terms of support or involvement in the project.

- Review project plan and project status—project leader reviews the status of the project and discusses any progress previously made. This is the opportunity to identify the point at which the project team will start in the project's life cycle.

- Identify challenges to project (issues, problems, risks)—project leader identifies and challenges to the project and discusses ongoing actions to resolve them.

- Question and Answer period—this is an opportunity for the project team to ask questions about the project. It is also a time to clarify any erroneous perceptions and dispel any rumors about the project.

- Obtain commitment from project team for the work—project leader asks if any member has any reservations about committing to the project and obtains commitment from all participants to make the project successful. The project leader must also make a commitment to the team and to the project.

Bringing in senior management, the project sponsor, and functional managers to acknowledge the project's importance and say a few words may be needed. Ensuring that the entire team and all team members are committed to the project is essential to getting a proper start.

6.3.6   Follow-On to Project Kickoff

Questions or comments that affect the success of the project must be resolved soon after the project kickoff meeting. Delayed answers and avoided questions raise doubts in the project team and will typically erode commitment.

Follow-on meetings may be required when the entire team is not at the initial kickoff meeting or when the composition of the team changes dramatically through matrix management or for other reasons. If there was less than full information about the project, a second meeting should be held.

Accomplishments that represent progress following the kickoff meeting should be celebrated. For example, when the kickoff meeting is prior to planning, the completion and approval of the plan should be celebrated.

This celebration is an acknowledgement that the team is working together and meeting its commitment to the project.

6.3.7   User Questions

1. Why is it important to have a project diary that is initiated at project startup?
2. What information should be covered in the project kickoff meeting and why?
3. Should senior management be invited to the project kickoff meeting? Why or why not?
4. Who is responsible for the success of the project as identified in the project kickoff meeting?
5. When should the project kickoff meeting be held?

6.3.8   Summary

Proper project startup is critical to the success of the project's goals and delivering the benefits of the project to the customer. Senior management, functional managers, the project leader, and the project team have important roles to play in getting the project started right. The project kickoff meeting is typically the time that all stakeholders converge to disseminate information and demonstrate their commitment to the project's success.

Planning the project kickoff meeting by reviewing and confirming the goals is the project leader's responsibility. Preparing an agenda and assembling information for presentation facilitates and make for an effective kickoff meeting. When appropriate, involvement of senior management to emphasize the importance of the project to the organization and their commitment to support the project can materially add to the meeting.

Follow-on meetings and celebration of achievements are important to reconfirm to the project team the importance of their role and commitment to project success. Changes to the project or project team may dictate a need to have more than one follow-on meeting.

A project diary is part of the initiation of the project and a log for informally recording actions and activities. It is the record of daily items that can affect project success and a source of information when additional discussion or action is needed on an item.

6.3.9   Annotated Bibliography

1. Archibald, Russell D., *Managing High-Technology Programs and Projects* (New York, NY: John Wiley & Sons, 1992), chap. 11. This chapter

describes the project start-up process and gives a case study of a project start-up in a telecommunications organization. It provides techniques to be used in a start-up workshop.

2. Stuckenbruck, Linn C., and David Marshall, *Team Building for Project Managers* (Drexel Hill, PA: Project Management Institute Press, October 1985, chap. I–VII. This monograph describes useful team building techniques and identifying roles in the project. There is guidance on holding a project kickoff meeting for project startup.

6.4   DEVELOPING WINNING PROPOSALS

6.4.1   Introduction

Proposals are the foundation for starting projects. Successful proposals are well planned, well written, cohesive, and competitively priced. Proposals may be either to an internal organization or external to a potential customer. Proposals to external customers are more formal and comprehensive.

All proposal efforts of any size have a proposal manager and a proposal team. The team is typically an ad hoc proposal team established for one specific effort. The temporary nature of the team requires that the proposal manager be able to quickly assemble and motivate the team. A kickoff meeting that communicates the need for the proposal and its importance to the organization is the best method.

> **Proposals tell the potential customers how you will satisfy their needs**

6.4.2   Planning the Proposal Effort

Winning a contract from any proposal requires a dedicated effort to develop the document for delivery to the potential customer. This development requires a disciplined approach to writing and assembling the proposal as shown in Fig. 6.2. The success of the proposal is determined by whether it conveys the proper message and commitment to performing the work.

Developing a Win Strategy

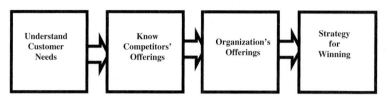

FIGURE 6.2   Strategy for winning the contract.

First, define a strategy for winning the contract through the proposal. What is the strategy and why will it work? This strategy results from the following:

- An in-depth analysis of the customer's stated requirements or needs
- An analysis of competitors' offerings
- An assessment of what your organization can offer
- A decision as to how to shape your proposal for the highest probability of winning

Once the strategy has been developed by a thorough assessment of the needs, competition, and your capabilities, the winning theme for the proposal is identified. The theme is the central idea, which provides cohesion to the proposal.

Responsibility for writing the proposal is assigned based on a task list and the available specialists. There is always a need for someone to ensure the cohesive theme is contained in all writings as well as addressing the technical aspects of the customer's needs. The proposal manager must ensure all tasks have a qualified person assigned to write a portion of the proposal.

A schedule for proposal work is needed to ensure all critical dates are met. Most schedules will include the times and durations for meetings, dates of critical milestones in preparing the proposal, and delivery date for the proposal. Timing is critical when the customer sets a "no later than" date for delivery or the proposal will be rejected.

Table 6.3 shows an example of a schedule shown with the general tasks associated with preparing a proposal. This matrix of tasks includes the start and finish dates for each task as well as the person responsible for performing the task.

The schedule is important to ensuring that the proposal is delivered to the customer on time. The above example shows discrete days for assign-

TABLE 6.3 Schedule of Proposal Activities

Task	Start of day	Finish of day	Responsible
Conduct background research	1	2	Proposal Manager
Develop proposal strategy and theme	2	2	Proposal Manager
Develop work plan with task list			Proposal Manager
Assemble and team kickoff meeting	3	3	Proposal Manager
Brief on theme and requirements			Proposal Manager
Assign tasks to team members			Proposal Manager
Brief on schedule of tasks			Proposal Manager
Set schedule of meetings			Proposal Manager
Write framework of proposal	6	15	Proposal Manager
Write task assignments	6	15	Technical Specialists
Review task writings and integrate	16	17	Proposal Manager/Technical Editor
Review complete proposal	18	19	Proposal Manager
Edit proposal	20	23	Technical Editor
Prepare in final form	23	25	Technical Editor
Obtain approval of proposal from senior management	26	26	Proposal Manager
Revise to address comments of senior management	27	27	Technical Editor
Prepare and produce final copies	28	28	Technical Editor
Deliver to customer	30	30	Proposal Manager

ment and delivery. However, assignments may be delivered early so the editing and integration can be started early. A technical specialist may write his/her portion in half the allotted time and deliver it to the technical editor. This levels out the work for the technical editor as well as releases the technical specialist to other work.

For large, complex proposals it may be necessary to make smaller steps in the writing portion. Experience levels of the writers also contribute to the ease or difficulty in preparing a proposal. Steps for a large proposal may include the following.

• Develop the strategy and theme for the proposal. The strategy and theme are based on the customer's requirements and the competitive offerings. Strategy and theme are designed to ensure a high probability of winning the contract through the proposal.

- Develop a detailed set of customer requirements and questions to be answered in the proposal. These detailed requirements and questions must be clearly addressed and readily identifiable in the proposal.
- Develop a detailed outline of the proposal similar to the table of contents for a textbook. Ensure any customer-prescribed format is followed when developing the outline. The outline must accommodate answering the questions and requirements identified in step 2.
- Develop a writing task list for all components of the proposal. This task list should identify the skills and knowledge required of the person writing the proposal section.
- Assign smaller tasks with shorter time frames to writers. Have each writer outline in bullet form his/her tasks to ensure consistency and following the proposal theme.
- Conduct frequent reviews of the writers' work and ensure the customer's requirements and questions are being answered.
- Using the detailed outline, prepare the framework and complete the general requirements of a proposal.
- Start integrating writers' contributions early and ensure the contributions are responsive to customer requirements.
- Conduct frequent team reviews to show progress and to identify written work that needs revision or additional work to clarify the proposal.
- Assemble all written work and conduct a comprehensive review with senior management to obtain approval for release. Senior management's approval is required because the proposal is a commitment of the organization.

6.4.3 Proposal Contents

Proposals address three areas for the customer. These areas address: What are you going to do? How are you going to manage it? How much will it cost. These primary components are:

- Technical—what is being proposed and how it will be accomplishment. This may be viewed as describing the work to be accomplished and the procedures used to perform the work. Another way of viewing this is "what problem is being solved and what is the approach to solving it."
- Management—what is the proposed method for managing this project work and what other information is required to establish credibility. The customer wants to be assured the work will be well—managed for successful completion.

- Pricing—what is the proposed bid price and what are the proposed terms and conditions for payment. The customer will typically require the price information to be in a specific format that may include detailed breakout of pricing for parts of the project.

Assembling the three modules for delivery to the customer may vary significantly. Delivery may be required in one single bound document or in three separate documents. Instructions for the format and assembly will usually be in the customer's solicitation document.

6.4.4   Technical Component of the Proposal

This component is concerned with the actual details of what is being proposed. The usual topics included are:

- Introduction—a non-technical overview of the contents.
- Statement of the problem—a complete definition of the basic problem that will be solved with the proposed service, product, study, or other work.
- Technical discussion—a detailed discussion of the proposed service, product, study or other work being offered.
- Project plan—a description of the plan for achieving the goals of the proposed project.
- Task statement—a list of the primary tasks required to achieve the project plan.
- Summary—a brief non-technical statement of the main points of the technical proposal.
- Appendices—detailed or lengthy technical information that contributes to the understanding of the technical proposal.

6.4.5   Management Component of the Proposal

This component is concerned with the actual details of what is being proposed for managing the project. The usual topics included are:

- Introduction—a non-technical overview of the contents of the technical and management components.
- Project management—details of how the project will be managed.
- Organization history—a brief discussion of the developmental history of the organization.

- Administrative information—a description of the organizational structure, management, staff, and overall policies and procedures that relate to the capability to perform the project work.
- Past experience—an overview of experience of a similar nature that contributes to the project performance.
- Facilities—a description of the facilities that will be used to provide the proposed service or build the product.
- Summary—a brief restatement of the main points of the management component.

6.4.6 Pricing Component of the Proposal

This component is concerned with the details of the costs for the project and the proposed contractual terms and conditions. The usual topics included are:

- Introduction—a non-technical overview of the main points of the technical and management components.
- Pricing summary—a summary of the costs for the proposed work.
- Supporting details—a breakout of the costs for items that appear in the pricing summary.
- Terms and conditions—a definitive statement of the proposed terms and conditions under which the product or service will be delivered.
- Cost estimating techniques used—a description of the methods used to determine the pricing in the proposal.
- Summary—a brief statement of the price of the proposed work and the major items being priced.

6.4.7 The Problem Being Solved

The most important aspect of any proposal is to identify and understand the problem that the customer is asking to be solved. Presenting your understanding of the problem and what is proposed to solve the problem is critical to being able to convince the customer that your proposal is the best one. Description of the problems usually involve:

- Nature of the problem
- History of the problem
- Characteristics of the optimal solution

- Alternative solutions considered
- Solution or approach selected

Describing the problem and the approach to solving the problem is the process of convincing the customer that you understand the situation. This area must be well stated both factually and convincingly to assure the customer that your proposal is the one to select. Identifying the wrong problem or providing subjective opinion will not convince the customer that your proposal gives the best solution.

6.4.8 Key User Questions

1. What is the first step in preparing a proposal and why?

2. Who is responsible for final editing of the proposal?

3. Why is there a strategy and theme needed to build a winning proposal?

4. What is the role of senior management in proposal preparation?

5. Why is there a need to look at the competition for the work?

6.4.9 Summary

Proposal writing is important to convey to a potential customer your understanding of the problem and that you have the best solution. The proposal states your technical approach to solving the problem, your management approach to the project, and the price at which you will perform the work.

Building a convincing proposal is a disciplined process that follows a general format with three components: technical, management, and pricing. The format provides the structure in which to describe your ability to meet the customer's needs for a product or service. Completing the format to accurately communicate your capabilities and desire to perform the work is in the detail.

Winning proposals are developed by a technically competent, motivated team. The proposal manager sets the tone for the proposal development and assigns the detail tasks. Proposal managers are often involved in the details of integrating the written tasks and working with the technical editor to build a smooth flowing document.

6.4.10 Annotated Bibliography

1. Freed, Richard C. (Contributors: Shervin Freed and Joe Ramano), *Writing Winning Business Proposals: Your Guide to Landing the Client,*

Making the Sale, Persuading the Boss (New York, NY: McGraw-Hill, 1995), 267 pages. This book presents an easy, systematic method to generate successful proposals. It shows how to develop key messages and themes for proposals.

2. Hamper, Robert J., and L. Sue Baugh, *Handbook for Writing Proposals* (Sumas, WA: NTC Publishing Group, 1996). This book is a concise discussion of the entire writing process. It uses a case study of a proposal writing team and has checklists to guide the reader.

6.5 PROJECT SELECTION—A REVISIT

6.5.1 Introduction

Selecting projects for an organization is important because of competing requirements and the need to bring the most value to the organization. Selecting projects for the business is critical to ensure continuity of the organization in a profitable mode. Random methods typically will not optimize the projects being pursued and will not enhance the organization's growth.

Internal project selection is often a random process that uses "ideas" generated by the marketing department or managers with needs for better capabilities. These ideas often do not have the payback capability or are optimistic in the outcomes and results. Developing projects may require that some brainstorming techniques are used, but the most benefit should be gained through a better selection process.

External projects, or projects that are a part of the organization's business, also require a structure for selection. Random bidding or accepting projects because it is work and may generate a profit are inherently counterproductive in the strategic position. A focused business plan that provides guidance as to project selection and criteria will enhance an organization's long-term position.

> **Selection criteria are needed for both internal and external projects**

6.5.2 Internal Project Selection Process

Selecting internal projects is as important as selecting external projects. The criteria are different and the selection process most often differs. Internal projects typically do not have as much weight on financial gain and more on enhancing the organization's capability.

Internal projects are typically an investment in the organization. Investments can come in many forms and the criteria for selecting internal projects may be unique to an organization. Some criteria that have been used in the past are listed below in no rank or specific order:

- Enhance organization's image. Example: hosting a picnic for handicapped children to demonstrate the organization's commitment to social issues.
- Enhance the productivity of a department. Example: streamlining a process for document management, both storage and retrieval.
- Optimize operations. Example: move a department to a location closer to its customer.
- Develop a new product or change an existing one. Example: upgrade software that is used internally for producing drawings.
- Develop a new marketing strategy. Example: change emphasis on new methods of doing business.
- Strategic prerequisite. Example: need to implement strategic project that depends upon information technology, such as E-Commerce.

The cost of each must be assessed for affordability, payback time, contribution to the organization's profit or cashflow. It may be difficult to assess the cost and benefit of each project, but the process for evaluating and selecting projects needs rigor.

6.5.3   External Project Selection Process

External projects are the source of revenue for many organizations. The organization's business is based on bidding for and executing projects that generate revenue streams and a profit. Selecting the appropriate projects is often critical to the growth of an organization because it is in the project delivery mode.

Organizations may start small and grow with the performance of projects over a matter of years. The bidding process is key to winning contracts, but first the type of contract being proposed is important. Winning contracts that an organization cannot deliver or detracts from the core business can cause an organization to fail to meet its goals.

Organizations should first determine the business that they desire to compete in and within that business line, identify the core competencies. Three to five core competencies may be the best for an organization to pursue because of the requirement for like skills, similar processes, and management capability. More than five core competencies may defuse the efforts of small to medium organizations.

General selection criteria for external, revenue-generating projects would be similar to the following.

- Matches the organization's core competencies
- Requirements are within the skills and knowledge of the staff or new resources that may be hired.
- Technology requirements are within the capability of the organization
- Management processes are capable of handling the project.
- The project contributes to the organization's public image and business image.

Other specific criteria may be applied to project selection based on the organization's business dynamics. For example, an organization may be extracting itself from certain types of projects or from a business line for lack of sufficient margin. These projects would not be bid or bids would be priced to make sufficient margin for continuation of the line.

6.5.4   Financial Considerations of Project Selection

All projects must contribute to the financial health of the organization, or a management decision made to pursue a specific project for good reason. There are many reasons for pursuing projects that either will show no profit or a loss. Some of these reasons include:

- The project is bid to obtain experience in that field. This builds on the organization's capability through some loss of profit.
- The project is bid to keep key technical staff employed. An organization may have a gap between work and a project that has no profit can be used to retain and pay the key technical staff. A temporary layoff could lose that capability.
- The project is bid because the organization knows the offeror will change his/her mind on the requirements. This project will be a profitable effort because the changing requirements will also result in increased expenditures. The project is initially viewed as a loss if the requirements are the same. The organization is at risk to provide the original requirements, but is betting on a change.
- The project is bid to become aligned with a major organization or to establish a basis for follow-on business. This project will typically establish the capability of the organization and position for new work that

is profitable. There may also be benefits in demonstrating an alliance with a well-known organization.

Financial return and profit are the typical indicators of success for selecting projects. Each project is reduced to a cost and the financial return that the project will bring. There are simple and complex methods of measuring the value of a project. It is recommended that the simplest form of value measurement be used and that the complex methods only be used when required.

Return on Investment, or ROI, is the most common approach to earnings analysis. The earnings before interest expense and taxes are related to the total investment are used in calculating ROI. ROI measures the return in money related to all project expenses. This is simply stated as:

Total project revenues divided by all costs to complete the project.

For example, (Anticipated) Project Revenues equal $1,500,000 and Total Project Costs are $1,200,000. The ROI is thus $1.5M divided by $1.2M, or $1.25. This measure that shows an ROI greater than 1 yields a profit while less than 1 is a loss.

ROIs may be used to compare similar projects as a potential measure of profitability and become a useful tool in the project selection process. It permits comparison of dissimilar projects on a common scale as it contributes to the organization's financial health.

Internal Rate of Return, or IRR, is another approach to calculating the rate of return for a project. IRR incorporates the concept of Net Present Value, or NPV. IRR more accurately reflects the value of a project when there is a long-term project and the payment is in the future.

NPV calculates the value of money today that is paid in the future. For example, $100 earned and paid today is worth $100. If the $100 is earned today and paid in one year, an inflation rate of three percent would decrease the payment value by $3, or a payout of $97.

The IRR approach on a project would look at the cost of the project at various time periods and calculate the net present value of each expenditure period. Then the payments would be calculated to reflect the NPV. A comparison of the period expenditures with payment schedules would provide the information to compute total IRR.

An example is shown in Table 6.4 for comparison of a project over five years demonstrates the concept. The non-discounted version shows the simple approach to determining the ratio. The discounted version incorporates NPV to obtain the IRR. Long-term project values may need to be calculated by the NPV method for IRR just to make a realistic comparison. Short-term project values may not have the need when inflation rates are low.

TABLE 6.4 Net Present Value for Projects

Project	Year 1	Year 2	Year 3	Year 4	Year 5	Project Value
Project Expenses	1,000	1,000	1,000	1,000	1,500	5,500
Project Revenue	1,000	1,100	1,200	1,300	1,900	6,500
Project Value Under NPV @ 5%	0	+100	+200	+300	+400	+1,000
Project Expenses	1,000	950	900	850	1,200	4,900
Project Revenue	1,000	1,045	1,080	1,105	1,520	5,850
Project NPV	0	+95	+180	+255	+320	+850

The table shows the difference between using a simple expense-payment approach and the NPV calculations with a five percent discounting per year. Note that the first year is not discounted in this example, although some organizations will use a discounted rate for each year.

NPV discounts expenses and funds received in future periods. The monetary value of a project is less than the computed amount when using the NPV and computing an IRR. The non-discounted version of Project A gives a ratio of 1 to 1.18. The NPV discounted version, or IRR, gives a ratio of 1 to 1.19. Higher discount rates, deferred payments, and a different expense or payment profile would distort the ratio significantly.

Cashflow analysis is another useful tool to determine whether a project should be selected. An organization with little capital operating in a period of high interest rates may need to perform a cashflow analysis. This will determine the amount of money required for expenses and the payment schedule for a project.

Cashflow is typically computed on a monthly basis where a project's expenses are estimated to determine the outward flow of money and the estimated payments each month from the project. A five-month project, for example, could have an expense to revenue profile as shown Table 6.5.

TABLE 6.5 Cashflow Analysis

Cashflow	Month 1	Month 2	Month 3	Month 4	Month 5	Remaining	Total
Expenses	12,000	12,000	12,000	12,000	12,000		60,000
Revenue	0	10,000	14,000	14,000	14,000	10,000	72,000
Difference	−12,000	−2,000	+2,000	+2,000	+2,000	+12,000	
Cumulative Deficit	12,000	14,000	12,000	10,000	8,000	−4,000	

The project will require financing between $8,000 and $14,000 during its life. This profile of more expenses than revenues throughout the life of the project dictates that financing be obtained to support the difference. Project selection must consider the need for additional funds and the cost of those funds during the project life.

6.5.5 User Questions

1. When would an organization bid on a project that is estimated to be financially a breakeven situation?

2. For short-term projects, what method would you use to assess profitability?

3. What does Net Present Value do for a project when used in conjunction with Internal Rate of Return?

4. Why would you want to know the Cashflow profile for a project and what does it tell an organization?

5. When, if ever, would an organization select a project to bid when it was not within the organization's core competencies?

6.5.6 Summary

Project selection is more than identifying an idea or finding a profitable venture that contributes to the organization's financial health. Projects must fit within the core competencies of the organization and bring benefits to the organization. Random project selection practices may deliver the wrong benefits and suboptimize the organization's project capabilities.

Using the simplest approach to select projects is best and ensures all pertinent factors are considered. Having criteria for selecting projects provides an objective structure for sorting through numerous choices. First, the project should fit the organization's strategic goals and bring benefits to the organization. Secondly, the project must bring financial benefits to the organization. Financial benefits should be assessed through one or more approaches such as ROI, IRR, and Cashflow Analysis.

6.5.7 Annotated Bibliography

1. Fabozzi, Frank, and Peter Nevitt, *Project Financing,* 6th ed. (London, England: Euromoney Publications, 1996), 380 pp. This book describes the criteria for obtaining financing for projects. The same criteria used by lenders applies to individuals selecting projects for implementation. It covers key criteria for success.

2. Marino, Joseph Paul, *R&D Project Selection* (Wiley Series in Engineering Management) (New York, NY: John Wiley & Sons, 1995), 266 pp. This book explains how to evaluate alternative projects and make selections based on advantages and disadvantages. The selection processes described use objective criteria and methods.

6.6 SELECTING AND USING PM SOFTWARE

6.6.1 Introduction

Project management software programs have grown out of a need for better management of information in projects. Perhaps the first paper system was created by Henry L. Gantt in the early part of the 20th Century with the bar chart to schedule and control production efforts. This was not improved upon until the 1950s when the US Navy and E.I. Dupont Company developed the Program Evaluation and Review Technique (PERT) and Critical Path Method (CPM), respectively, as the foundation of modern schedule networks.

The advent of the computer and its growth since the 1980s has brought more and better software programs that meet the needs of project teams. Today, desktop computers with project management software give team members a powerful capability for planning and controlling projects. The software capabilities are dependent upon the sophistication and complexity of the programs.

Project management software programs are available for projects with a variety of features. The cost of the programs range from $25 for the low end, basic scheduling tool to several thousands of dollars for the high end, mainframe hosted systems. The most popular programs, that is those with the greatest sales volume, range in price from $300 to $600 for hosting on desktop computers.

Managing multiple projects or large projects that have many tasks and hundreds of resources requires the next step up in project management software. Software that can accommodate projects or multiple projects with ease ranges in cost from approximately $2,000 to $10,000. Selecting capable software will materially enhance the organization's project management capability.

6.6.2 General Functions of Project Management Software

Project management software initially started with a scheduling function, or the capability to layout tasks over time and track the progress of work.

Later, the cost of project work was integrated into the software programs and most recently the resource management functions.

Project management software for the small to medium projects is currently designed to handle scheduling, resource management, and computed costs of resources. These general functions are addressed here, but not the more sophisticated functions, such as changing costs of resources over time.

> **Project management software generates critical decision-making information on time, cost, human resources, and material resources**

Described below are the general functions that are the most common, and that provide the basics for all software programs. A short description is provided for each function:

- Time management—the capability to develop schedules that depict the work in tasks and summary activities. This is a planning requirement and then converts to a tracking function. Time management entails the capability to set a baseline and apply actual accomplishments into the schedule for measuring progress during the execution and control phases.

- Cost management—the capability to develop a project budget with detailed costs assigned to each task. The individual costs are summed to obtain the total anticipated cost of the project. Because costs are associated with each task and each task is placed in a time frame for execution, the budget is readily available as a time-phased plan for expenses. During execution and as resources are consumed, the system should be capable of accumulating expenditures for comparison against the baseline budget, both in detail and at the summary level.

- Human resource management—individuals with the requisite skills to accomplish project work on tasks are assigned to those tasks. This assignment makes the schedule a "resource-loaded" schedule with all identified people needed to complete the project. Because the resources are assigned to tasks, the time that they are required on the project is identified. The system should be capable of adding or changing the human resources during execution and control phases, as the situation dictates.

- Resources (other than human)—materials, equipment, vendor services, contracts, travel, rentals, and other costs may be required to complete the project. These resources can be assigned to tasks and the price of each entered into the schedule. This permits ordering and receipt of general resources for the project to meet the requirement in time for the project continuation. The system should be capable of adding or changing the resources during execution and control phases, as the situation dictates.

These four categories are the minimum for software other than some low end products that plan and track only schedule functions. These categories provide the capability to manage projects from extremely simple and of short durations to major projects with long durations.

6.6.3 Considerations for Software Selection

An organization must view its needs in terms of managing different components of the project and identify the needs for planning, executing, and controlling projects. One of the major drivers in considering purchase of new software is the number of users. If an organization has distributed planning, execution, and control functions, the software will typically be provided to several individuals. The capability of such software would be in providing only time, cost, and resource management on the low end of the scale.

An organization with centralized project management planning, execution, and control functions may select more capable software. There would be additional functions and capability, such as earned value management—the cost-schedule integration function for measuring project progress. A more capable software system may also be considered for managing multiple projects or for managing projects through a master schedule.

Training is an expensive and time-consuming process for an organization. Although not apparent many times when the software selection is being made, training has a dramatic affect on the organization's implementation of new software across several users.

Compatibility with current software in the organization is another consideration. The transfer of data between the project management software and the corporate accounting system, for example, may be difficult. Transfer of data between different project management systems may also be required if there is more than one type of software.

Reporting of project information is a significant area for assessing and determining whether any new software will produce the reports required to fill the management functions at all levels. Ideally, information from the project management system will meet the needs of the project leader to track progress and report status, the project sponsor for summarized data, and senior leaders for decision-making data. The project management software should format the information in the proper reports for all consumers.

Computers must have the capacity and capability to host the project management software. Desktop computers will typically host the midrange of project management software, but there needs to be sufficient hard disk space and enough random access memory to ensure optimum performance.

Printers and plotters are needed for producing paper reports or graphics to meet the needs of all consumers. Many of the project management software programs will produce the reports if there are printers and plotters available. Printers and plotters may not be available to support the reporting functions that the software is capable of producing.

Price of project management software is the last item to consider. The purchase price is only part of the cost. Training can be extensive and expensive. Additions or changes to the hosting computer systems and printers/plotters can also be a large cost. Special costs such as on-site support of software or additional hardware should also be included in the overall price.

6.6.4 Detailed Software Requirements

An organization buying software must consider the projects being managed and the details of the project requirements. Figure 6.3 shows the minimum areas for consideration.

Detailed discussion of the basic considerations for software selection include:

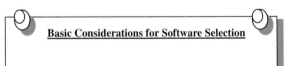

Basic Considerations for Software Selection

- Size of projects
- Number of resources
- Number of cost categories
- Task dependencies
- Project consolidation into master schedule
- Resource leveling

FIGURE 6.3 Basic considerations in project management software selection.

- Size of projects—the size of the projects is measured by the number of tasks that can be handled by a software program. Most small projects will be less than 250 tasks. Large projects, multiple projects, or a master project may exceed 50,000 tasks.

- Number of resources—assigning and tracking resources on tasks may be constrained by the software. If the organization's methodology requires numerous resources to be assigned to one task, the software needs to have that capability. The number of resources that may be placed in a resource library is important as well.

- Number of cost categories—the budget will require a number of cost categories for an organization to obtain the granularity on the cost. Software should be able to accommodate all cost categories.

- Task dependencies—the number of dependencies, or connections between tasks, will be defined. The organization's planning and project management methodology will dictate the number of dependencies.

- Project consolidation into a master schedule – the ability to consolidate several projects into a master project may be desirable. This permits schedules to operate from a single resource library. It does require that the total volume of tasks can be accommodated.

- Resource leveling—the capability to level resources either through use of available float or through changing the project's end date. Resource leveling reduces or eliminates conflicts when one resource is required to perform two tasks simultaneously.

The human side of selecting new project management software must be considered to ensure user acceptance. Areas that must be considered for all participants' satisfaction are:

- Ease of use—the software should be easy to use and consistent with other software being used. There should not be different conventions for using the software and difficulty in the population of data, for example.

- Reliability of software—the software should be consistently reliable and not have any major flaws or failures. It should produce consistent, uniform results.

- Reporting functions—reports should be easy to format and produce. The data for headings and legends should be saved for projects.

- Training time—the time to learn a new software program should be reasonable and cover most of the routine features. Advanced features, if used, may require some extra effort and training time.

- Transfer of data—data from other sources should be easily transported and imported into the schedule. Data formatting should only require simple and easy manipulation.

- Report formats—standard or customized reports should meet the needs of all consumers without further manipulation. The range and type of reports must serve the team members, project leader, project sponsor, and senior leaders without overly complex manipulation.

6.6.5 Selecting Project Management Software

Selection of project management software involves several people. A team approach is best to review all the requirements and to consider all functions and features. The following people must contribute to the process and information required:

- Senior leaders and project sponsors—information requirements from the software.
- Project leaders, project planners, and project controllers—capability to plan and control projects; capability to permit number of dependencies, resources, tasks, and consolidation of projects.
- Project team members—ease of use for reports to inform and to update progress; accuracy of information generated in reports.
- Software user—ease of use, easy to learn, and easy to import or export electronic data; easy to generate and produce reports; speed of calculations for updates or recalculations; ability to make changes and to record progress.

The software selection team should develop criteria for the software based on the requirements of the consumers, validate that the computers are capable of hosting the software, validate that software is compatible with report production devices, and coordinate any electronic data transfer requirements with the owners of other systems, such as accounting systems.

6.6.6 Key User Questions

1. Why should anyone other than the user of project management software have requirements for selection of a new system?
2. What requirements for a project management software system would be appropriate for senior leaders?
3. If the majority of projects have only a requirement for 300 tasks, what would be a legitimate reason for selecting a software program with 50,000-task capability?

4. If there is a need to transfer time card hours from the accounting system to the project management system, should this be an absolute requirement? Why or why not?

5. New project management software is being evaluated for the organization. As the project planner, what are your needs?

6.6.7   Summary

Selection of project management software is a complex task that has to consider several aspects of the organization. First, there needs to be an assessment of the organization's needs in terms of project management capability, the project needs to be met, requirements of all levels of leadership, and user acceptance of the software.

Software selection has to consider the capabilities of existing systems to host the program and the ancillary devices for generating reports. Electronic data transfer, either import or export, features are critical to the smooth operation with other systems within the organization. Training requirements on the new software must be considered in terms of ease of use, number of users, and cost to train.

The price of the software is the last to be considered. This price must include the purchase of the software with any support requirements, upgrades or replacements to existing hardware, and training for users. When all costs are identified, a cost-benefit analysis may be conducted to determine whether there are similar competing systems that should be selected.

6.6.8   Annotated Bibliography

1. Research and Management Technicians, Inc., Document, *Decision Support System Specification,* Gaithersburg, MD, March 7, 1991. This was a detailed study of the requirements for a project management software system to support a $500,000,000 a year program. It considered report functions that meet client needs.

2. Sample Project Management Software Vendors

   - Microsoft Corporation (*www.microsoft.com*)
   - Primavera Systems, Inc. (*www.primavera.com*)
   - Scitor Corporation (*www.scitor.com*)
   - Timeline Solutions Corporation (*www.novel.com*)
     These vendors sell project management software that is under continual development to enhance the products. Their Internet sites are listed for ease of consulting the most current information on capabilities and contacts. There are other vendors as well who may meet the needs of project management software users.

6.7   PROJECT CONTRACT NEGOTIATIONS AND ADMINISTRATION

6.7.1   Introduction

It makes good sense for a project manager to understand some of the basics of how to negotiate and administer the contracts involved in a project. In some cases, a project manager takes over a project where the contract terms have been negotiated, and the project manager's main concern is in the administration of the contract. In this section, some of the basic ideas about the project manager's role in contracts are explained. The reader is cautioned to remember that expert contract negotiation and administration is a specialized activity that supports a project. The project manager is advised to seek appropriate expertise when needed.

6.7.2   Some Practical Guidelines

Most project managers are not lawyers, but they are involved in the process of negotiating and administrating contracts. Some basic principles include:

- Seek orientation through the organization's legal office to become acquainted with the general aspects of contract negotiation and administration.
- Remember that at times during the management of a project, the project manager can become the *de facto* contract manager.
- Recognize the legal aspects and responsibilities that a project manager has regarding the project.
- The project manager must understand (1) the basic negotiation of the contract; (2) the clauses that exist in the contract that limit liability and risk sharing; and (3) the administration of a contract in a non-negligent manner.
- The project manager must also understand the limits of his or her knowledge in contracting, and know when to seek specialized support from contract specialists or legal counsel.

6.7.3   Contract Negotiation

Negotiation is a process whereby parties with differing interests reach an agreement through a process of communication and compromise. In any negotiation process it is necessary to:

- Separate people's emotions from the contract issues.
- Remember that each party to the negotiation is a person just like yourself, who may get angry, frustrated, hostile, and offended when their personal views regarding the project do not prevail.
- Remember that any people problems that exist must be dealt with outside of the substantive issues of the contract that are being negotiated.
- Put yourself in the other person's position—which can be helpful.
- Know that both sides to the negotiation must work hard at maintaining active communication.
- Focus on the common interests in the process rather than the opposing positions of the negotiating parties.
- Remember that behind the conflict issues, common interests and motivation can often be found.
- Ensure that basic conflict over substantive issues is dealt with so that both parties win something.
- Be careful of committing to a specific position, and be caught in the need to defend this position without any hope or pretext of compromise.
- Acknowledge the interests of the other parties as part of the negotiation process—this will help to bring about a successful negotiation.
- Be firm in the negotiation process, yet be open and supportive of the human side.
- Attack the substantive issues—not the people involved.
- Carefully identify and prepare potential creative solutions before the negotiation process starts—this is essential.
- Insist that the negotiating process will be based on objective criteria.

6.7.4   The Contract

During the negotiation process, ensure that the basic contract is understood along with clauses that will be integrated into the final contract. Other considerations are:

- Seek to limit the legal liability of the project manager and the organization.
- Seek an understanding of how to use warranties, indemnification, liquidated damage clauses, and other clauses which limit liability.

6.7.5 Warranties

The concept of a warranty is that the seller's verbal or written commitment means that the deliverables of the project will meet certain standards. Key provisions include:

- The warranty imposes a duty on the seller who can be held liable by the buyer if this commitment is breached.
- The buyer can bring legal action to recover damages or rescind or cancel the contract.

Two basic types of warranties exist:

- *Verbal or written warranties* which pledge a specific commitment to perform on the contract.
- *Implied warranties* which are assurances or promises that are a matter of law and practice, rather than a specific promise made in the contract. Implied warranties arise from specific laws or from what is by precedent expected in the product or services delivered to the buyer. The implied warranty of a product or service holds that such deliverables must be reasonably suited for the ordinary purposes for which they are used.

6.7.6 Project Manager Actions Regarding Warranties

- Be cautious in making statements regarding performance or design of the project's product or service.
- Be prudent in putting information into the contract that could have some warranty repercussions.
- Remember to state what a warranty covers as well as what it does *not* cover.
- Remember that inserting warranty language into contracts for performance of services can have the result of raising the standard of performance and customer expectations.

6.7.7 Indemnification

Indemnification is the act of protection to guard against legal suit, or body injury to a person, or organization for a loss incurred by that person or organization. There are two types of indemnification:

- Common-law
- Contractual

Indemnity provisions vary considerably from contract to contract as to the extent of the liability transferred. These provisions are generally of three types:

- *A broad form,* the most severe, which obligates the indemnitor to indemnify and hold harmless the indemnitee against all loss arising out of the performance of the contract.
- *An intermediate form,* which holds the indemnitor responsible for all claims or suits arising out of the contract except those arising out of the sole negligence of the indemnitee.
- *A limited form,* where one party agrees to indemnify the other only for the claims arising out of the indemnitor's negligence.

6.7.8   Project Contract Administration

Contract administration includes (1) supervising the work to be done under the terms of the contract; (2) preparing and processing the changes that come about; (3) providing interpretation of contract language and forms; and (4) approving invoices as the work is performed. In performing these responsibilities, the project manager needs to be aware that conduct that is less than reasonable can give rise to charges of negligence against the project manager, the organization, or team members. Such unreasonable behavior can also lead to allegations of breach of contract. Company executives must establish policies and procedures describing the manner in which contract administration is carried out in the organization. Several key principles of behavior involved in contract administration include:

- All required signatures, comments, and approvals have been obtained and documented before the contract is issued.
- No work should be performed before the final contract is issued, or pending the finalization of the contract, a formal letter of authorization is provided.
- Excellent contract files and documentation regarding the administration of the contract to include the recording of events by the project manager that may have a bearing on the contract and the quality with which the contract is administered.
- Assurances that the cumulative services and billings for the organization's services do not exceed the scope and budget for the contract.
- Develop and document a contract change control process so that there is a record of why, when, and how contract changes were reviewed and approved.

- Seek legal or contract counsel before getting into difficulty in the contract negotiation or administration process.

6.7.9 Key User Questions

1. Does the culture of the organization clearly establish the importance of project contract negotiation and administration? If not, why not?
2. What policies and principles have been established in the organization concerning the project manager's responsibilities regarding the contracting process?
3. Has adequate training been provided to project managers and the project team on how best to become involved with project contracts?
4. Do the members of the project team understand the difference and nature of expressed and implied warranties, and the applicable types of indemnity provisions?
5. Do the project team members understand the conditions under which they might be held liable for negligent behavior involving the project?

6.7.10 Summary

In this section, some of the basic notions involved in project negotiation and administration were presented. This presentation was made with the cautionary advice that the field is a large and complex one in which much legal consideration predominates. Accordingly, the project manager should seek legal and contract counsel before the contract is in trouble.

6.7.11 Annotated Bibliography

1. Speck, Randall L., "Legal Considerations for Project Managers," in David I. Cleland, ed., *Field Guide to Project Management* 2nd ed. (New York, NY: John Wiley & Sons, 2004), chap. 32. This chapter provides a summary of the legal considerations about which a project team should be knowledgeable.
2. Cleland, David I., *Project Management: Strategic Design and Implementation,* 2nd ed. (New York, NY: McGraw-Hill, 1994), chap. 15, "A Project Manager's Guide to Contracting." This chapter provides a summary of some general considerations for a project team to be aware of in their project contracting activities.

6.8   QUALITY IN PROJECTS

6.8.1   Introduction

Quality in projects is defined as meeting the customer's requirement. This entails meeting the customer's technical requirements and assuring that the customers are pleased with the results. Building quality into the project's products is a rigorous process of first understanding the customer's requirements and continuously working toward the end result.

Quality in projects is neither an accident nor is it achieved by chance. Quality is a concerted effort by all project stakeholders to focus on the customer needs and to work toward satisfying those needs. The project team leader is the linchpin for determining those customer needs and ensuring that all participants converge on the solution.

Senior management can support or adversely affect the product quality be demanding that shortcuts be taken in the project or technical tradeoffs reduce product functionality. Supporting building product quality requires some understanding of the customer's needs and how that will be achieved. This should be detailed in the project plan.

The project team must understand their roles in doing the work right and following the specifications. Deviations or shortcuts on building the products will typically result in less than satisfactory quality. To build to the specification, the project team has to be technically competent and proficient in their work.

6.8.2   Quality Principles

A quality program will follow principles that ensure the best results are achieved through a disciplined process. Deviations from these principles will typically result in less than a satisfied customer because his/her needs are not being met. Deviations also create waste in time, materials, money and profit through rework.

Quality principles consistent across all industries are as follows.

- Doing the work right the first time saves time and money.
- Quality is a prevention process.
- Quality is conformance to the requirements or specifications.
- Quality is built into products through dedicated attention to customer needs.
- Testing is a validation of functionality, not a defect identification process.

Deming's 14 Points

1. Create a constancy of purpose for improvement of product and service.
2. Adopt a new philosophy.
3. Cease dependence on mass inspection.
4. End the practice of awarding business on the price tag alone.
5. Improve constantly and forever the system of production and service.
6. Institute training.
7. Institute leadership.
8. Drive out fear.
9. Breakdown barriers between staff areas.
10. Eliminate slogans, exhortations, and targets for the work force.
11. Eliminate numerical quotas.
12. Remove barriers to pride of workmanship.
13. Institute a vigorous program of education and training.
14. Take action to accomplish the transformation.

FIGURE 6.4   Deming's 14 points.

- Testing does not identify all defects in products.
- Quality is the responsibility of everyone, not just the quality assurance/ quality control personnel.
- Quality is a continuous improvement process.

The best known guidelines for project quality were developed and published by Dr. W. Edwards Deming, a recognized expert in quality in Japan and the US. Dr. Deming's 14 points are applicable to any quality program. They are shown in Fig. 6.4.

6.8.3 Quality Management Components

Quality for projects comprises three components: (1) quality planning; (2) quality assurance; and (3) quality control. These three components define the approach to a quality program for a project and determine the extent that quality instructions are needed for the project.

Quality planning comprises the identification of relevant standards for the project and the degree to which they apply to the workmanship, product, and processes. For example, an industry standard for workmanship may be identified as the proper one for a specific project. This sets the level of workmanship that is expected and provides a basis for any inspection or examination of the project in subsequent quality control measures.

Quality assurance establishes performance criteria for the project to assure the process is working and to give confidence that the project will meet the standards identified in quality planning. This function entails the overall project quality approach and the measures that will ensure a quality product. Quality improvement efforts are included within this function.

Quality control is measuring work results against the standards to ensure there is conformance to attribute, characteristic, and functionality elements. Measuring the work results in a validation process and defects will be identified through the efforts of quality control inspectors. Defects identified during inspection are opportunities for improvement in the process.

6.8.4 Cost of Quality

The cost of quality is determined by functions in five areas:

- Prevention—the cost to establish and maintain efforts that preclude defects from entering into the process or product. Prevention centers around such areas as training of performing participants, establishing and proving processes, and continuous improvement efforts.

- Appraisal—the cost of all efforts to examine, test, inspect and demonstrate product compliance with requirements. Quality control efforts typically are the majority of costs in this area.

- Internal failure—the cost of fixing items that fail prior to delivery to the customer. These are often the result of quality control tests and inspections.

- External failure—the cost of fixing items after they have been delivered to the customer. These costs include the cost of repair and other costs associated with such items as replacement, delivery, services, and complaint handling.

- Measurement and test—the cost of equipment and tools to perform measurements and testing of processes and products. This also includes calibration cost for precision measuring devices.

Projects have shown a significant cost for all these items and in one instance the cost of rework, materials, and labor added 46 percent to the cost of the project. Typically, the cost of quality is 12 to 20 percent of the cost of goods sold. This adds a significant cost to products for no added value. The estimated cost of quality is 3 to 5 percent of the cost of goods sold.

Table 6.6 demonstrates the need to shift to a new way of quality management and reduction of costs in waste of materials, labor, and time. The actual cost and the projected cost show dramatic improvement to profitability as well as organizational image as a producer of high grade products.

TABLE 6.6 Areas of Cost for Quality Products

Cost area	Actual cost of goods sold 12 to 20%	Should cost of goods sold 3 to 5%
Prevention	10	70
Appraisal	35	15
Internal Failure	48	10
External Failure	7	5
Measurement & Test	< 0.1	< 0.1

This demonstrates that more emphasis needs to be placed on prevention and less on appraisal. The reduction of actual quality costs (12 to 20 percent to 3 to 5 percent) has a dramatic effect on profitability as well as customer satisfaction. This chart shows that quality costs can be reduced by 75 percent (low end 12 percent to 3 percent), which can yield both lower prices (competitiveness) and increased profit margins.

6.8.5 Continuous Improvement

Continuous improvement of quality is achievable within large projects and in a series of projects. Large projects have the interest and emphasis of senior management to plan and execute the work in an efficient and effective manner. This planning phase gives the project team time to identify the most effective process for execution. Small projects can capture lessons for both repeatable activities and for those that should not be repeated.

Continuous improvement is based on fundamental concepts that must be implemented and improved over time for projects. These concepts include:

- Clear requirements—customer requirements are clearly and succinctly defined in the statement of work, specifications, product description, and illustrations.

- Project processes are well-defined—processes used in project planning, execution, and control are designed to bring forth the best results for the project. Processes are proven and have rigor.

- Process is capable—processes used in projects must be proven capable through audits and validation. Training must be conducted on new processes or processes that are changed.

- Process is in control—controls for processes and the check points must be in place prior to use. During implementation, controls must be effective to demonstrate the process is in or out of control.

- Policies are in place—quality policies must be in place and enforced to be supportive of the quality goals. Policies must be focused on defect prevention.

6.8.6   Problem Solving Process

Quality programs need a process for solving problems or defect correction as the problems and defects are identified. There is a need for a consistent approach and a rigorous process that ensures problems and defects are corrected in an efficient manner. The following six-step process outlines an effective approach:

- Define the problem—defining the problem looks for what is wrong and the immediate effects of the problem.

- Fix the problem—correct the defect or problem as identified by step one.

- Identify root causes—determine what caused the problem or defect, not just the symptoms of the problem or defect.

- Correct the process deficiency—identify the weakness in the process and correct the process to ensure the root causes are resolved.

- Evaluate the corrective action—examine the process to ensure the corrective action is effective and that it removes the root causes of problems or defects.

- Follow-up—review the corrective action to ensure the new problems or defects do not result from changes to the process.

6.8.7   Quality Team

The quality team is comprised of all persons who have a role in ensuring the product of the project meets customer requirements and anyone who can affect the outcome of building the product. This team, working together to continually work toward the technical solution, can easily affect the end product and the project's capability to design, plan, build, and evaluate the required product.

The project quality team consists of seven groups of people:

- Senior management—sets the tone for quality through policies and directions to the project team.
- Project leader—implements quality policies and develops quality plans for projects; enforces quality initiatives within the project.
- Functional managers—implement quality policies as well as the project quality plan for any work accomplished for the project.
- Suppliers/vendors—provide pieces, parts, components, assemblies, and materials of the specified grade for the project.
- Contractors—meet the quality standards specified in the contract and to meet customer requirements.
- Project team—meets quality standards in workmanship and specification requirements.
- Customer—sets the requirements for quality and stabilizes the requirement to permit building to the requirement.

6.8.8 Key User Questions

1. Quality is the responsibility of the entire quality team. How can the customer affect the quality of the product when he/she is not involved in building the product?
2. Project team members require skills, knowledge, and abilities to perform the work to the required specifications or standards. How does motivation become involved in quality products?
3. The project leader is the focal point for designing quality into the project plan and implementing that plan. Who should review the quality plan prior to its being implemented and why?
4. Senior management may not become involved in the daily performance of the project work. How can senior management affect quality, either positively or adversely?
5. Dr. Deming's 14 points are universal to most projects and organizations. Which of the 14 points are being implemented in your organization?

6.8.9 Summary

Quality in projects is the responsibility of all stakeholders. Anyone may adversely affect the quality of the product through lack of understanding of the requirement, lack of appropriate skills to perform functions, or lack of motivation to do the work right the first time.

Quality products from projects are the result of disciplined efforts to design accorrding to requirements and to build to those requirements by using proven processes. Assurance is provided through planning and actions that build confidence in the capability to deliver the product. Control of the quality is performed to ensure conformance to requirements and to validate the product's attributes, characteristics, and functionality.

Process management is used to provide consistent, proven methods of performing the work. These processes can be adjusted or changed when there are problems or defects in the product. Processes may also be changed to make continuous improvements to the method of performing on the project. Processes must always be proven prior to implementation to be supportive of the quality program.

6.8.10 Annotated Bibliography

1. Walton, Mary, *Deming Management at Work* (New York, NY: Perigee Books, 1991), chap. 1–7. This book describes the Deming philosophy for quality in an organization and gives several examples of organizations implementing quality programs.

2. Ireland, Lewis R., *Quality Management in Projects and Programs* Project Management Institute Press, (Drexel Hill, PA: 1991), chap. II–VI. These chapters describe the quality process and the people involved in quality. The quality team is defined and described.

6.9 PROJECT TERMINATION

6.9.1 Introduction

In general, projects are terminated for one of two reasons: project success or project failure. Project success means that the project has met its schedule, cost, and technical performance objectives, and has an operational or strategic fit in the organization. Project failure means that it has failed to meet its cost, schedule, and technical performance objectives, or it does not fit into the organization's future. Both success and failure are relative factors.

Projects that are failures typically need to be terminated as soon as feasible to save the organization's resources. Too often, projects are continued without a reasonable probability of successful completion.

> **Terminate projects that have a
> low probability of success**

6.9.2 Reasons for Project Termination

- The project results have been delivered to the customer.
- The project is overrunning its cost and schedule objectives and/or is not making satisfactory progress towards its technical performance objectives.
- The project no longer has an operational or strategic fit with the owner organization's future.
- The project owner's strategy has shifted, reducing or eliminating the need for further expenditure of resources on the project.
- The project's champion has been lost, putting a lower priority on the anticipated outcome of the project.
- A desire exists to reduce the resource cost of a project whose ultimate outcome is unknown.
- The project time-to-market window was exceeded or your competition beat you to it.

6.9.3 Evaluating a Project's Value

By asking and seeking complete and candid answers to the following questions, valuable insight can be gained regarding a project's value, both prior to its beginning and during its life cycle:

- Does the project promise to have an operational or strategic fit in the organization's future product, service, or organizational process strategies?
- Will the project results complement a strength of the organization?

- Does the project avoid a dependence on a weakness of the organization?
- Will the project results help the organization to accomplish its mission, objectives, and goals?
- Will the project results lead to a competitive advantage to the organization?
- Is the project consistent with other projects and programs of the organization?
- Can the enterprise assume the risks likely to be associated with the project?
- Are adequate organizational resources available to support the project?
- How well can the project be integrated into the operational and strategic initiatives of the organization?
- What would happen to the well-being of the organization if the project were cancelled?

By seeking and discussing answers to the above questions organization managers can gain an understanding of whether or not a project should be terminated.

6.9.4 Continuation of Projects Whose Value is in Doubt

There is a tendency to continue to expend resources on projects whose continuation and expected results may not make sense. These reasons usually include:

- Tendency to view serious project problems as normal and to be corrected, given enough time and resources.
- Estimated and anticipated high project termination costs.
- Project managers are motivated to persevere and "stay the course", even in the face of serious problems impacting the project.
- An uncanny ability on the part of the project manager and team members to see the project only in accord with preconceived ideas of value and success.
- Hanging on to a project even though it is in serious trouble, fearing the loss of power, loss of a job, or adverse impact on the image of the project manager and the team members.
- Organizational "pushes" and "pulls" as well as the inertia that impedes withdrawal from losing projects.

- Organizational politics that prevent a project termination.
- Failure to accept "sunk costs"—"We can't cancel it, we've got too much invested."
- The idea that projects are expected to succeed, not fail.

The above reasons may well have an emotional basis, but can be real limitations to an objective evaluation of a valuable outcome of the project results.

6.9.5   Consideration of Termination Possibilities

An ongoing consideration of termination possibilities should be carried out regarding the portfolio of projects in the organization. Figure 6.5 provides an overview of major factors involved in such a consideration.

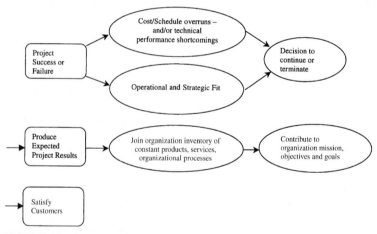

FIGURE 6.5   Considerations of termination strategies.

In this model, project success or failure is dependent on how well the project is meeting its cost, schedule, and technical performance objectives as well as how the project fits into the operational and strategic purposes of the organization. A decision to terminate or continue will impact the expected project results, which if continued and carried to completion will join the inventory of products, services, and organizational processes of the enterprise. The ultimate outcome of a successful project is one that contributes to an organization's mission, objectives, and goals—leading to satisfied customers. A listing of considerations that can provide insight into whether to terminate or continue follows:

- Review the project and its potential operational and strategic context on a regular basis.
- Be aware of the psychological and social forces that can cause project managers, and team members to "stay the course" in continuing the project beyond any rational basis.
- Recognize that there are often prevailing beliefs that the commitment of additional resources will ensure project success "just around the corner."
- Have available policy guidance that defines what project "success" and "failure" means to the organization.
- Listen for stakeholder concerns about the project.
- Is it perceived that termination of the project may adversely impact the career of key people in the organization?
- Consider having an independent audit of a project that could provide more objective insight into the project's value.
- Develop and maintain an organizational culture in which both good and bad news is welcome.
- Always keep in mind that projects are building blocks in the design and execution of organizational strategies, and as building blocks, should be constantly evaluated to see if they make sense from the organization's future.

6.9.6   Key User Questions

1. Are projects regularly reviewed for assessment of total progress, as well as the prudence of continuing to expend resources on such projects?
2. Does the cultural ambience of the organizations support the philosophy that projects can both succeed and fail, and that failure should be no stigma?
3. Have project termination policies and procedures been established in the organization?
4. Are the members of the organization aware of the emotional issues that can influence project termination?
5. Who has responsibility in the organization for making project termination decisions?

6.9.7 Summary

In this section, the concept and processes of project termination were presented. The point was made that all projects are terminated, either when they have delivered their technical performance objectives on time, within budget, and satisfying some operational or strategic need, or when further expenditure of resources on a project no longer makes sense. Organization senior managers should work to create an organizational culture that facilitates an ongoing review of project progress, to include the explicit assessment of whether a project should be continued or terminated.

6.9.8 Annotated Bibliography

1. Cleland, David I., and Lewis R. Ireland, *Project Management: Strategic Design and Implementation,* 4th ed. (New York, NY: McGraw-Hill, 2002), chap. 15, "Project Termination." This chapter describes the concept, processes, and techniques to use in assessing under what conditions a project should be terminated. A model, to include a work breakdown structure, of a typical project termination is provided.

2. Staw, Barry M., and Jerry Ross, "Knowing When to Pull the Plug," *Harvard Business Review,* March-April 1987. This article addresses by example and explanation how projects can be continued beyond a point in which termination made sense.

SECTION 7
PROJECT PLANNING AND CONTROL

7.1   PROJECT PLANNING

7.1.1   Introduction

Project planning is the process of having a vision for the project and thinking through the strategy for how a project will be implemented to achieve the desired goals. The finished plan will describe what is to be accomplished and how it is to be accomplished as well as stating what will not be done.

Project planning is perhaps the most important part of a project in that it defines goals and strategies by sorting through several courses of action to select the most favorable roadmap. Once the goals that will lead to client satisfaction are defined, all planning leads to their achievement through concentrated effort to describe the roadmap and application of resources to various tasks or work packages.

Project planning is a top-down approach to the continual elaboration of details that explicitly documents the objectives, goals, and strategies necessary to accomplish the project. A successful plan guides the project team to satisfaction of the project's cost, schedule, and technical performance goals.

7.1.2 The Need for Project Planning

No single element has more impact on the success of a project than project planning. There are numerous examples of poor planning in projects that caused major replanning efforts during critical times during project execution. Goals have been poorly defined, deliverable products have been redefined, customers have not understood the process, and the project team has been unsure of the next steps. Weak project planning will nearly always yield poor results.

Successful planning leads to successful projects

There are many reasons for weak project planning. Some of the more common are:

- Senior management and the customer emphasize starting the work quickly to see some progress in building the product.
- Concepts and principles of project planning have not been taught in US schools of higher education.
- Project team members have little experience or training in project planning
- Project planning and project execution are different skill sets.

7.1.3 Organizational Context for Project Planning

Project planning is done within the context of the organization—relying heavily on the organization's strategic and business goals. All planning must flow down from the organization's high-order plans to ensure that project planning is accomplished within the context of the organization's

- Mission or purpose
- Strategic and business objectives and goals
- Project management capabilities

Projects must have a strategic and operational fit with the organization to be successful. When projects are randomly started or selected without regard for the business, industry, or technology, it is highly likely that they will not meet their full potential for the organization.

7.1.4 Project Plan Contents

A project plan for a large project requires extensive description of project management components and the functions that are required. A project plan would include the following items.

- **Scope statement**—the description of the project, its goals, and the extent of the project in terms of technical, time, and cost.
- **Work breakdown structure (WBS)**—the division of the project into manageable parts for work assignment and control. The WBS also will be used in such areas as communication, cost estimating, scheduling, and work authorization.
- **Schedule**—the work laid out over time, which will typically be at two levels, detail and summary. The schedule will also contain milestones for the control of project progress and validation of work accomplishment.
- **Project budget**—the cost estimate converted into a time-phased expenditure plan that gives an expected rate at which funds will be used to complete the project.
- **Risk assessment**—the detailed analysis of the project's work and associated risk. The assessment will give visibility into areas of risk for managing throughout the project.
- **Interface plan**—the description of external interfaces for the project's product. This may be physical, electrical, hydraulic, digital, or other external connections for the project's product.
- **Work authorization plan**—the process and practice for authorizing the release and completion of work packages. It is a part of the control mechanism for regulating and validating work accomplishment.
- **Logistic support plan**—a document that describes the means for supporting the project's product after the project is complete. It may describe such items as the repair and maintenance of the product once it is commissioned.
- **Communication plan**—the description of the process and practice of communicating project information to internal and external participants. This would typically include a list of who is to receive what information and the frequency. Team meeting management would also be descibe in this document.

- **Procurement plan**—the description of the goods and services that will be obtained for the project, to include details on when to request them and when they are needed for the project.

- **Quality assurance plan**—the description of actions to be taken to build confidence that the project's product will meet client needs.

- **Human resource list or plan**—the list or description of human resources required for the project and when they are needed. Typically, an organization will name individuals, but often there will only be a skill identified for a period of time.

- **Stakeholder management plan**—the description of the stakeholders and how they will be managed in a proactive manner to ensure the project progresses as planned.

- **Project closeout plan**—the process and practices to be used to close all activities and reassign resources upon completion of the project. This may include transition planning for handover of functions or activities that must be continued, such as transfer of a building that was used by the project team.

- **Product commissioning plan**—the process and practices to be used to commission a product, such as testing an electrical generator for the client and operating the generator for a period of time.

   Project plans consist of several sections or chapters, depending upon the size of the project and the extent of project planning. Planners may scale down the following list of sections to fit the project size and need for detailed instructions. All projects regardless of size will have the following—some formally stated in a project plan and others informally stated.

- **Project statement of work (SOW)**—a succinct description of the work to be accomplished and how it will be accomplished. The SOW may include blueprints, drawings, and other illustrations that describe the work.

- **Technical specification**—the technical parameters for a product that will result from the project. Typically, this document describes the details of a product to be built. It should be noted that a part of the project may be to write the technical specifications for client approval.

- **Technical goals for the project**—a list of the features, attributes, and functions that the project must achieve to be successful. These may be included in the SOW and technical specification.

- **Schedule goals for the project**—a list of dates and accomplishments that must be achieved. This may include dates for project phases, milestones for critical dates to be met, and project review dates. Project planners will typically focus on the end date or the product delivery date.

- **Cost goals for the project**—a single or phased list of expenses for the project. Short projects typically have a single dollar value stated, whereas long-term projects may have time-phased expenditure requirements.

7.1.5   Sequence of Planning Actions

**Planning is a building block process—
sequence is important**

Project planning should follow a sequence that builds on prior planning and document development. The detailing of tasks and actions is a logical flowdown from higher-order documents. This hierarchy of documents requires that one be completed before another is started. Thus, the sequence of planning is important.

- **Project charter**—this is typically the first project document and follows project selection. It describes in general terms the project's purpose, general parameters such as time and cost objectives, appointment of the project manager, definition of the roles and responsibilities for all participants on the project team, identification of the client, and any other authorization or limitation placed on the project manager.

- **Requirements definition**—the process of converting general statements into measurable parameters for the project's product and service. Technical parameters and descriptions must be documented to establish the description of the end product—the objective of the project.

- **Project planning sequence**—the sequence of building the project plan, like a schedule, aligns the order in which planning will be accomplished. First in planning is to define what is the end product of the project and describe it fully to promote understanding about the product's scope. This is typically the technical specification and statement of work. Next, one can schedule the work, followed by the pricing of work. Once these functions are accomplished, other planning will use pieces and parts to develop plan components.

7.1.6   Project Work Packages

Critical to project success is the process of defining project work packages. Project work packages are the lowest level of management for project work. Whereas the detailed WBS defines the lowest-level part of the projects, the natural work package may be a combination of these parts. These are combined in to work packages for assignment and tracking of progress to one work unit—whether it is a single person, a small team, or a vendor.

Uses and consideration for use of work packages are many. Some of the more important items are as follows:

- Decide which work packages will be done in-house.
- Identify project work packages that will be subcontracted.
- Obtain the commitment of the responsible functional work managers.
- Plan for the allocation of appropriate funds through the organizational work authorization system.
- Develop procurement specifications and other desired contractual terms for the delivery of the goods and services to be provided by outside vendors.
- Develop the master work package schedules.
- Identify the strategic issues that the project is likely to face.
- Estimate the project costs.
- Develop the project budgets, funding plans, and other resource plans.

7.1.7   Competencies for Project Planning

Planners must know what can be translated from a plan to a working situation. Simple concepts and straightforward work approaches in plans are more easily put into practice than complex approaches. Planners should not make the project plan any more complex than needed to describe the requirements.

Planners need to visualize the overall requirement and describe in detail how tasks are to be performed. Planning is accomplished from general to specific. Planners must identify the general requirement and continue to elaborate on the work until it is in sufficient detail for another person to understand.

Planners also must understand the fundamental principles of planning. Planning is taking a mission, developing goals, assembling facts, supplementing facts with assumptions, and describing the process. Treating an area lightly or omitting facts or assumptions can lead to a flawed plan.

7.1.8 Planning Responsibilities

The primary responsibility for planning the project typically rests with the project leader. The project leader prepares the project plan, using the project team and other available resources, for presentation to senior management. The plan must be complete and describe the entire process for performing the project work and delivering the product to the customer.

Project team members are responsible for contributing to the development of the project plan and ensuring that the work described can be achieved with the listed resources. Team members contribute their technical knowledge of the product and the process by which it is built. Working cooperatively, team members ensure the project plan is factual and realistic for the performing team.

Senior management is responsible for approving a complete project plan that is success-oriented. Senior management also has responsibility for rejecting a weak plan or one that does not meet the criteria for describing the project's work. Weak plans result from poor planning and inadequate review by the approval authority.

Functional managers are also responsible for project planning when they have elements to prepare. For example, an engineering function may prepare the drawings that graphically describe the product. Any weakness in the drawings will transfer to the plan and ultimately the project.

7.1.9 Adverse Affects on Project Planning

There are many different items that can have both positive and negative impacts on project planning. These items need to be considered from the applicability and potential negative impact. Remember that the success or failure of the project may hinge upon one or more items.

A list of items that can adversely affect project planning and the quality of the resulting project plan follows. This list is not inclusive, but it provides a general identification of areas for consideration.

- Need for clear customer requirements and understanding with the customer
- Need for well-defined goals for quality/ technical, schedule/time, and cost/price
- Need to identify facts and assumptions regarding the project
- Need to identify issues related to the project and resolution prior to project execution
- Need for a capable tracking and control system to be in place prior to project execution

- Need for a risk analysis to be conducted prior to project execution and mitigation or completion of contingency planning
- Need to identify the right skills and resources during planning and ensure that they are available through normal procedures (either internal to the organization or contracted)
- Need to identify interfaces and dependencies and coordinate same
- Need to identify major issues during project planning
- Need to ensure work is planned to be accomplished by phase and that there is an approval process to move to each new phase.
- Need to see closeout of the project is planned and delivery or transition of the product is defined

These top-level items are essential to ensure the detailed planning is conducted. Detailed planning for the project follows the organization's project methodology and defines the individual requirements. This detailed plan provides the guidance to team members on how to execute the project.

7.1.10 Common Errors and Excuses of Project Planning

There is a tendency to make excuses for poor planning. This is a natural reaction when the person feels that planning has little value to the project. Some examples used in different projects reflect the types of excuses.

- *Actual situation was poorly defined requirement.* "I use rolling-wave planning. I plan a little and work a little." This excuse is an attempt to say the planning for the project can be done in phases or "waves" without the requirement being well thought through. Incomplete planning during the planning phase will not converge on the technical solution. Any plan that does not include all phases and the delivery of the product is flawed. Delivery of the product in the plan demonstrates an understanding of the technical solution.
- *Improper use of assumptions.* "I have an assumption that covers that part of the project." This is an attempt to divert attention away from a critical aspect of the project and avoid doing the required research or coordination of that area. To continue with an assumption when the matter should be resolved is to have a flawed plan.
- *Lack of documented facts.* "I don't list facts because they never change." Facts must be listed because they form the basis for the project and are supplemented by assumptions where facts are not available.

Listing facts also ensures that the plan approving authority has all information regarding the project and agrees to those facts.

- *Mixing facts and assumptions.* *"I* don't differentiate between facts and assumptions. They are all the same to me." Facts and assumptions are uniquely different in that facts change only when there has been a misunderstanding. Assumptions are anticipated future positive outcomes that must become true statements or the project will be impacted. Assumptions require tracking and managing to ensure a successful project.

- *Inadequate planning.* "Plans never come true anyway. So I just wing it and please the boss with something that looks good." Actual performance seldom exactly replicates the project plan, but in many cases is very close to the plan. Planning provides a direction and ensures most areas are thought through and options are weighed. Good plans will ensure most situations are covered and then the few variations can be managed.

- *Complex plans.* *"My* plans are so sophisticated that most people cannot understand them. Planning is one of my strengths." Any plan that is complex or unduly complicated will probably be misunderstood or misinterpreted. Plans should be simple and complete. Planning is to make things clear and avoid confusion.

- *Planning projects from technology orientation.* "It takes someone familiar with the technology to do the planning." Understanding the technology may be required in some areas of preparing a project plan. However, most of the planning is around the business aspects of the project, and the planner needs to understand how the organization works, how schedules will work, where resources are located, how much the project will cost, and how to communicate with all stakeholders.

There are many examples of errors and excuses used in planning. These are typically used when the planner is unsure of what is required in the plan or an area seems to be intuitively easy. When the plan-approving authority acts on an incomplete plan, he/she becomes a part of the weakness in planning.

7.1.11 Benefits of Proper Project Planning

Significant benefits are derived from thoroughly planning projects. Benefits have not been documented because of inadequate measures of effectiveness for comparison between projects. Industry, project size, and technology involved dictate the parameters that should be used to plan a project. Some rules of thumb that have been gleaned over several years of planning and implementing projects give some guidance.

Good project planning yields benefits

The range of these rules of thumb is dependent upon the planning knowledge and skills of the project team as well as the willingness of senior managers to enforce proper planning. It is obvious that planning, as the foundation for future project work, should be in the range of 10 to 15 percent of the time and budget. More complex projects may require more time and funds.

Benefits of proper planning show that good plans lead to projects that can be more easily executed in a consistent fashion for better end results. Plans that guide project execution to the proper technical solution are of significant benefit to an organization in terms of productivity and profitability.

Experience shows that significant savings in time can be achieved by proper planning. Some rules of thumb have been developed based on more than 20 years of work on projects. The duration and complexity of the project play significantly in the savings. There is an increased savings when proper planning is performed on more complex and longer-duration projects.

In high-technology industries, the rule of thumb is a ten-to-one ratio for execution savings to planning for small to medium-sized projects with moderate complexity. For each hour spent planning, it is expected that ten hours will be saved in execution. This approximation assumes minimum planning skills by the project team. However, the planning process must be proven and adequate to result in a complete project plan.

Rule of thumb for project planning benefits	
Project characteristics	**Savings in labor hours for each hour of planning**
Short-duration, simple project	3 to 5 hours
Short-duration, complex project	5 to 7 hours
Medium-duration, simple project	6 to 9 hours
Medium-duration, complex project	9 to 11 hours
Long-duration, simple project	10 to 12 hours
Long-duration, complex project	12 to 15 hours

7.1.12 Key User Questions

1. What is the difference between a fact and an assumption? Give examples of both for a project plan.

2. Why do good project planners use assumptions in project plans and what is the impact on the plan because assumptions are used?

3. When the customer's requirements for the project are unclear, what is the best method of ensuring an understanding between the customer and project leader?

4. Have all of the planning work packages for projects in the organization been accomplished before work is started in your organization?

5. Have organizational policies, plans, and other strategies been developed in the organization to support projects that are underway?

7.1.13 Summary

Project planning is hard work that requires the commitment of several levels within an organization to ensure completeness and validity. The planners must do their best to write instructions for project execution at the proper level of detail. The project leader must provide guidance and develop the project plan that is a continual elaboration on the customer's requirements. Most important, senior management must approve for implementation only plans that provide the project the best possibility of success.

Several items mitigate against good project planning. Both the customer and senior management often place emphasis on starting the project work (execution) before the planning is complete. This view of project work assumes more progress when the team is doing the project's work. The importance of planning is played down.

Proper project planning has a beneficial effect on the success of the projects and reduces the stress on the executing project team. Proper project planning gives the customer and senior management the best prediction of the outcome of the project. Proper project planning also compresses the execution time for the project and reduces waste through rework.

7.1.14 Annotated Bibliography

1. Weiss, Joseph, and Robert Wysocki, *5-Phase Project Management: A Practical Planning and Implementation Guide* Boulder, CO: Perseus, 1992), 121 pp. This book covers project planning fundamentals and contains templates to use in planning. The principles and tools encompass the full range of planning.

2. Angus, Robert B., and Norman A. Gundersen, *Planning, Performing, and Controlling Projects: Principles and Applications* (Upper Saddle River, NJ: Prentice Hall College Division, 1997), 320 pp. This book

gives the theory and practice of planning technical projects. It outlines a specific four-phase systematic approach and explores its implementation in a real-world case study.

7.2   ESTABLISHING PROJECT PRIORITIES[1]

7.2.1   Introduction

In today's dynamic work environment, it is often a question of setting priorities for our work and doing those first tasks that have the most urgency of need. If Parkinson's Law is correct that *work expands to fill the time available*, we need to identify and meet our obligations to the important work. Too often, we find ourselves inundated with tasks and without a plan or priority for the work.

> **Getting important work done first
> is critical to business**

One individual stated the obvious when met with conflicting tasks: "I'm so busy with so many tasks that all I have time for is to make excuses why I am not getting anything done." This is a situation where he should stop worrying about tasks until there is some order of importance assigned—either self-assigned priority, or let the boss sort the tasks as to urgency of need.

Another individual stated, "My boss keeps giving me more work and I don't have time to complete the tasks that he has given me before." When advised to make a priority list that ranked the urgency of need for the work efforts, he stated that the boss would just say, "Get all of the tasks done." This individual was unable to either get his boss to prioritize the tasks or to put them into an urgency-of-need ranking. He did not have the time or people to perform all the tasks.

Projects are no different than individual work efforts—everything needs some ranking as to what is important and urgent. Projects established without a priority, or ranking as to what is needed first, create situations that are disruptive and conflicting for the person charged with leading the effort. Too often, everything is considered "priority one" by senior management.

7.2.2 Importance of Project Work

Most people will agree that projects are important because they were selected by the organization as a worthwhile venture. The degree of importance may be debated based on how one sees the contribution of a project to the organization. Some examples of the importance of projects are listed below.

- CEO—project makes a large revenue contribution to the organization.
- Vice president of projects—project is a showcase of what a project should be and an example for future project work.
- Director of finance—project requires a low investment of funds to perform and has a good cash flow into the organization.
- Project manager—project is doable with the available resources and gives a lot of satisfaction to the project team.
- Director of engineering—project has a challenging product design that will stretch the ability of our engineers.
- Director of marketing—project will meet the expectations of our clients and result in a lot of sales.

As seen from these examples, different people in the organization can assign unique definitions of importance to projects. These definitions of which project is most important rely on one's organizational position rather than the organization's goals. It is essential that the importance of work be defined through the goals, both strategic and operational, for the organization to ensure growth. Importance for a projects, viewed from the organization's goals, could be a combination of all the factors listed above.

The most important projects should receive first consideration for resources

---

"Urgency" relates to time and how soon the project's product is needed. Priorities for projects show the relative order of need and the emphasis that must be assigned for allocating resources to perform the work. A "priority one" would receive resources before a "priority two" project, and so on, if there were conflicting resource requirements. A priority system guides managers in allocating resources as well as tracking the schedule for delivery.

"Urgency of need" is important for managing projects to meet the goals of the organization. When a priority system is flawed or not present, decision-makers are often without the guidance to consistently emphasize and support projects that should be first in line for resources. Uninformed decisions can materially affect the organization's business and impact client satisfaction through late deliveries.

7.2.3   The Project Priority System

Often the size of a project is identified as the priority for a product or service. Large projects are assumed to have the highest priority, and the resultant priority for resources. This type of priority system is flawed in that many times a small project has a significant impact on the organization's business and is needed immediately. Resources should be given first to that project that must be delivered first.

> **Project priority is based on the contribution to the organization's goals**

The urgency of need for a project, or its priority, should be based on the project's contribution to the business. Personal criteria should be set aside when establishing a priority for a project because those criteria may conflict with supporting the organization's goals. A rigorous system of assigning priorities to projects is needed to ensure congruence with the organization's purpose.

7.2.4   Rationale for Prioritizing Projects

Organizations should develop a list of reasons for assigning a high priority to a project. These reasons would flow down from the strategic and business goals as well as be supportive of marketing efforts. Some reasons for assigning a high priority to a project are listed for consideration.

- Promised project deliverables to client on a specific date
- Early delivery of products to increase revenue stream
- Project's product feeds another important project

• Project's product improves company image
• Project has a high potential for follow-on work if accomplished early.

> **Senior managers must decide which projects are more important—or the project manager and team members will.**

When there is no priority ranking of projects, all take on equal status and importance. Senior managers with the responsibility for assigning a priority, or order of importance, to a project pass the decision to lower levels. Project managers working without a priority will typically assume that their project is the most important for the organization and seek resources on that basis.

Project team members working in a matrix organization often have latitude as to which project they will work on to accomplish specific tasks. When the priority is either "all is priority one" or there is no priority, the team member must make the decision as to what is to be accomplished. Experience shows that team members will sort on the following items to select the work when given no distinction between projects.

• Work that is most interesting or challenging
• Work supporting the project manager they like
• Work that is easiest to accomplish to show progress
• Work where friends are working

7.2.5   A Model Project Priority System

> **A model is a guide to a better solution**

A typical project priority system should have from three to five priority categories. More categories diffuse the system, while fewer than three may not have the individual characteristics with which to differentiate between projects. The following may be a start for an organization in establishing a project priority system.

Priority	Sample Criteria
1	• Project is needed earliest to correct a situation that affects organizational performance. • Project is needed soonest to satisfy a major client's needs. • Project is needed soonest to solve a problem affecting a client. • Project's product is needed soonest to meet a marketing need. • Project's product is needed soonest to establish a new product or service line.
2	• Project is needed quickly to meet a client's needs. • Project is needed quickly to maintain a cash flow. • Project is needed quickly to take advantage of environmental situations. • Project develops a new product for near-term sales. • Project is used to change the organization to optimize the use of resources.
3	• Project is needed to conduct routine work for the organization. • Project is needed to deliver a product, service, or organizational change to a client on a routine basis. • Project is needed to explore long-term product development options. • Project is needed to meet routine commitment to client efforts.

The above generic priority system is just a start at ranking the importance of projects to the organization. One needs to first identify the strategic and operational goals to ensure alignment of priorities with the organization's purpose. Too often, "pet projects" serve the interests of a person rather than the organization. Linkage to the goals in a priority system are therefore critical to organizational success.

The drawbacks to selecting and implementing projects that do not have a priority based on organizational goals are many. Some of the more common issues are as follows.

• Project is unique and receives resources that should be used elsewhere.
• Project may contribute to an individual's agenda without benefit for the organization.
• Project is not a building block for organizational success.
• Project detracts from the "real work" of the organization.

7.2.6 Implementation of a Priority System

Once a priority system is defined and established in policy, there is a need for rigorous implementation to ensure the benefits are realized as soon as possible. The above three-level priority system is an example of what might be implemented, but with greater detail for how priorities are assigned and how they may be changed on a project.

Organizations without a priority system need to establish a working group to review all active and candidate projects for their contribution to the organization. Using a guide and understanding the strategic, business, and marketing goals of the organization, the working group would classify all projects into groups. An example of the classification groups might be as follows.

- Continue the project and change its priority to a higher ranking.
- Continue the project and change its priority to a lower ranking,
- Continue the project and retain its priority.
- Terminate the project as being inconsistent with the strategic and operational goals of the organization. Harvest all product components or scrape all components.
- Modify the project to align with the organization's strategic and operational goals. Assign a priority consistent with the project's contribution to the organization.
- Place project on hold until a review is conducted to determine its value to the organization.

Project priority can change during the execution phase to either a higher priority or lower priority. If the organization's goals change, there is a need to ensure the project is still a viable entity within the changing direction. Changes to the strategic goals, for example, could cause a project to be cancelled if the organization was no longer pursuing a particular product line. Another situation that could cause a change to priority is a shift in the market or a client's needs.

7.2.7 Advantages of a Priority System

A project priority system, when properly used, makes the decision process a rigorous process based on facts and organizational needs. Random ranking of project priorities results in less than optimal performance for the organization through poor resource allocation and utilization. Some rationale for establishing and using a disciplined process is suggested:

- Better utilization of resources by placing the resources where they will do the most for the organization
- Reduced conflict between projects for resources by establishing the order of importance for projects
- Possible termination of projects that are low yield and low priority
- Possible expedited projects to gain early benefits for important projects

Priority systems establish the basis for selecting projects and communicating the importance of each project to the organization. When consistently applied, a priority unifies the efforts of everyone in the organization toward success—i.e., the best allocation of resources to meet the organization's goals. Without a system of priorities, allocation of resources becomes a random process.

7.2.8   Key User Questions

1. Does your organization use a rigorous process for assigning priorities to projects? If not, why?
2. Size of a project is important because a large project should provide more benefits to an organization than a small project. Why would a small project often be given priority for resources over a larger project?
3. What impact does a priority system have on the utilization of human resources? Explain.
4. Who sets the priority for a project and who can change it in your organization?
5. What should be done when a project is the last in order of priority for resources?

7.2.9   Summary

Priorities for projects establish the basis for allocating resources to meet the organization's strategic and business goals. When senior managers insist that all projects are "priority one," the decision as to what is important is passed to the lower echelons of workers to decide what to do first. These individuals, typically, do not have the information or knowledge of the organization's operations to make a valid selection of "important work."

Randomly selected and prioritized projects typically do not apply the proper emphasis on delivering products through projects in an efficient manner that is supportive of organizational goals. Random selection of projects increases the risks against achieving organizational goals through an arbitrary process that may or may not be consistent with the project management process.

Developing and implementing a prioritization system for projects that are unique to the organization should be tailored around the strategic and business goals. Any change to these goals should trigger a review and adjustment to the priority allocation system. Further, there should be a review of the system if the environment or client's needs change.

7.2.10 Annotated Bibliography

1. Archibald, Russell D., *Managing High-Technology Programs and Projects*, 2nd ed. (New York, NY: John Wiley & Sons, 1992), pp. 139–141. Archibald discusses project priorities in a combination of technical and schedule activities and suggests a three-level ranking of project priorities. He also suggests establishing a project priority review board for the purpose of weighing the relative importance of projects and assigning them a priority.

2. Ireland, Lewis R., "Managing Multiple Projects in the Twenty-First Century," *Proceedings of the Project Management Institute Seminar/Symposium 1997*, Chicago, IL (FU-15). Ireland discusses some basic approaches to categorizing and prioritizing projects based on contributions to the organization. He differentiates between "size" and "priority" to show the need to assess both for the health of the organization.

7.3 PROJECT SCHEDULING

7.3.1 Introduction

Scheduling of projects is considered one of the basic requirements of project planning. Over the past 50 years, scheduling has matured and the tools associated with scheduling have improved significantly. Scheduling tools are available to nearly all projects for developing time lines.

Through the use of automated tools, almost any project team member can accomplish scheduling. These schedules vary in sophistication and utility based on the understanding that the preparer has of scheduling practices. There is a tendency to either put too much in a schedule or too little. The right balance is often not known by the preparer.

Another challenge to proper scheduling is the definitions of terms used by different tool manufacturers for product differentiation. Some refer to the schedule as the project plan and others mix the terms for the lowest level of work in the schedule as task, activity, or work package. Standardization of the terms is needed to ensure consistent communication of the schedule components.

7.3.2   Key Scheduling Definitions

The key definitions used in scheduling include:

- Task—the lowest level of work in a project schedule.
- Milestone—the non-resource-consuming task that is used to signify a critical control point in the schedule. Milestones use neither time nor resources to be achieved.
- Critical Path Method (CPM)—a method of network scheduling that determines the longest time path through the project. The longest path is called the critical path and there may be more than one critical path. CPM, in its original form, depicted the tasks by the arrows and the nodes showed logical connectivity.
- Program Evaluation and Review Technique (PERT)—a method similar to CPM that uses three time estimates to calculate the longest time path. PERT and CPM are used interchangeably by some individuals.
- Precedence Diagram Method (PDM)—a CPM derivative scheduling method that uses the nodes for tasks and the arrows for logical connectivity.
- Work Breakdown Structure (WBS)—a disciplined process of defining the product within a project. It decomposes the project's work to the required visible level and defines those items. (See Section 7.12.)
- Scheduling Relationships for Tasks—the logical relationship between two tasks. There are three recommended relationships: (1) Finish to Start when one task is completed and the next one starts; (2) Start to Start when two tasks start at the same time; and (3) Finish to Finish when two tasks are completed at the same time. These relationships are often modified by lead (prior to) or lag (following) time constraints.

7.3.3   Project Schedule Purposes

The basic purpose of the schedule is to describe the work that will be accomplished over time. This laying out of the work on a time line provides the plan for the work sequence and at what time to start and finish tasks. Sequencing the work ensures that it is all in the time frame for the project and that the project completion is identified.

A second purpose of the schedule is to communicate to all stakeholders that all work has been included for the project and that tasks will be accomplished at a specified time. The communication of this information ensures confidence is conveyed for the project's activities. This coordination of the scheduled work supports a united effort in completing the project.

The third purpose of the schedule is to establish the benchmark for comparing the actual performance with the plan. The schedule depicts what is expected and actual performance shows how closely the work results match. Schedule validity is achieved when there are no significant, unexplained variances.

7.3.4   Preparing a Schedule

Preparing a schedule requires using the WBS, which defines the project work. This work is then converted to tasks and summary tasks that depict work to be accomplished. The WBS defines the product or functions required for the complete project and should be used to provide discipline to the process.

Some organizations use a task list to develop the schedule items. This method has application in research projects and projects without the formal structure required of a disciplined process such as the WBS. In these projects, the outcome or product of the project is unknown. Using a task list also gives a random level of detail for tasks as well as opens the possibility that some tasks are not identified.

Once the tasks are identified, place them in a Gantt chart or bar chart format. Scheduling tools permit this and allow manipulation of the information to continue the process. Use a standard task duration as an initial solution and connect the work into its logical sequence for performance.

Evaluate each task to determine whether it is a "fixed duration," requires a fixed amount of time regardless of the number of resources, or if it is "effort driven," may expand or contract based on the assignment of resources for completion. Assign resources to the tasks and develop a critical path.

Figure 7.1 shows a simple Gantt chart that has tasks with different durations and relationships. This chart also includes lines showing the relationship between tasks. The chart does not show whether the tasks are fixed duration or effort drive, have a fixed start or the links drive the start, and if the tasks are resource loaded or not.

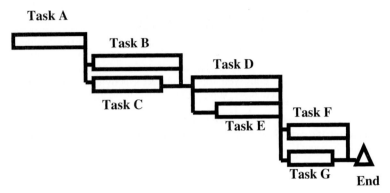

FIGURE 7.1 Simple Gantt chart.

Milestones are inserted in the schedule at critical points to signify completion of a phase or transition to another work group. In the above example, the one milestone used is for completion of the project. Other milestones may be used to control components of the work and for reporting purposes.

Simple relationships between tasks are best to schedule the work when one task is complete and the next one is started (finish to start). There are occasions that dictate task relationships such as two tasks beginning at the same time (start-to-start) and two tasks completing at the same time (finish to finish). These may entail delays or advances on one or the other task to make the relationship more complex.

As a rule of thumb, the "finish to start" for tasks should be used for about 90 to 95 percent of the relationships. "Start to start" should be

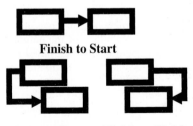

FIGURE 7.2 Task relationship chart.

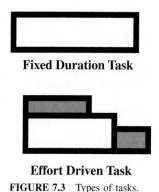

Fixed Duration Task

Effort Driven Task

FIGURE 7.3   Types of tasks.

approximately 3 to 7 percent of the relationships. "Finish to finish" should be 2 to 3 percent of the relationships.

Figure 7.2 shows the relationships of tasks. The upper left tasks show the "finish to start" relationship. Below and to the left is the "start to start" relationship with the finish to finish relationship just to the right.

A fixed duration task is shown in the upper area of the next illustration. This task cannot be reduced in duration regardless of the number of resources assigned. An example of this is a four-hour meeting with 10 people. If eight attend or if 12 attend, the meeting is still four hours in duration.

An effort-driven task is shown in the lower area of the illustration. The shaded areas in Figure 7.3 depict change in duration. The longer shaded area indicates that less than one full resource is assigned to the task, say for example, one person is assigned 50 percent of the time. Less than a full resource causes the task to become longer in duration. The shaded area above the task indicates more than one resource is assigned and the task is therefore shorter in duration. For example, assigning two people to a task reduces the duration to half the estimated duration.

7.3.5   Resource Loading a Schedule

Assigning resources to a schedule is accomplished by matching the proper skill sets with the task requirements. Resources are only assigned to the lowest level tasks and no assignments to milestones. At least one resource is required for each task, while some large tasks may require several resources with similar or different skills.

The number of resources assigned determines effort driven task duration. Each additional resource decreases the duration of the task. There are, however, limitations on the number of resources that can effectively work on a single task. Large tasks may need to be divided for better control and resource utilization.

Some principles for assigning resources to tasks include:

- Assign one resource per task as a minimum and make that resource full time on the task.
- Assign similar resources to tasks for the same amount of time to ensure best effort.
- Ensure resources are assigned by name rather than skill to identify specifically who will be working on tasks.
- Update the schedule with the resources at least 30 days in the future (some resources will leave the organization and not be available for the project).

7.3.6   Schedule Tracking and Control

The resource-loaded schedule is implemented after being approved by the project leader or senior management. Implementation is the allocation of tasks to be started and ensuring the resources are informed when the work is to start. This initial effort starts the clock running on the project toward the completion date.

Project status and progress are collected on a periodic basis, typically weekly. The performing individuals report tasks completed, percent complete of tasks, and percent remaining on started tasks. Actual progress is posted to the schedule and variances are plotted. Variances represent work that is ahead or behind schedule.

Variances in the schedule may result from inadequate estimates of the amount of work required on tasks, poor control over the work being accomplished on the tasks, lower productivity levels on the tasks, or late starts on tasks. Variances must be analyzed to determine the impact on the schedule and costs. Major variances may require significant changes to the schedule while minor variances may only dictate the situation be more closely monitored.

Variances on the critical path have the potential for negatively impacting the project completion date. These variances may require additional resources be placed on tasks on the critical path or that the logical relationships between tasks on the critical path be changed. Any change to the schedule will probably be less efficient than the original schedule unless there is better information for predicting the future.

7.3.7 Key User Questions

1. What is the simplest task relationship and why is it used approximately 90 to 95 percent of the time?

2. What is the impact of having a "finish to start" relationship for two 10-day tasks with a negative lag of five days?

3. When a significant schedule variance of 10 days occurs on the critical path, what is the impact on the project?

4. During analysis of the schedule, it is discovered that two three-day tasks in parallel are on the critical path and they can only be accomplished in series; i.e., one follows the other. What actions can be taken to ensure the project is not delayed by three days?

5. During the weekly reporting, one person gives the status of an eight-day task as 50 percent complete and five days remaining on the work. What are the implications of this report?

7.3.8 Summary

Scheduling is a mature technique and practice. Tools and techniques have been perfected that support the rapid development of schedules and ease of change. These tools and techniques, however, use subjective data, such as estimates of task duration and estimates of labor skills that can materially affect the outcome of the project's delivery on time. Scheduling skills are familiar to many members of the project team and each may assist in developing or managing the time line.

Schedules should be as simple as possible for ease of development and ease of understanding by all the stakeholders. Relying on the fundamental procedures and practices of scheduling will provide a solution that is superior to complex task relationships and resource assignment by less than 100 percent of effort. The more complex the schedule, the more complex the management of the schedule.

Tracking and controlling the schedule status and progress provide the measurement of the effectiveness of the schedule as a plan for the project. Significant variances and discovered work are indications of weak schedule planning. Small variances and explainable variances give assurance that the schedule is workable and represents the best solution for the project.

7.3.9 Annotated Bibliography

1. Duncan, William R., *Guide to the Project Management Body of Knowledge* (Upper Darby, PA: Project Management Institute Publications,

February 1996), chap. 6. This chapter describes the sequential activities required to build a schedule. The process is laid out in detail for ease of understanding.

2. Frame, J. Davidson, "Tools to Achieve on-time Performance," in David I. Cleland, ed., *Field Guide to Project Management,* 2nd ed. (New York, NY: John Wiley & Sons, 2004). This chapter describes current scheduling techniques and practices. There is a complete coverage of network diagrams that promote understanding and comprehension.

7.4   *PROJECT MONITORING, EVALUATION, AND CONTROL*

7.4.1   Introduction

In this section, information will be provided concerning how project performance can be monitored, evaluated, and controlled. Control is the process of maintaining oversight over the use of the resources on the project to determine how well the actual project results are being accomplished to meet planned project cost, schedule, and technical performance objectives. There are four key elements in a control system as depicted in Fig. 7.4.

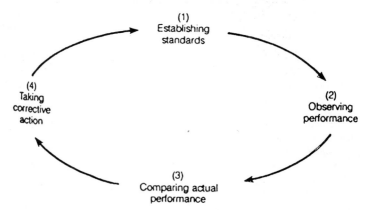

FIGURE 7.4   The project control system. (***Source:***   David I. Cleland, *Project Management: Strategic Design and Implementation,* 3rd ed. (New York, NY: McGraw-Hill, 1999), p. 325.)

In order to properly assess project progress, several conditions and understandings are required. These are:

- Team members must understand and be committed to the importance of the process of project monitoring, evaluation, and control.
- Information derived from the Work Breakdown Structure is required to measure project progress.
- The work package is the basic project unit around which progress on the project can be measured and evaluated.
- Information used for project control purposes must be relevant, timely, and amenable to the plotting of trends in the use of project resources.
- Measurement of project results must start with an evaluation of the status of all of the work packages on the project.
- Information collected and compiled concerning the status of the project must be tempered by the judgment of the project team members and executives concerned.

7.4.2 Elements in the Control System

These elements are drawn from Figure 7.4.

1. *Establishing standards* from the project plan typically include:
- Scope of work
- Product specifications
- Work breakdown structure
- Work packages
- Cost estimates
- Master and supporting schedules
- Technical performance objectives
- Financial forecasts and funding plans
- Quality/reliability standards
- Owner satisfaction
- Stakeholder satisfaction
- General/senior manager satisfaction
- Project team satisfaction
- Vendor/subcontractor performance metrics

- Project team effectiveness
- Other performance standards derived from the project plan

2. *Observing performance* involves receiving relevant, sufficient and timely information about the status of the project, which comes from many sources:

- Formal reports
- Briefings
- Informal conversations
- Review meetings
- Miscellaneous documentation, such as customer perceptions; letters; e-mail; memoranda; audit reports; "walking the project"; observations; talking with stakeholders; listening; and listening some more

During the project review meetings, asking and seeking answers to the following questions can give valuable insight into how well the project is doing. These questions include:

- Where is the project with respect to schedule, cost, and technical performance objectives and goals?
- Where are the project work packages with respect to schedule, cost, and technical performance objectives and goals?
- What is going right on the project?
- What is going wrong?
- What problems are emerging?
- What opportunities are emerging?
- Does the project continue to have an operational or strategic fit with the organization's mission?
- Is there anything that should be done that is not being done, and are we doing things we shouldn't be doing?
- Are the project stakeholders comfortable with the results of the project?
- How is the project customer image; is the customer happy with the way things are going?
- Has an independent project evaluation been conducted?
- Is the project being managed on a total "project management systems" basis?
- Is the project team an effective organization?

- Does the project take advantage of the strength that the organization possesses?
- Does the project avoid a dependence on the weakness of the organization?
- Is the project making money for the organization?

Ongoing monitoring of the project performance should be carried out at the following levels:

- The work package level
- The functional manager's level
- The project team level
- The general manager's level
- The project owner's level

3. *Comparing planned and actual performance* in the use of resources on the project to determine how well such use has contributed to the fulfillment of the project objectives. Project performance must be evaluated on a regular basis to identify variances from plans.

An important part of the control process is to take preventive action to solve actual or anticipated problems. The major preventive actions include strategies for:

- Overall change control that involves coordinating project changes across the entire project
- Scope control
- Schedule control
- Cost control
- Technical (product) control
- Risk control

The functioning of an effective project control system in the context of comparing planned actual performance includes the opportunity to find answers to several questions about the project:

- How is the project performing?
- If there are deviations from the project plan, what are the causes of these deviations?
- What can be done to correct these deviations?

- What can be done to prevent these deviations in the future?
- How many open issues are there compared with our history of projects at this point and size?

4. *Corrective action* can take many strategies to bring the planned and actual use of resources back into harmony. Some of these strategies include:

- Replanning
- Reallocation of funds
- Reallocation of resources
- Reassignment of authority/responsibility
- Rescheduling
- Revising cost estimates
- Modification of technical performance objectives
- Change project scope
- Change product specifications
- Terminate the project
- Stopping work on the project and replanning/redirecting the entire project

7.4.3   Project Audits

Project audits provide an opportunity to have an independent and expert appraisal of where the project stands. The basic purposes of an audit include:

- Determining what is going right, and why
- Determining what is going wrong, and why
- Identifying forces and factors that have prevented or may prevent achievement of cost, schedule, and technical performance objectives
- Evaluating the efficiency of existing project management strategy, including organizational support, policies, procedures, practices, techniques, guidelines, action plans, funding patterns, and human and non-human resource utilization
- Providing for an exchange of ideas, information, problems, solutions, and strategies with the project team members

A project audit should cover key activities depending on the nature of the project. A sample of these key activities are included in Table 7.1.

TABLE 7.1   Project Audit Activities

---

1. Engineering
2. Manufacturing
3. Finance and accounting
4. Contracts
5. Purchasing
6. Marketing
7. Human resources
8. Organization and management
9. Quality
10. Test and deployment
11. Logistics
12. Construction/Production

7.4.4   Post-project Reviews

Post-project reviews can serve to identify and pass on to other project teams the efficiency and effectiveness with which a particular project has been managed. Such reviews provide insight into the degree of "success" or "failure" of a project, as well as a composite of lessons learned from a review of all the projects in the organization's portfolio of projects. The results of such reviews can provide insight into such matters as:

- Accuracy of estimating project costs
- Better means of anticipating and minimizing risk
- Evaluation of project contractors
- Improve future project management

7.4.5   Key User Questions

1. Do the project manager and the other project stakeholders understand the concepts and processes involved in monitoring, evaluation, and controlling the use of resources on the project?
2. Have appropriate project performance standards been established through the planning process for the project?
3. Are adequate feedback mechanisms and review processes in effect for the project?
4. Are review processes carried out at the appropriate level to facilitate a fuller understanding of where the project stands with respect to cost, schedule, and technical performance objectives?

5. Are remedial strategies in place to facilitate the use of project resources to better accomplish project ends?

7.4.6 Summary

In this section, the basic control cycle was presented as a model on which to base a philosophy and process for the monitoring, evaluation, and control of a project. The key elements in this control cycle were explained, along with suggestions on what could be done by way of corrective action to get the project back in harmony with the project plan. Finally, basic ideas were presented about project audits and post-project reviews as a means of evaluating a particular project, and passing on the "lessons learned" to other project teams.

7.4.7 Annotated Bibliography

1. Cleland, David I. and Lewis R. Ireland, *Project Management: Strategic Design and Implementation,* 4th ed. (New York, NY: McGraw-Hill, 2002), chap. 13, "Project Monitoring, Evaluation and Control." This chapter.provides a general framework and process for the design and execution of appropriate monitoring, evaluation, and control of resources to support project purposes. Included in this chapter are suggestions for what information is required for use in each of the elements of the control cycle.

7.5 RISK MANAGEMENT

7.5.1 Introduction

Risk in a project is the probability that some adverse event will negatively impact the project's goals. Project goals are the baseline for measuring the risk to the project and the risks are typically against the cost, time, and technical goals. There may be other goals, such as customer satisfaction, that will be considered.

All projects have some risk or there would not be projects. Projects are initiated when there is some element of risk and management wants the sharp focus of a project plan and team to perform the work. Too much risk is sometimes assumed without the requisite understanding of the elements that can cause the project to fail. Too little risk means the project is not pushing the thresholds for cost, time, and technical performance.

7.5.2 Project Uncertainty

Uncertainty is a major contributor to project risk. Complete uncertainty is the total lack of information while certainty is the totality of information. Projects will typically not have all the information to plan and execute the work. Sometime a project leader may have as little as 40 percent of the required information and must proceed because of commitments to customer or market conditions. It is estimated that project leaders will have between 40 and 80 percent of the required information during the planning phase of most projects.

Figure 7.5 shows the complete range of available information and gives a perspective of the spectrum of complete uncertainty to complete certainty.

Uncertainty is often accounted for in project assumptions. When there is insufficient information available to make a decision or plan a project, assumptions bridge the gap. Assumptions are reasonable, but not without the probability of failure if they do not come true.

Project uncertainty has been identified in several different areas within an organization. Some of the areas that have been most visible are as follows:

- State of the art for the technology used
- Organizational capability to perform repeatable project management processes
- Availability of technical and project management skills

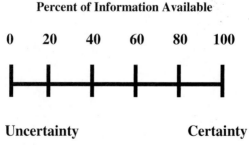

Percent of Information Available

0   20   40   60   80   100

Uncertainty   Certainty

FIGURE 7.5 Uncertainty to certainty spectrum.

- Equipment availability for the project
- External project interfaces
- External project suppliers
- Technical impasses
- Test results for products of the project

7.5.3   Internal and External Project Risk

Risk may be divided into two categories: internal project risk and external project risk. See Fig. 7.6 for a conceptual view of the two categories of risk. Internal project risk is that risk inherent in the project that the project leader has control over and can reduce through direct actions, such as developing contingency plans. External project risk is that risk which is outside the control of the project leaders. Examples of this would be project interfaces that are unknown and the interface definitions are being accomplished by another party.

Internal project risks are part of the constraints placed on the project through establishing goals. The delivery date for the product may be optimistic and the plan has to reflect that date. Planning will focus on the delivery date and schedule of work to ensure the delivery is feasible. Cost is typically an area that is constrained. Planning will drive the type and quantity of resources that will be used on the project, even when the

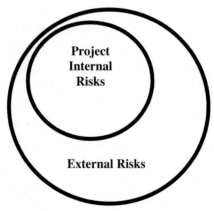

FIGURE 7.6 Project internal and external risks.

budget is less than desired. Technical solutions will be placed in jeopardy when the delivery time is optimistic and the funds are limited.

External project risks may be influenced by the project leader and anticipated, but there is no direct control over the risk events. These risks are influenced through agreements and contracts with other parties. The degree of influence exercised by the project leader is determined by the identification of external risks and the cooperation extended by the other party. Some examples of other parties could be other project leaders, functional managers, vendors, and contractors.

7.5.4 Risk Identification

Risk events are the potential adverse affects on the project. Identification of these events should be first made during project initiation and then in project planning. During project initiation, the project leader should be identifying the project interfaces and the degree of difficulty in achieving those interfaces. There may also be early identification of internal project risks through such activities as anticipating using new technology or a new workforce.

Identifying internal project risks during planning starts with the goals and asking whether the plan will deliver the desired results. Starting with the technical aspects of the project, ask the question as to whether the solution is feasible within existing technology. Record any events that may fail and the probability of failure. Some guides to identifying risk are shown Table 7.2.

TABLE 7.2 Risk Identification Practices

- Checklists for project risk areas
- Lessons learned from previous projects
- Resource availability lists
- Resource training records for applicable skills
- Peer review of project plans
- Senior management reviews of plans
- Organizational project management capability audit

Identify the schedule failures as planning is being accomplished. The schedule must have assigned and available resources to perform the work or there is risk associated with the lack of qualified personnel to perform the work. The delivery of the product is always subject to evaluation for feasibility of finishing the work. Milestones may be dictated that would also be subject to evaluation for feasibility.

Cost risk is the total of the materials to build the technical solution and the human resources to perform the work. Other costs may be in the project, such as travel, shipping, and duties, that may or may not be major cost drivers. The total cost is a budget that represents the time-phased expenditures for the project. Risks may be from total cost or the rate of expenditure if there is a cash flow constraint.

7.5.5   Risk Quantification

Risk quantification is a means of weighting the risk events so that risks may be ranked for mitigation. Figure 7.7 shows probability on a scale of 0 to 1.0 and the consequence of a risk event. Consequence for cost is in dollars and for schedule is duration of extra time required. Technical risk is the failure to achieve satisfactory functionality or performance. Technical risk is typically translated into additional cost or schedule duration.

A less precise quantification of a risk event is accomplished on a two-dimension matrix that has probability of occurrence on the vertical scale and consequence if the event occurs. The simple matrix of Fig. 7.8 may use a color scheme for management purposes.

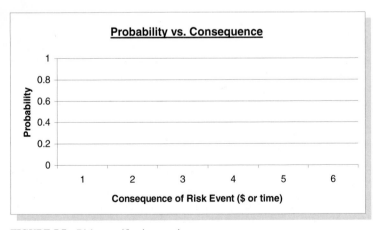

FIGURE 7.7   Risk quantification matrix.

Probability	Risk Rank + Weight				
Very High	5	6	7	8	9
High	4	5	6	7	8
Moderate	3	4	5	6	7
Low	2	3	4	5	6
Very Low	1	2	3	4	5
	Consequences or Outcome ←⎯⎯⎯⎯⎯⎯⎯⎯→				
Legend: Green – 1, 2, 3 Yellow – 4, 5, 6 Red – 7, 8, 9					

FIGURE 7.8   Risk ranking by intersection.

Managing risk is accomplished in order of the magnitude of impact on the project. Green, for example, would require only that the risk event be monitored to ensure it did not increase. Yellow would indicate there was a need to actively monitor and reduce the risk where possible. Any increase would trigger a response. Red would indicate that some action is needed to either lower the risk event or take a different approach.

7.5.6 Risk Mitigation

Mitigating or reducing risk is accomplished through changing the plan, adding resources, using a different technical approach, or other actions. As indicated in the quantification of risk matrix, all Red events would be mitigated to either a Yellow or Green status prior to continuing. This reduction of risk, either through the probability of the event happening or the consequences of the outcome, improves the risk picture for the project.

Risk mitigation may not be possible for a risk event and senior management may be asked to decide whether a high risk item is acceptable or whether the project should be cancelled. High risk items may dictate some other action or a delay until better solutions can be found. The decision on risk events that have major consequences is typically reserved for senior management.

7.5.7   Contingency Planning

Contingency planning is a part of risk reduction through identifying and documenting alternative solutions if a risk event happens. When the decision is made to continue in spite of a major consequence, contingency plans are prepared. The contingency plan may be an alternate technical solution for technical problems, contracting out part of the work to a more qualified organization, reducing the scope of work, or using different technology.

Contingency plans for cost or schedule failures could include reducing the technical scope for less work, eliminating part of the project, changing the human resources, or other cost-cutting and work reduction efforts. It is important that the technical scope of the project be determined and whether there are changes to that area first. Cost and schedule changes will typically occur when there is a change to the technical scope.

7.5.8   Contingency Reserves

With several risk events, there will be some occur that have both cost and schedule implications. Any technical risk event that occurs will add cost to the project and increase the duration. Therefore, it is best to have some contingency reserves for both cost and schedule. Contingency reserve represents a percentage of the cost and a percentage of the schedule.

A cost contingency can be computed to some degree of accuracy by multiplying the probability of occurrence (probability in the range of 0 to 1.0) by the consequence of the risk event (in dollars). Thus, a risk event that has a probability of 0.6 of occurring and a consequence of $10,000 would have a negative value of $6,000. This would be assigned a time for probable occurrence and plotted against the schedule.

A schedule contingency can be computed similarly to the cost contingency. Multiply the probability of occurrence (0 to 1.0) by the additional time that would be consumed (consequence) if the risk event occurred. This provides the negative value for additional units of time that will be required. A contingency of schedule in a high risk project is a necessity.

Contingency reserves for cost and schedule are not required throughout the duration of the project. Once a risk event has been avoided and the project is continuing, any contingency reserve may be reduced by the amount of the negative value. For example, a cost risk that is avoided would permit release of any funds in a contingency reserve for that event to reduce the amount of money held for future use.

7.5.9 Risk Control Responsibility

The project leader is responsible for managing the project's risks. Each risk event should be put on a schedule of when it may occur. In this manner, the project leader can monitor risks until they either occur or until the period of time has past. Risks that do not occur can be removed from the active tracking and any contingency can be removed from the project.

By placing the risk in the project schedule, one can maintain the time relationship between risk events. New or emerging risks can be inserted into the schedule while past events can be marked complete. One method is to use a milestone for each risk event and identify each with a specific risk.

Risks that do not occur are easy to manage. Risks that occur will require intensive effort to maintain the forward progress of the project. Anticipating the risk event and the corrective action that may be required if it occurs will minimize the disruption.

7.5.10 Key User Questions

1. Uncertainty induces risk into a project because there is a lack of information for proper planning. What can a person do to reduce risk associated with uncertainty?
2. Risk is inherent in every project. Why are projects initiated with high risk?
3. Quantifying risk gives an order of magnitude for the risk events. Why is an order of magnitude important in managing risk events?
4. If a project has many high risk items that cannot be mitigated, senior management must decide what is the appropriate course of action. Why does senior management get involved?
5. A high risk technical event is identified and quantified. Where can the risk event be changed to reduce the overall risk?

7.5.11 Summary

Risk management requires a rigorous process to identify, quantify, mitigate, and manage risk events. High risk items should be mitigated first and reduced to the level deemed necessary by senior management. Moderate and low risk events may be accepted without mitigation if their negative impact has little effect on the overall project.

In mitigating risk, the first area to address is the technical functions. Any risk in a technical area will typically extend into cost and schedule areas. Once the technical risk events are addressed, then schedule and cost may be addressed.

Schedule risk events will most likely affect cost. Any schedule slippage will affect the cost of resources and any fixed costs for the project. Assessing schedule risk events is the second area to be addressed after technical functions.

Cost is the last area to address and is affected by both technical and schedule. Cost is always affected by technical function rework and schedule delays. The resources required for the rework or new work as a result of risk events occurring will drive cost increases.

7.5.12 Annotated Bibliography

1. Duncan, William R., *A Guide to the Project Management Body of Knowledge* (*PMBOK® Guide*) (Newtown Square, PA: Project Management Institute, February 1996), chap. 11. This chapter sets forth a process for identifying and quantifying risk. It also describes procedures for developing a risk response and risk control.

2. Wideman, R. Max, ed., *Project and Program Risk Management: A Guide to Managing Risks and Opportunities,* (Newtown Square, PA: Project Management Institute, 1992). This book is an easy-to-understand approach to managing risk. It places risk in the project context and defines risk areas with clarity.

7.6   PROJECT AUDITING

7.6.1   Introduction

Project audits are conducted to determine whether the project is performing according to plan. There is an implied purpose of the project being successful, but success is contingent upon the project plan and its ability to guide the project team to a successful conclusion. Audits are a comparison of what is being accomplished versus what was planned to be accomplished.

The project audit is not a random assessment of the project. It is a planned activity that may be scheduled at a time in the execution of the project or when a certain threshold triggers a need for an audit. Planned audits may be conducted at different stages of the execution phase of a project. Triggered audits may be conducted when, for example, a project milestone is not achieved.

An audit validates whether a project is or is not proceeding according to plan

---

7.6.2   Types of Project Audits

Project audits will vary according to the need for comparing the plan to the actual execution practices. Planning and conduct of the audit focus upon the purpose and expected outcome. Typical project audits are as follows:

- Progress Audit—the review of project progress in terms of the three primary goals for schedule progress, budget expenditures, and technical progress. The outcome for this type of audit would be a comparison of the planned progress in each of the three areas against the actual accomplishments. This would provide senior management a report on the effectiveness of the project's execution.

- Process Audit—the review of practices by the project team to ensure they are following the prescribed process and that the process is effective in meeting the planned goals. One example of a process audit could be a test of a new piece of equipment to determine the functionality. The test process would provide assurance that the testing was producing the correct results. The results of the audit would provide confidence as to whether the process is capable of producing the desired outcome.

- System Audit—the review of a technical or administrative system to ensure it is functioning according to the documented guidance. The system would typically be a project support operation or function such as Issue Management, Change Management, or Communication Plan. These areas would often be supporting functions for the overall project and areas that are critical to ensuring project accomplishment. The result of the audit would be a report on the adequacy of the system to support the project's work.

- Product Audit—the review of the technical accomplishment of the project in building the product. It is a comparison of the technical achievement according to the plan (e.g., product description, statement of work, specification) and compared to the actual. The outcome of the audit gives confidence as to the convergence of the technical parameters and the work. The result of the audit is a report stating the degree of convergence on the technical solution.

• Contract Audit—the review of work on the project and a comparison with the contractual requirements to determine compliance. This audit provides confidence that the project team is performing the work required by the contract and that all specific requirements, such as workmanship or process usage, are being met. The result of the audit is a report of the degree of compliance with the contractual requirements.

• General Audit—the review of all aspects of a project and a comparison of those planned versus actual accomplishments. This type of audit encompasses all areas of the project and is used to identify noncompliance with the plans, practices, processes, procedures, and requirements. The audit may vary as to the depth of examination and time spent in any one area. The result of this type of audit is a report on the degree of compliance with all requirements for conducting projects.

• Special Audit—the review of specific parameters of a project to determine the progress or status of a project. This type of audit is typically triggered by a loss of confidence in the project's achievements in a specific area and senior management desires to determine the actual progress or status of that area. The result of a special audit is a specifically focused report on the progress or status and, perhaps, recommendations for improvement of conditions.

7.6.3 Audit Teams

Audit teams are formed to conduct the various audits and to report the results of audits. The composition of the team is often dependent upon the type of audit and any special purpose. The success of the audit team is dependent upon the skills, knowledge, and ability of the individuals and the needs of the project. It is not necessary for a person auditing a project to be technically qualified in project work.

Figure 7.9 depicts the three types of audit teams and the overlap between internal and external team composition.

There are three basic types of audit teams: the internal team, the external team, and the composite team of both internal and external team members. The functions for each type may be similar, but the degree of independence in the audit is the primary difference.

• Internal Project Audit Team—a select group of individuals from a project team that comprise an audit team. This team conducts self-audits of the project functions and identifies compliance issues. The team has credibility with the total project team and team members are familiar

Audit Teams

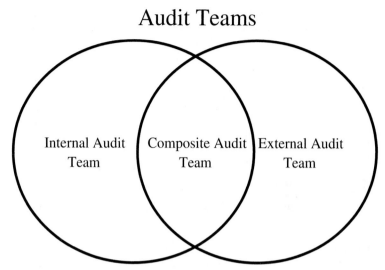

FIGURE 7.9   Audit team types.

with the work. There is a rapid start-up and completion of the audit. Using project team members, however, detracts from the work that was to be performed and does not provide an objective assessment of the audited area.

- External Audit Team—a select group of individuals with specific skills, knowledge and abilities that comprise an audit team, but have no member from the project team. This audit team has a high degree of independence and perceived objectivity in making the assessment. It does, however, require time and effort to form, become familiar with the project, and collect information. A dedicated external audit team can minimally impact the project's progress while reaching objective findings.

- Composite Audit Team—a select group of individuals with specific skills, knowledge, and abilities that comprise an audit team, and are a mix of both internal and external resources. This composite arrangement obtains critical resources from the project and from other sources to build an audit team that can balance objectivity of assessment with the knowledge of the project work. A composite team has the strengths of being knowledgeable in the project work and the audit function while having minimal impact on the objectivity of the outcome.

There are many variations of the audit team and the source of people to form the team. Constraints on availability of technically qualified people and the urgency of need for the audit may drive a unique solution. One must recognize the strengths and weaknesses of the different variations and how the variations may negatively impact the desired results for the audit.

7.6.4 Planning Project Audits

Project audits must be planned to ensure the pertinent areas are examined and a comparison made of each area with the project plan, project standard, process, procedure, or practice. Planning provides the guide to how the audit will be conducted and what will be examined. Corrective actions are typically not part of the audit, but recommendations made by the project audit team are used to initiate required remedial actions.

A project audit plan should consider the following items:

- Purpose of audit. This is the reason for the audit and what it is to accomplish. An example of a purpose statement could be: "To audit the cost and schedule functions of Omega Project to determine the extent of compliance with the project plan."

- Scope of audit. This would entail the amount of time to be spent on the audit as well as the areas to be examined. An example of a scope statement could be: "The audit team will have five days to audit Omega Project in the areas of cost and schedule at Newtown Plant. The team will review documents, discuss information with the project team, and compare the information against the baseline. A report will be rendered to Ms. Jo Smith within five working days following the audit."

- Resources and Assignments. This is a detailed list of the audit team and their skills. It would also give general assignments to each person for the audit, to include writing the report of findings.

- Methodology. This is a general approach as to how the audit will be conducted. It may describe the pertinent procedures for the conduct of the audit. For example, it may have detailed instructions as to documents to be reviewed or it may describe the function to be reviewed. This would include any preparation required prior to the audit, such as a meeting to discuss assignments and the depth of auditing services.

- Report. This describes the report that the audit team will render upon completion of the audit. It may range from a simple narrative report to detailed reports with briefings to selected audiences. The report is the deliverable product of the project audit team and represents its efforts in conducting the audit.

Random or quick fire audits do not have the same effectiveness as the planned one. Project audits must be conducted in a formal manner that is understood by the project team as well as senior management. Random and quick fire audits convey a message that the audit team is focused on finding fault. All audits should be to identify those areas in need of improvement and confirm that other areas are performing well.

7.6.5 Conducting the Project Audit

The project audit team, whether an independent or internal team, must always conduct itself in accordance with a set of guidelines that allow them to set a professional atmosphere. The audit team has an obligation to the project team as well as senior management to perform all duties in a professional manner.

Some guidelines that are helpful in conducting the audit are listed.

- Be prepared to perform your part of the audit and use as little time of the project team as possible.
- Brief the project team on your mission and the time frame that is required to conduct the audit.
- Always ensure the project team members know your role and the mission of the audit team.
- Collect and record information in a formal manner; avoid misleading a project team member to obtain information.
- Allow project team members to initiate corrective action, but never direct any action.
- Do not promise any project team member anything; refer questions about the audit to the audit team leader.
- Be accurate in recording information and confirm the accuracy by obtaining feedback from the project team members.
- Immediately report any unsafe or illegal activity to the project audit team leader.
- When appropriate, give an informal debriefing to the project team prior to completing the audit.

A well-informed and prepared project audit team can quickly and professionally collect the required information to make a comparison between the requirements and the actual results. This professional approach ensures completion of the audit's mission to identify what is progressing well and what needs attention. Audit reports are based on how well the audit team performs its functions and meets its professional requirements. The value of the audit is directly related to a professional job.

7.6.6   Key User Questions

1. What are the purposes of project audits and what does each type of audit do for the organization?
2. What is the composition of a project audit team if one is seeking an independent assessment of the project's progress?
3. What skills are useful in conducting audits within your organization?
4. When would you use an independent audit team on one of your projects and why?
5. When an audit team member is asked questions about the audit, is it appropriate to give a project team member emerging results? Why or why not?

7.6.7   Summary

Project audits serve useful purposes for a variety of reasons. The type of audit will focus attention on the specific area of interest and require specific skill sets to ensure completion of the audit. Audits, however, are all similar in that they are designed to determine the compliance and differences between the requirements and the actual work.

Compliance with the variances from the project plan, the schedule, the budget, the work standard, or other baseline document is the purpose of the audit. It is as important to identify those activities that are in compliance with the baseline document as it is to know where there are variances.

All successful audits will have an audit team that performs the work in a professional manner that demonstrates the audit team is focused on the mission. The audit team will be well prepared and knowledgeable of the requirements so as to perform the audit functions in the least time and with the least interruption of the project team's work. Reports of audits will be based on facts professionally collected and accurately recorded.

7.6.8   Annotated Bibliography

1. AT&T Bell Laboratories, *Current Best Practices: Project Management Audit* (December 1993). This document describes the practices and

procedures for conducing project management audits. It details the work required to prepare for the audit and to report the findings.

2. Bittel, Lester R., *Encyclopedia of Professional Management* (New York, NY: McGraw-Hill, 1978), pp. 64–71, 93. This book describes the purpose for conducting management audits. It describes different types of management audits and the participants.

7.7 *SCHEDULING STANDARDS*

7.7.1 Introduction

Scheduling standards are defined as the rules to be applied in the development and maintenance of schedules for projects. Project scheduling is the distribution of tasks over time to permit the best accomplishment of the project's work.

Laying out tasks and applying resources to the tasks is fundamental to the project planning. It is often done quickly and without the rigor to ensure all tasks are included in the schedule. A random development process relies heavily on the prior experience of the planner to construct a comprehensive schedule.

All planning activities need standards by which the components are uniformly developed for implementation. Standards bring about consistency of work practices. Consistent practices also permit adjustment of the standard when errors occur. Randomness in planning does not provide the practice stability with which to improve on the process.

The standards outlined here are consistent with many of the automated project scheduling tools and follow the general sequence that one would use with an automated project scheduler. Sequentially, the application of the standard guidelines builds and adds to prior items.

7.7.2 Scheduling Standards

Each organization will build its scheduling standards to meet the level of detail in planning as well as the control processes needed by the business. Managerial control requires different standards applied to schedules for large and small projects. Large projects need the detailed definition of work to ensure the convergence on a visible technical solution. Small projects, however, can often be general because the technical convergence can be visualized without a lot of details.

The 18 items depicted in Fig. 7.10 are derived from an actual scheduling standard. These items highlight the areas to be considered and the practices accompanying the item.

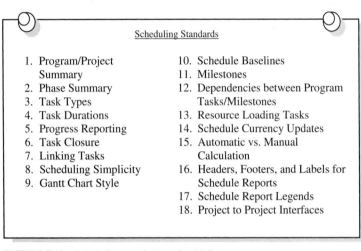

Scheduling Standards

1. Program/Project Summary
2. Phase Summary
3. Task Types
4. Task Durations
5. Progress Reporting
6. Task Closure
7. Linking Tasks
8. Scheduling Simplicity
9. Gantt Chart Style

10. Schedule Baselines
11. Milestones
12. Dependencies between Program Tasks/Milestones
13. Resource Loading Tasks
14. Schedule Currency Updates
15. Automatic vs. Manual Calculation
16. Headers, Footers, and Labels for Schedule Reports
17. Schedule Report Legends
18. Project to Project Interfaces

FIGURE 7.10   Scheduling standards and guidelines.

Descriptions of the standards guidelines are as follows:

1. *Program/Project Summary.* Start building each project schedule with a summary line. The first line becomes the capstone for time duration of all phases, tasks, activities, milestones, and summary tasks. This single summary line permits collapsing the project to a single line and provides a summary of start and end dates.

2. *Phase Summaries.* Major subdivisions of the program/project are phases. Below the program/project line and indented are Concept, Definition, Development, Introduction, and Deployment Phases. Schedules will have all these phases as the first level of indenture. (Note: the project methodology may differ for an organization and the project may have other titles for phases.)

3. *Task Types.* Two types of tasks will be used in the schedule: Effort Driven tasks and Fixed Duration tasks. Effort driven tasks change as the amount of labor is applied while duration driven remain unchanged regardless of the number of labor hours (people) applied. Examples: effort driven "software coding" task of 10 days (80 hours) duration. One person can complete the task in 10 days while two

people can complete it in five days. (Linear relationship assumed for application of labor.) Duration tasks, however, are fixed in the number of days. A one-day conference with three people will still take one day if only two attend or if ten people attend.

4. *Task Durations.* Tasks should not exceed 80 hours for positive control of the project. If the task exceeds 80 hours, then subdivide it. There may be situations where the project manager will want to make that as little as eight hours for extremely tight control and measurement of progress. The least duration for a task should be ½ day or 0.5 day. (This 80 to four hour range appears to be adequate for the type of work being accomplished and for the degree of control exercised over the tasks. There are exceptions and these should be handled on a case-by-case basis.)

5. *Progress Reporting.* Progress reporting should be in 10 percent increments on the schedule. Avoid using the auto update/status change feature in some computer programs unless it is a certainty that the project is on schedule. Most of the time it will not be a perfect match between progress and the "now" line.

6. *Task Closure.* Once a task is closed (i.e., 100 percent complete), the task may not be opened again for additional work. Any task requiring a review will not be completed until the review is complete. (This gives the person reporting closure greater incentive to ensure the work is truly complete.) Discovered work is placed in a new task to show the growth of the scope of work.

7. *Linking Tasks.* Logic or connectivity between tasks:

- Connect all tasks at the detail (lowest level) (Summary level connections lose the control over the details.)
- Connect tasks in a "finish-to-start" relationship. Use "start-to-start" and "finish-to-finish" only when necessary. Never use "start-to-finish" relationships.
- Never use negative lag on connection/relationship. It gives a "start date" based on an unknown date.
- Use positive lag as necessary to indicate a delay between two or more tasks.

8. *Scheduling Simplicity.* Keep scheduling as simple as possible. Sophisticated spins are difficult to understand and scheduling software does not handle complex relationships well during schedule calculations.

9. *Gantt Chart Style.* Use the Gantt chart to show progress. Print in landscape mode. Fit the Gantt chart to one page. If the Gantt chart exceeds one page in the landscape mode, then determine if there is a better way to present the information. Two methods to use are "summary view only" or "view only the current period of work."

10. *Schedule Baselines.* Do not save schedules in baseline mode until the project manager approves the schedule for release. (Note: If uncertain, you may want to save an electronic file to a disk prior to baselining. Once saved in the baseline mode, the schedule is difficult to change and the calculations often are erroneous for changes.)

11. *Milestones.* Use milestones to indicate a significant event or control point in a schedule such as approval of a phase's product (e.g., report, specification, design). This example is the transition milestone that provides the "go ahead" to the next phase. Other milestones may be used for control purposes and may be "owned" by senior management. Typically, these milestones require the project manager to report achievement, slippage, or potential slippage. The senior management owner may "fix" milestones and only permit movement of tasks in the time frame between the milestones.

12. *Dependencies Between Project Tasks/Milestones.* Each task, except the first and last ones in the program, should have a dependency relationship for its start and finish. Independent starts (i.e., fixed starts in the middle of a project) remove the dynamics from the schedule as does other fixed dates, e.g., milestones with external ownership. The last task or milestone in the project should be the delivery of the product or service to the customer.

13. *Resource Loading Tasks.* Tasks are accomplished by people and each task must have at least one person assigned. Personnel will be assigned 100 percent of the time to a task, whenever possible, to take advantage of "continuous flow" work. When two or more people with the same skills are assigned to a task, all will be assigned for the duration of the task. This precludes leaving one individual on the task to deliver while others have performed "level of effort" work.

14. *Schedule Currency and Updates.* Schedules will be maintained and updated for customer on a weekly basis. The update cycle and the cutoff on data (currency of information) is the prerogative of the project manager. Updated information will be delivered to the customer on (day, time) each week. The project manager will specify who is to get copies of the schedule. (Note: Use electronic mail for delivery, if possible.) Always check to ensure the schedule has been calculated prior to printing and delivery.

15. *Automatic vs. Manual Calculation.* When working with large schedules, set Calculation to Manual to permit faster working rather than a calculation with each operation. Remember to reset it to Automatic after the schedule is developed.

16. *Headers, Footers, and Labels for Schedule Reports.* Labeling schedule reports is important for identification and tracking the dates the reports are produced. In the Header (under page setup), place the Program/ Project Name in the center. Use Ariel 14, bold for consistency in all schedule reports. In the Footer, place on the left side the file name, system date, and system time (i.e., date/time report is printed). In the center of the Footer, insert the page number as "Page _ of _ Pages." On the left side, place the Program/Project Manager's name and telephone number. Use the default font for all entries in the Footer.

17. *Schedule Report Legends.* Legends are the key to the symbols used in the Gantt chart. Print schedule reports without the Legend as the default position. Print the Legend only upon request of the user.

18. *Project to Project Interfaces.* There are typically two types of project interfaces that need to be identified and placed in the schedule: Dependencies on other project's products (i.e., internal to organization) and dependencies on external sources (e.g., vendor products and facility availability for installations). Subcontractors and vendors working within the project are not normally considered interface areas. These folks are managed on a contractual basis and their products/ services are included in the schedule is the same as an employee.

7.7.3 Dependencies and Interfaces

Dependencies and interfaces are often missed in the planning of projects. These are critical to the completeness of the planning and to establish responsibilities for managing each area. The responsibility may be defined as another party, but coordination assures an agreed upon linkage.

- *Internal Dependencies.* Dependencies (interfaces), suggested in Fig. 7.11, with other projects must be identified and placed in the schedules of both projects. This establishes a point in time for the hand-over of a product or service that is mutually agreeable by both project managers and consistent with the delivery requirements of the receiving project. Frequent tracking and agreement that the delivery is going to be made

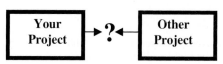

FIGURE 7.11 Interface responsibilities between projects.

on time is required to ensure there is no impact on the receiving project. (Note: Internal interfaces are best managed by mutual agreement between the project managers, but may require escalation to a decision body when there is a difference of judgment on when the delivery should be made.)

- *External Dependencies.* Dependencies (interfaces) with external sources, such as vendors delivering products, must be identified and tracked on a continuous basis to ensure any potential slippage is addressed early on. The interfaces are placed in the schedule at the point of expected receipt, but the management of those interfaces is continuous. (Note: External interfaces shown in Fig. 7.12 usually are outside the management arena and rely on persuasion to affect change to the advantage of the program.)

FIGURE 7.12 External interfaces dependencies.

- *Schedule Symbols for Interfaces.* Use milestones as the Interface Points, i.e., zero duration. Label interface milestones similar to the following examples: ">>>Deliver Invoice Engine to Program XYZ" for the delivery. "<<<Receive Invoice Engine from Program ABC" for the receiving program. (Note: When schedules are combined into a Master Schedule, these will be electronically linked.)

- *Interface Management.* Interfaces must be actively managed to ensure the best possible solution or the earliest identification of a mismatch in meeting the time requirements. Once there is agreement between the delivering and receiving parties, the Interface Point Milestones should be fixed in the schedule. A schedule of contacts between the parties should be established for updating or confirming the validity of the date. Any indication that the delivery will negatively impact the receiving project should be immediately addressed and contingency actions planned.

The scheduling standard is an example that may be used to develop one that fits the needs of another organization. The details contained in this standard were adequate for large-size projects on software development. It is recognized, however, that the standard is directly linked to any automated scheduling tool and this would need to be tailored into any new standard.

Scheduling standards can improve the quality of planning while reducing the planning time. Having a standard to guide the planner is useful to ensure all items are included in the schedule as well as uniformity in how the schedules are built. Schedules that follow a standard result in consistency for reporting information to the project team, and senior management, and posting project internal updates.

7.7.4 Key User Questions

1. What benefits do you see in your organization when using scheduling standards?
2. What disadvantages are there to not having scheduling standards?
3. Who should develop the scheduling standards for the organization?
4. What size projects should use scheduling standards in your organization?
5. How would scheduling standards affect managing interfaces within your organization?

7.7.5 Summary

Scheduling standards are needed in all organizations to ensure consistency of development of schedules. The advantages of a standard is that uniformity is achieved in an organization's schedules during development and implementation as well as consistent reports being delivered to senior management. The standard will also identify the need for interface management between projects and facilities.

Standards improve schedules developed during the planning process and make them easier to maintain in the implementation phase of the project. Uniform schedules between projects permit better interface planning and coordination. Consistency is required in schedule planning and implementation to effectively manage across several projects.

Standards consider all areas of scheduling, to include interfaces with organization's internal projects where there is a dependency and with external parties that can affect the success of the project. Establishing interfaces and managing to the documented agreement is critical to the organization's project management capability.

7.7.6 Annotated Bibliography

1. Frame, J. Davidson, "Tools to Achieve on-time Performance," in David I. Cleland, ed., *Field Guide to Project Management,* 2nd ed. (New York, NY: John Wiley & Sons, 2004). This chapter describes proce-

dures for developing a schedule and computing the schedule duration. The fundamentals of scheduling can be easily understood and used.

2. Duncan, William R., *A Guide to the Project Management Body of Knowledge* (Upper Darby, PA: Project Management Institute Publications, February 1996), chap. 6. This chapter describes sound schedule development practices and procedures. It is process driven and covers all the fundamentals of scheduling except managing interfaces.

7.8   OUTSOURCING PROJECT MANAGEMENT

7.8.1   Introduction

Outsourcing, or out-tasking as some organizations call it, is a growing field. Historically, the soft areas such as custodial services, food services, and landscaping have dominated the outsourced work; nearly 50 percent of recently surveyed organizations identifying these areas. Maintenance of fleet automobiles and aircraft are also examples of areas for outsourcing.

Outsourcing is contracting for services that could be provided by the organization, if the organization had the capability and desire to perform those functions. Outsourcing has many advantages in that it is typically more economical to buy the services than provide them in-house. There is no investment in the function in terms of people, equipment, or maintenance for the outsourcing organization.

Outsourcing of core competencies is not recommended because an organization loses its capability to function effectively. Core competencies must be nurtured and grown rather than contracting for them through another organization. A providing organization has little incentive to improve on another organization's core competencies as long as the products and services are being purchased.

Outsourcing support functions is good business

7.8.2   Project Management as an Outsourced Service

Project management services have been outsourced in several instances and the future looks promising for more outsourcing. The benefits of outsourcing project management includes improved and more economical operations.

Project management services can be improved by transferring the functions to an organization specializing in project management. This specialized company hires the right skills and uses the best of breed practices because it is their core competency. They have resources that are focused on providing these services and have the in-depth expertise to perform at high-performance levels.

Outsourcing relieves the parent company of the burden of managing project management services. Through contractual relationships, the parent company states its needs and then manages the contract and delivery of services. There is less effort required to receiving the products and services from outsourced project management than to manage it in house.

Outsourcing companies are concerned with loss in the project management function. Organizations outsourcing will lose visibility into the details of work, but there is no need to see detail when the end product meets the requirements. Outsourcing companies should only be concerned with the delivered product and less concerned over the details of production.

7.8.3 Outsourcing Trends

The trend today is to outsource that work which an organization does not want to perform, work that can be accomplished better by an organization specializing in the functions, or work products or services that can be purchased for less cost than in-house production. This trend is good business and makes for a better approach to producing the product or service to complement core competencies. Outsourcing continues to grow with new business approaches.

Project management outsourcing has been primarily one of providing services to organizations by one person at a time. It is referred to as "body shop" or "hired gun" because the project management professional is typically working on-site with the organization's staff. This type of arrangement is profitable for the organization doing the work, but is not outsourcing.

The "body shop" or "hired gun" approach mixes different levels of proficiency and competency. The lower level of project management proficiency and competency is typically the resident employee of the organization and the outside person brings the expertise. This places the resident employee as the driver and the expert as the follower. There is considerable waste in talent, time, and money with this approach.

Project management services outsourcing will, like other professional services, continue to be used to fulfill the needs of organizations. This has a high-growth potential because of the ability of project management service providers to deliver better products in a timely manner at lower cost. Project management service providers will have the expertise and skills to build better products than a part-time effort in-house.

7.8.4 Selecting an Outsource Provider

Successful selection of an outsource provider is accomplished in four sequential steps, shown in Fig. 7.13. These steps give a high degree of assurance that the best provider is selected and that the outsourcing relationship is satisfactory.

**Four Steps for
Selection of an outsource provider**

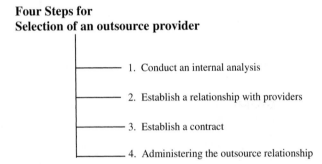

1. Conduct an internal analysis

2. Establish a relationship with providers

3. Establish a contract

4. Administering the outsource relationship

FIGURE 7.13 Selecting an outsource provider.

- Conduct an internal analysis—identify those functions that can be outsourced, assess the tactical and strategic impact of outsourcing each identified function, evaluate the total cost of each function selected for possible outsourcing, and determine the advantages and disadvantages of outsourcing. Classify each function as "no outsourcing," "possible outsourcing," and "definite outsourcing." Use criteria for each category such as listed in Table 7.3.

- Establish a relationship with providers—issue a request for information (RFI) that solicits interest in providing products/services, select two or three respondents and ask for due diligence. Conduct due diligence survey to validate the capabilities of the potential providers.

- Establish a contract—from the three respondents meeting the due diligence survey criteria, negotiate a contract with one. Establish scope and boundaries for the contract, describe the following:
  - Resources to be used to produce the products and services
  - Key deliverables and the schedule for delivery

TABLE 7.3   Criteria for Outsource Provider Selection

Category	Criteria
No Outsourcing of Function	• Part of core competency • Negative benefit or value added to outsourcing • Impacts strategic goals
Possible Outsourcing of Function	• Cost and other benefits show no advantage or disadvantage • Neither contributes to strategic goals nor impacts strategic goals • All factors equal for in-house or outsourcing
Definite Outsourcing of Function	• Definite cost savings to outsource • Reduces complexity of management for in-house work • Better product/service results from outsourcing • More responsiveness to needs of organization • Outsourcing uses better practices and technology than available in-house

- Performance measures and other quality metrics
- Invoice and payment schedule to include provisions for timing of payments
- Change and termination provisions of the contract
- Administering the outsourcing relationship—establish the contract management process, establish the technical review process, establish a change order process, and establish a steering committee or oversight committee. Involve the users or consumers of the products and services in the steering/oversight committee.

Identifying providers who can meet the organization's needs involves more than establishing a contract and working to enforce the provisions and clauses. A good contract is the basis for promoting understanding with the provider of products and services. It is also essential that the provider have a record of meeting contract provisions.

References should be checked during due diligence and references should be asked the following questions.

- Did the contractor deliver the products and services called for in the contract?
- Were the products and services usable by the consumer as delivered?

- Did the contractor demonstrate flexibility in minor changes to the contract or was each minor change an issue?
- Were change orders to the contract performed on a fair and equitable basis?
- Do you recommend the contractor for these products and services (identify)?
- Would you award a contract to this contractor again?

7.8.5   Outsourcing Project Management Services and Products

From an organization's perspective, there is a need to determine what may be outsourced and to what project management service provider. First, the areas of project management that may be outsourced need to be identified and, second, identify the best provider.

Project management products and services that may be outsourced depend upon the organization's structure for project management, the degree of maturity in project management, the number of projects and the management style. The organization must know what is needed in terms of project management products and services. An organization with a loosely structured project management capability may not know what is needed.

The entire project may be outsourced. This frequently happens within such fields as Information Technology or Information Systems where the organization only wants to be concerned with software solution and not the challenges of designing, developing, testing, and delivering software releases. When only components of project management are outsourced, Table 7.4 may be helpful.

7.8.6   Project management outsourcing guidelines

Outsourcing for any products and services needs to consider all aspects of the relationship and how it is managed. Transition of the responsibility for products and services must be coordinated and performed according to an agreed upon plan. The following items are helpful in guiding an organization through outsourcing of project management services and products:

- Products and services—identify the products and services to be transitioned (or initiated if there are no current comparable in-house). Describe the products and services in detail and conduct a mutual understanding meeting to ensure both parties have full agreement on the products and services.

TABLE 7.4 Outsourcing by Project Management Component

Project management component	Outsourcing potential
Project planning	High—project planning expertise is typically not resident in companies. A centralized planning effort with the project team can provide many benefits by getting the projects started from a solid basis.
Project scheduling and maintenance	High—scheduling and schedule maintenance skills are often resident in-house, but the work is time consuming for highly qualified technical people.
Project cost estimating	High—cost estimating is an art that requires specific skills sets that are typically not resident in an organization. It can be time consuming for technical people who are better employed on other tasks.
Project progress reporting	High—status and progress reporting are part of the schedule maintenance. Consistent reporting to senior leaders will be achieved as well as standard reports across projects.
Change control • Product • Project	High for both categories—this is the administration of the change control process. Initiation of changes and decisions on changes are typically performed in-house.
Issue tracking	High—tracking issues by a standard procedure and collecting information on the status is a routine matter that should not burden the project manager.
Problem tracking	High—tracking actions by a standard procedure and collecting information on the status is a routine matter that should not burden the project manager.
Risk assessment and risk tracking	Medium—conducting risk assessments and mitigation actions can be performed by an outsource provider. There must be good coordination with the project manager for any project being assessed.
(Other components) Identify	(This open item allows tailoring to meet individual organizational requirements.)

- Standards and formats—identify the standards and formats that are considered a requirement. Standards and formats provide a consistency to the delivered products, but will change the current in-house products.
- Frequency of delivery—determine the frequency of delivery for products and time required for preparing products. Typically, reports are provided on a weekly basis, but there may be requirements for rapid turnaround for some products.
- Method of delivery—products may be delivered in hard copy, electronic copy, or a combination of these. Determine the most appropriate and effective means of delivery for all products.

7.8.7 Key User Questions

1. In selecting project management functions to be outsourced, what would be one of the major considerations?
2. Selecting a contractor to provide project management services requires due diligence. What are some items to look for in due diligence?
3. As a project management service provider, you have been contracted to provide schedule development and maintenance services. Can the outsourcing organization change the frequency of delivery on updated schedules?
4. As a project management service provider, you are asked to sign a contract that will be defined later as to the scope of work. Should you sign an open contract and what is the potential impact?
5. In seeking a project management service provider to outsource selected project management functions, you ask to place some of your people in the provider's organization to oversee the production of deliverable products. Is this reasonable? Why or why not?

7.8.8 Summary

Outsourcing project management products and services has the potential for significant gains for an organization. Outsourcing relieves the outsourcing organization of a technical area for which they may not have the trained resources to perform and to perform at a lower cost. Outsourcing can place many of the project management functions in the hands of project management professionals.

Entering into an outsourcing arrangement requires some background work to identify and select the provider with the best qualifications and reputation as a contractor. The time and effort spent on selection of the provider will save effort in managing the relationship.

Outsourcing requires thoughtful assessment of the areas to be contracted and establishing a relationship that is cooperative. There also needs to be a thorough mutual understanding between the two parties as to what, when and how services will be provided. A contract can describe the requirements, but it takes communication with the provider to make the outsourcing arrangement work to the satisfaction of all parties.

7.8.9 Annotated Bibliography

1. Raines, Christy, "Outsourcing Should Be 'Marriage of Convenience for Both Parties,' " *Unisys World,* Austin, TX, January 1999, p. 1. This article discusses the relationship and its duration for outsourcing. It addresses how to select an outsource supplier and how to manage that supplier.

2. Hudson, Katherine M., "The Role of Trust in Outsourced IS Development Projects," Association for Computing Machinery, *Communications of the ACM,* New York, NY, February 1999, pp. 80–86. This article discusses the need for balance in performance, structure, and trust in outsourcing relationships. Trust is considered vital to the relationship and plays a large part in outsourcing services.

7.9 DECISION-MAKING IN PROJECT MANAGEMENT

7.9.1 Introduction

Decision-making is a daily affair for project managers and functional managers. Understanding decisions and the decision-making process is important to assure the best courses of action are selected with the information available. Following a rigorous process of sorting through the alternatives gives the best chance for a successful project

7.9.2 Decision-Making Considerations

A decision entails the act and process of selection of a course of action leading to desired objectives and goals. In the management of projects, decisions have to be made and implemented. The decision process in project management involves: *first,* the act of selecting a course of action regarding the commitment of project resources, and *second,* the use of resources through a strategy to accomplish project objectives and goals. Some additional basic notions about the decision context of project management include:

- A decision removes the uncertainty on how the resources will be committed and used in the furtherance of project purposes.
- All decisions involve risk and uncertainty.
- Adequate information is required to make decisions on a project, but at some point it becomes necessary to cut off the gathering of information, and move ahead with making a decision.
- A project manager must have the knowledge, skills and attitude to make a decision on behalf of the project and the organization to which the project belongs.
- Decision making in the matrix organization can require extraordinary patience since consensus of the team members should be sought as well as advice from relevant stakeholders.
- Decisions can have operational (short-range) and strategic (long-range) implications.
- A project manager has the responsibility for *making* and *implementing* decisions involving the use of resources on a project, even though a "limits of authority" may restrict their decision power. In these circumstances they are responsible for finding where the decision authority exists.
- A decision is a judgment involving a choice between alternatives.
- Before a decision is made the counsel and opinion of the project team members, and other project stakeholders, should be obtained.
- A decision on a project should not be made until adversarial opinions have been found and evaluated.
- An alternative always available to the decision maker is to do nothing.
- A decision has to be made if a condition on the project would degenerate if nothing is done.
- Policy and procedures should exist to refer appropriate decisions to higher level managers as appropriate.
- When a decision is made there are likely reverberations throughout the stakeholder community.
- An effective decision on a project must be followed by the commitment of the project team members to carry out the decision.

TABLE 7.5 Classification of Decisions

- *Routine/programmed decisions,* which can be made according to, established policies, procedures, established methods and techniques. These decisions can also be normally classified as short-term operational decisions. A decision concerning the selection of vendors to support a project would be an example.
- *Strategic/Non-programmed decisions* for which there may be little precedence which involved high degree of risk, uncertainty, and ambiguities, such as decisions involving the cost, technical performance, and schedule of a project whose life cycle extends into the future for many years.

- A project manager is involved in a wide variety of decisions. One way of classifying these decisions is shown in Table 7.5 which are discussed in the material that follows.

7.9.3   The Decision-Making Process

The starting point in the decision-making process involves the determination of whether a decision needs to be made. Consideration needs to be given to the issue of what would happen, if anything, if a decision was not made. There have been many models suggested for the steps involved in the decision process. A couple of examples follow.

The common steps for rational decision making usually suggested for a decision-making process includes:

- *Recognize need:* Recognize the need for a decision to be made.
- *Gather relevant information:* Gather background information needed to provide insight into the forces and factors that suggest the need for a decision.
- *Develop alternatives:* Consideration and development of alternatives to be considered in the decision process.
- *Evaluation of alternatives:* Develop criteria for evaluation of decisions.
- *Select alternative*: Selection of best alternative(s).
- *Implement alternative:* Develop and implement strategy on how the decision will be executed.
- *Feedback:* Evaluation and feedback concerning how well the decision is being implemented.

The decision-making process can also be viewed as a sequential progression from the identification of the problem or opportunity, through the making of a decision concerning the use of resources, and then on through the implementation of the decision and the monitoring evaluation, and control of the use of resources that support the decision. Figure 7.14 portrays this total process.

The notion of the decision-making process is more applicable to non-programmed decisions then to programmed decisions, many of which are routine and procedural. The details of Fig. 7.14 are discussed below:

* *Problem or opportunity*—A necessary condition for a decision is a problem or opportunity. For both a problem and an opportunity there is a gap between what is being done—or in the case of an opportunity—what should be done. For non-programmed decisions there is a direct relationship between planning and the decision-making process which should lead to the establishment of specific and measurable objectives.

* *Developing alternatives*—In developing alternatives to solve the opportunity or exploit the opportunity an early step is to address the issue of what can be done about the problem or opportunity. Flexible alternatives are potential solutions; and the decision maker must consider the potential consequences of each alternative. Developing alternatives is an extremely difficult and important act, since the decision maker will probably end up choosing one of the alternatives. Many decision makers have often wondered if any important alternatives have been left out of the process.

* *Evaluating Alternatives*—There are many mathematical models drawn from the Operations Research field that can give insight into the relationships that exist between the alternatives and the possible outcomes. These include:

  * Conditions of certainty—such as adequate knowledge of the organizational, environment, and competitive melded into the decision-making process.

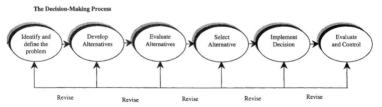

FIGURE 7.14   The decision making process.

- Conditions of risk—possible or probable risks in obtaining resources, weather, dependence on other people or political expediencies to provide services to the enterprise. A project that is pushing the technology would probably have associated technological, schedule, and cost risks.
- Condition of uncertainty—such as having absolutely no idea of the probabilities associated with the various alternatives. In such situations the judgment, intuition, experience, and personality of the decision maker plays an important role. In the construction industry, the uncertainty of the weather is also a consideration.

- *Selecting An Alternative*—The purpose of selecting an alternative is to solve the problem or exploit the opportunity in order to achieve the predetermined objective, goal, or condition. The decision is not an end in itself but only a means to an end. The decision maker should not forget the factors or conditions that led up to the decision or the factors that follow the decision, such as implementation and evaluation. When the decision is made, a commitment to a course of action usually involving the use of resources has been set out. When a decision has been made, the sequence in events that are set out usually affect other elements of the project. For example, a decision to approve an engineering change on a project will have reverberations on the project cost, schedule, and after-sales support. The decision-maker should remember that a good decision can be hurt by poor implementation, such as when people fail to support a project manager's decision through procrastination, avoiding responsibility, or even ignoring the decision.
- *Evaluation and Control*—In this final phase of the decision process, an assessment is carried out concerning how well the decision is producing the results that were expected. The ideas put forth in Section 7.4 concerning the role of monitoring, evaluation, and control of the project have application here.

There are many different models of how to make decisions on a project. Project managers should select the best model, which supports their particular needs.

7.9.4 Team Decision Making

Although we have examined the decision-making process as though individuals made decisions, decisions of the project team are more often made by the team members through an interaction process. There is evidence that a team will make better decisions on the project than an individual—even the project manager. However, there are at least two common tendencies that can interfere with effective team decision-making.

- *Individual dominance*—where a strong member of the project team dominates the discussions and the decision-making process. Often described as "Group Think," in which there is a deterioration of the participation of the team members when confronted with a strong individual who may "outrank" the team members. "Group Think" includes several symptoms:

  - Feelings of superiority on the part of the "Group Think" leaders.
  - Willingness to take unusual risks based on the feelings and rationalizations that discount potential risk.
  - Stereotypical and unrealistic thinking about project conditions.
  - Strong suppression of dissent within the group because dissent means disloyalty.
  - A belief that everyone feels the same about the issues—silence means consent.
  - Self-appointed "mindguards" who keep bad news about the project under cover.

  How can the project manager counteract the "Group Think" risk on the project?

- Encourage every team member to express doubts and criticisms of proposed solutions.
- Demonstrate an ability to accept criticism.
- Break the project team into subteams to analyze the decision, and then have the subteams confront one another to examine their differences.
- Have other project stakeholders participate in the discussion.
- Develop insight into some of the signals of the formation of a "Group Think."
- Have the team discuss how they feel about how the decision-process is being carried out.
- When the project team takes part in the decision-making involving the project, there is breadth with advantages and disadvantages.

7.9.4.1 Advantages

- A greater amount of knowledge is available.
- A broader range of alternatives will emerge, depending on the individual team member's perception.
- Broader perceptions and viewpoints are available.
- The team will probably be willing to accept more risk than an individual.

- When the team members participate there will likely be a greater support and motivation to support the decision.
- Greater creativity will likely result from the interaction of the individuals.

7.9.4.2 Disadvantages

- Because the team cannot be held responsible, the risk occurs that no one is responsible and there will be a lot of "buck passing" results.
- Because time is usually a valuable resource, team decisions are costly.
- If a decision is to be made quickly, working it through the project team may not be practicable.
- Team decisions can be the result of compromise and/or the operation of the "Group Think" phenomenon.
- If senior managers are present the team members may be reluctant to be frank, or if one member has a dominant personality, the team decision may not be in reality a team decision.

7.9.5 Key User Questions

1. Does the project manager have sufficient basic and workable knowledge and skill to make, or cause to be made, key decisions on the project?
2. In approaching the decision-process involving the project, is adequate consideration given how to best make decisions?
3. What mechanisms exist for referring decisions to upper management for which there are "limitations of authority" for the project manager?
4. Has a process model of how decisions are made in the enterprise been developed as a general guide for decision makers?
5. Do the project managers discuss the pending project decisions with the members of the project team?

7.9.6 Summary

In this section, the subject of "Decision-Making in Project Management" was presented. A few guidelines were provided that, if followed, can strengthen the project manager's ability to make decisions, or to refer those decisions on a project which should be sent to upper management.

The field of decision making is quite broad and supported with abundant literature. The field of Operations Research addresses the decision making process, and how to evaluate risk and uncertainty in decision making.

7.9.7 Annotated Bibliography

1. Stoner, James A. F., *Management,* 2nd ed. (Englewood Cliffs, NJ: Prentice-Hall, Inc., 1982), chap. 6, "Problem Solving and Decision Making." This chapter is a classic in the sense of giving advice on a protocol on how to solve problems and make decisions.
2. Milosevic, Dragan Z., *Project Management Toolbox* (New York, NY: John Wiley & Sons, 2003), chap. 9, "Risk Planning." In this chapter the author presents risk planning tools which have a strong quantitative base in both concept and processes. The principal tools provided are Risk Response Plan, Monte Carlo Analysis, and Decision Trees. The material on Decision Trees provides a graphical device for analyzing project situations under risk. A careful perusal of this chapter will heighten a project manager's awareness of the important role of quantitative analysis in decision-making in the management of a project.

7.10 ESTABLISHING A PROJECT MANAGEMENT SYSTEM (PMS)

7.10.1 Introduction

Effective project management requires that an organization follow a methodology that includes a project management system tailored to the business being conducted. Organizations using projects as a strategy must follow a rigorous process to achieve its goals and objectives in the most effective manner. Random process application yields random results.

An organization's commitment to using project management is also a commitment to establishing and tailoring a system that works best for that organization. This section provides the basic guidelines for the project management system.

7.10.2 The Project Management System

Once a decision has been made to institute the use of project management in the organization, a strategy needs to be developed to establish a framework on how project management will be carried out. The use of a PMS provides for a "way of thinking" as well as an effective way of portraying how a "systems approach" can facilitate the management of projects as building blocks in the design and execution of organizational strategies. The principal "subsystems" of a PMS include the following:

- The *Facilitative Organizational Subsystem,* which is the organizational design used to provide a focal point for the use of resources to support the project purposes. The key characteristics of such a subsystem include:
  - Superimposing the project teams on the existing functional structure of the organization.
  - The creation of a "matrix" organization which provides a paradigm on how complementary authority, responsibility, and accountability are assigned and allocated among the project team, the supporting functional elements, the work package managers, and the general and senior managers of the organization.
  - A definition of how individual and collective roles are assigned in the "matrix" organization.
- The *Project Control Subsystem,* which provides for the philosophy and process for selection of performance standards for the project, the design of feedback mechanisms, the comparing of planned and actual performance, and the initiation of corrective action as required to keep the project on track. The key requirements for such a control system include:
  - Regular and ongoing review of project progress by relevant stakeholders to provide intelligence regarding how well work is going on the project.
  - The commitment of all project stakeholders to provide accurate and timely reporting of key matters involving the use of project resources.
- The *Project Management Information Subsystem,* which contains the information essential to effective planning and oversight of the project. Both formal and informal information is useful in managing the project to include, but not necessarily restricted to:
  - Information required to develop project plans and relate those plans to the strategic management initiatives of the organization.
  - Formal and informal information required to provide intelligence for the ongoing review of project progress.
  - The identification and assessment of information required to make and implement decisions in the management of the project.
- *Techniques and Methodologies Subsystem,* which contains the techniques and methodologies such as PERT, CPM, and related scheduling techniques as well as specialized techniques for estimating project costs,

technical performance assessments, and other management science methodologies. The use of quantitative methodology to evaluate risk and uncertainty in the use of project resources is included in this subsystem.

- The *Cultural Ambience Subsystem,* which is the general cultural environment in which project management is practiced. This subsystem typically entails:

  - The social context of the perceptions, attitudes, prejudices, assumptions, experiences, values, mores, and behavior patterns of the people associated with the project.
  - The cultural ambience influences how people act and react, how they think and feel, and what they do and say—all of which affects the behavioral norms of the people associated with the project.
  - Education, training, team building, and similar techniques to enhance the interpersonal skills of the project stakeholders can, if effectively developed and applied, improve the cultural ambience of the project.

- The *Planning Subsystem,* which provides for the means to identify and develop strategies on what resources will be required to support the project, and how these resources will be utilized during the course of the project. Some of the principal elements of project planning include:

  - Development of the work breakdown structure which shows how the project is broken down into its component parts.
  - Project schedules and budgets are developed, technical performance goals are selected, and the organizational design for the project is stipulated.
  - The creation of an integrated project plan which can become the performance standard against which project progress can be monitored, evaluated, and controlled.

- The *Human Subsystem,* which includes most of everything associated with the human element. This subsystem includes:

  - Some knowledge of sociology, psychology, anthropology, communications, semantics, leadership, and motivation theories and applications.
  - A management and leadership style on the part of the project and supporting managers that engenders trust, loyalty, and commitment among the managers and professionals associated with the project.
  - The artful management style that the project manager develops, demonstrates, and encourages with the project team will have a marked impact across the project stakeholders.

Figure 7.15 depicts a PMS in the context of a public utility organization. It shows how the existence and operation of a PMS approach cuts across the strategic management and the functional levels of the organi-

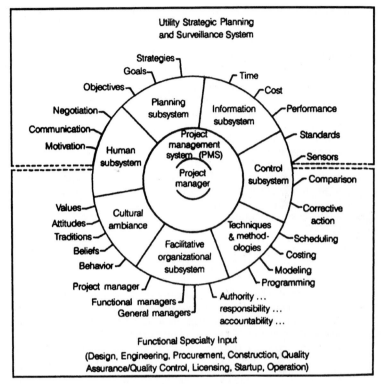

FIGURE 7.15 The project management system. (*Source:* Adapted from D. I. Cleland, "Defining a Project Management System," *Project Management Quarterly*, vol. 10, no. 4, p. 39.)

zation. As such it can be used as a fundamental philosophical model of how a project should be managed.

7.10.3 Key User Questions

1. How can the use of a PMS help to ensure that a total "systems" approach to the management of the project is carried out?
2. Are the projects within the organization being managed using a PMS as a paradigm for such projects?

3. If inadequate attention is being given to any of the subsystems of a PMS for the projects in the organization, have any project management inefficiencies been noted?

4. Has the idea of a PMS been accepted and practiced in the organization? If not, why not?

5. If a PMS philosophy is not being used in the management of projects in the enterprise, what philosophy is being used to ensure a total "systems" approach to the management of the project?

7.10.4 Summary

In this section, the concept of a PMS was offered as a basic guide to how to think about and manage projects in the organization. The point was made that if all of the subsystems of a PMS are "up and running", the chances for the successful management of projects in the organization are enhanced.

7.10.5 Annotated Bibliography

1. Cleland, David I, and Lewis R. Ireland, *Project Management: Strategic Design and Implementation,* 4th ed. (New York, McGraw-Hill, 2002), pp. 123–125. "The Project Management System." A fuller explanation of the PMS is contained in this reference together with its components.

2. Kast, Fremont E., and James E. Rosenzweig, *Organization and Management* (New York, NY: McGraw-Hill, 1970). This book is considered a classic in the management literature in that the text incorporates some of the emerging concepts in the systems approach into the traditional principles of management and organization theory. The authors portray the evolution of management in such a manner that the reader can understand how a systems approach can provide an effective philosophy on how to manage the multiple systems found in contemporary times. The book set an intellectual historical standard in the subsequent and continuing evolution of systems theory and practice in modern times.

7.11 MANAGING COSTS IN PROJECTS

7.11.1 Introduction

Managing costs in projects requires a disciplined approach to estimating, budgeting, and controlling expenses. This process starts during the planning phase when estimates are made to determine the most likely total project cost and establishing a project budget. The project budget is the document that lists all planned expenditure categories and the amounts for each category.

Variances between the budget and the actual cost affect the project's total cost. Changes to the scope or duration of the project also have implications that usually increase costs. Inefficiencies in productivity and discovered new work negatively impacts the budget.

Controlling costs is achieved by tracking expenditures and comparing them with the budget. Tracking expenditures must be a proactive approach to collect expenditures in real-time for comparison and analysis. The result of the analysis forms the status and progress report for the weekly review of expenditures.

A major challenge for controlling costs is the timeliness of capturing expenditures. Using corporate accounting systems will often delay the accumulation of expenditures for as much as 45 days. For example, it is in the interests of a company to delay paying contract costs. A delayed payment of 30 days following invoice from the contractor does not give real time information. Projects typically work in a seven-day time frame and act upon variances in that window.

7.11.2   Planning for Project Costs

Project costs are managed differently by different organizations. The reason for the unique management approaches is the need for detailed project costs versus the need for total project cost. Some organizations manage at the project level while others manage at the work package level. Figure 7.16 shows the difference between an organization's approach and the project's approach to cost accumulation.

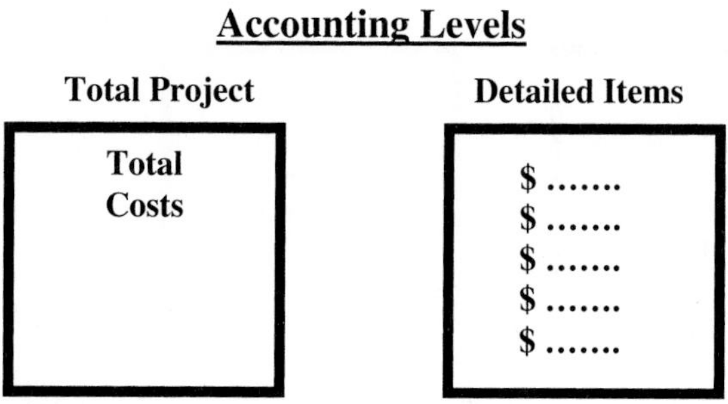

Accounting Levels

Total Project   **Detailed Items**

Total
Costs

$ ....
$ ....
$ ....
$ ....
$ ....

FIGURE 7.16   Cost management levels.

A second consideration in planning for project cost management is the ability to collect detailed and summarized costs. Many organizations use their corporate accounting system to collect project costs. Others use a project level system to collect cost information. The difference between the two systems is the timeliness of cost information. Projects need near real-time cost information to manage the budget on at least a weekly basis.

Corporate accounting systems are designed to delay costs and the recording of expenses. Contractor invoices are typically paid 45 to 60 days following the completion of work. Materials ordered for a project may have a billing and payment date as much as 90 days following receipt of the goods. This delayed accounting system serves the organization well because of deferred cash flow. For a project that is six months duration, these delays are counter to any cost management efforts.

Projects must collect all costs and analyze them no later than every two weeks if there is to be any corrective action or redirection of the project based on cost. Projects may have an informal accounting system that is used, but it typically will differ in the amount of cost from the corporate system as much as 15 percent. This difference may be attributed to such areas as discounts for early payment, changes in price for materials, change in labor rates for personnel, and errors in charges to the project.

7.11.3   Types of Project Costs

Project costs typically are in two categories, direct and indirect costs as suggested in Fig. 7.17. Direct costs are charges to the project that have a direct relationship to the amount of services or goods received. Indirect costs are often computed costs based on a percentage of the direct costs or some other factor that has been determined to be equitable.

Examples of direct costs to projects are:

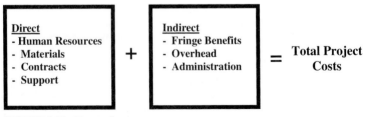

FIGURE 7.17   Types of costs.

- Human resource costs—the price of labor to perform the tasks on the project and any project related expenses, such as supervision.
- Material costs—the price of goods and services that are consumed in the project and become a part of the end product.
- Contracts—outsourced work for a part of the project. This may be a part of the project's work or it may be support to the project, such as independent analysis of an area.
- Support costs—the price of miscellaneous items such at travel, entertainment, and other small expenses.

Examples of indirect costs to projects are:

- Fringe benefits—the cost of human resource benefits that do not directly contribute to projects. These include holiday pay, jury duty pay, military training duty pay, and health benefits.
- Overhead—the cost of facilities such as building space, telephone services, heat, lighting, facility maintenance, and general supervision.
- Administration—the cost of administering the human resource function. Typically these areas are payroll department and human resource department.

The above general categories may have duplicate items under the subcategories. For example, training can be either a direct or indirect cost. The test of where it belongs is: Is it required by the project to complete the project? If it is directly related to the project's completion, then it is a direct cost.

Residual benefits are usually not considered in training programs if a project is first to require the training. An example would be training for racial sensitivity. It is an organization program and should be paid by the organization rather than the project. A second example would be training in a new software programming language that is being used on a project, but may benefit the organization with follow-on projects. This is a project cost for training.

7.11.4 Developing a Budget

A budget represents the time-phased expenditure plan for the project by expense item. The budget is the aggregate of all project costs, both direct and indirect, and may include the profit for the project. It is time-phased to show the cash flow profile for expenditure rate and by category.

Budgets are the accumulation of all expenses and placed in an expense category. This is a control mechanism that authorizes expenditures on the project only in those identified and funded categories. Expenses exceeding the budget or in a category that are not identified must receive approval from senior management.

Costs are estimated from the task items developed in the work breakdown structure and transferred to the schedule. The work identified in the schedule requires resources to accomplish, usually human resources and materials or services. These items are estimated at the detail level and totaled to summary levels. For example, the cost of planning a project may be a summary cost category and the components of planning the detailed level of cost.

Budget development must be consistent with the cost. These are the following categories of cost that must be addressed:

- Raw cost—costs that will be actual cost of the goods or service.
- Burden or indirect cost (also called overhead and administration)—the percentage of the raw cost that is distributed by the organization as the project's fair share.
- Burden cost—costs that include the indirect costs as a percentage of the raw cost.
- Total cost—costs within category for the project.
- Profit—typically a percent of total project cost.
- Total price—the price of the project to the customer, which includes all costs and profit.

The following illustration in Table 7.6 is an example that shows the categories and how costs are computed. Note that different categories of costs will have different burden costs. Labor, for example, has a high burden because of the fringe benefits while materials will often have only a handling charge.

Computing costs on projects should ensure the estimating process retains either raw costs throughout the process or burdened costs. Organizations set their standards for computing project costs and a person must follow those standards to be accurate with the budget development. Many organizations use a burdened cost for labor and raw costs for materials, contracts, services, and other direct costs.

7.11.5 Labor Costs

In labor-intensive environments, the cost of labor can be the most significant part of a project's budget. Labor costs are estimated based on the number of hours required for each labor category.

TABLE 7.6 Cost Categories in Budget

Raw cost and categories	Raw cost	Burden or indirect cost loading	Burden cost	Total cost	Profit	Total price
Labor						
• 1202 hours @ $45/hour	54,090.00	0.87	47,058.30	101,149.17	0.24	125,424.97
Materials						
• Supplies	2,100.00	0.09	189.00	2,289.09	0.24	2,838.47
Services						
• Cleanup	200.00	0.57	114.00	314.57	0.24	390.07
Contracts						
• Computer Rental	450.00	0.17	76.50	526.67	0.24	653.07
Project Totals	56,840.00	N/A	47,437.80	104,279.50	N/A	129,306.58

Organizations have used average costs and nominal costs in computing the budget estimate rather than actual wages paid. This has two purposes; it permits charges to a project without revealing a person's actual salary, sensitive information in some companies, and it is convenient for computing future costs based on pay raises. Some categories of labor costs for budget estimates are as follows:

- Actual labor rate—the rate of pay per hour for a specific individual or category of labor. This does not include fringe benefits or overhead expenses.
- Burdened labor rate—the rate of pay per hour for a specific individual or category of labor. This includes all fringe and overhead expenses.
- Average labor rate—the arithmetic average of the rate of pay for individuals in a project, in a category, or in an organization.
- Nominal labor rate—the organization's stated rate of pay per hour for individuals or category of labor. This rate may or may not be an accurate representation of the actual cost of labor. It may or may not include the burdens for labor.

There is a need for consistency by using only one type of labor rate through the project. When the information on the amount of labor hours consumed during a period is reported, ensure computation of costs with the proper formula. If labor hours are reported through the corporate time-card system, then it is important that the reported costs match the costs used to prepare the budget.

7.11.6   Budget Execution

The project budget is approved upon completion of the planning phase. Typically, the costs of planning the project are not recorded in detail, but may be brought forward in a lump sum. These costs are true project costs and represent the investment in front-end work on the project.

An approved budget is executed by a statement that project work may begin and charges will be to the project's budget or through a series of work package releases. Releases have the effect of only permitting charges against specific work packages, whereas the general release authorizes individuals to work on any part of the project.

As work is accomplished and expenses are collected, the budget is reduced by the actual amount of expense. Variances between planned and actual expenses are analyzed and corrective measures taken, where appropriate. Changes to the budget as a result of new work or scope change are recorded and managed.

An important aspect of cost management is the projection into the future to determine the total actual cost of the project based on the accuracy of the plan. Often, the cost baseline, i.e., the budget, is optimistic and a trend of excess costs will indicate the budget goal is less than actual costs. The project leader must obtain additional funds from senior management or reduce the scope of the project.

7.11.7 User Questions

1. A budget is a time-phased plan for expenditure. What additional costs would be charged to the project if the schedule was extended by six months and no additional resources were required?
2. What is the difference between an average labor rate and a nominal labor rate?
3. What are some of the reasons for using a nominal or average labor rate over the actual labor rate?
4. If a project uses the organizations automated accounting system to track its costs, what are some of the potential impacts on proper project cost management?
5. If a new company policy is put into effect that grants all employees an additional week of vacation, will this affect your cost? Why or why not?

7.11.8 Summary

Managing costs for a project is dependent upon the accuracy of the cost estimate and resulting budget. Accuracy in the application of cost categories will also determine the validity of the budget and whether the budget represents all costs in the project. Any variation from accepted cost estimating practices or any erroneous methods used to compute the costs will impact the validity of the budget.

Both the development of the budget and its execution are disciplined processes that rely on the rigor applied to the processes and the information collected for estimated and actual work. Projects must use the processes to ensure costs are managed to the plan and variations are explainable.

In both planning and managing budgets, the project leader must be aware that any change to either the technical product description or the duration of the schedule affects project cost. Increases in product functionality may have a penalty of both time and cost because of more materials, more labor, and additional costs to the project through overhead charges, for example.

7.11.9   Annotated Bibliography

1. Ward, Sol A., *Cost Engineering for Effective Project Control* (New York, NY: John Wiley & Sons, 1991). This book provides an integrated overview of methods for compiling cost, schedule and quality data. It covers basic estimating and analysis patterns.

2. Bent, James A., and Kenneth K. Humphreys, eds., *Effective Project Management Through Applied Cost and Schedule Control* (New York, NY: Marcel Dekker, 1996). This book is a reference to advanced applied cost and schedule controls for a project. It provides techniques procedures and methods for managing costs in projects.

7.12   PROJECT WORK BREAKDOWN STRUCTURE

7.12.1   Introduction

The work breakdown structure (WBS) is a fundamental consideration in planning, organizing, and executing a project. This family tree division of products, services, and functions defines and graphically displays the elements of a project that are to be managed. Decomposition of the work into manageable components is a major contributor to understanding the project, getting customer buy-in on the scope of the project, and being able to manage the project.

The WBS divides the overall project into work packages representing distinct elements for assignment to someone on the project team or to other stakeholders, such as vendors, functional managers, and contractors.. A work package is a unit of work required to complete a specific project, such as a report, a piece of hardware, an element of' software, or a service which can become the responsibility of' one operating unit within the organization or with stakeholders, such as a supplier or customer. Other characteristics of' work packages include:

- A work package is negotiated and assigned to a specific individual who executes the work to he accomplished or who may be performing as a *work package manager.* Such a manager is held responsible and accountable for the results to be accomplished on the work package.

- A work package has a specific objective to be accomplished at it a designated price on a defined schedule to support the project's purposes.

  Developing a WBS for a project provides the means for the following:

- Summarizing all products and services comprising the project, including support and other tasks
- Displaying the interrelationships of he work packages to each other, to the total project, and to other functional activities in the organization
- Establishing the authority-responsibility matrix organization
- Estimating project costs
- Performing risk analysis
- Scheduling work packages
- Developing information for managing the project
- Providing a basis for controlling the application of resources on the project
- Providing reference points for getting people committed to support of the project

7.12.2 Characteristics of a WBS

The WBS is a means for dividing a project into easily managed increments that helps ensure the completeness and continuity of all work required for successful completion of the project. The project is divided into major groups, and then the major groups are divided again. This process is continued until the project's increments are understandable, manageable, and assignable. The lowest level of decomposition is the task level, which makes up the work package that can be assigned to a single work unit.

One division of a hardware product is defined as follows in a hierarchical arrangement:

- System
- Assemble
- Subassembly
- Component
- Subcomponent
- Part or piece

Each element has a discrete number that is not duplicated again in the project. This numbering system shows both the level of division and hierarchical relationship. For example, an aircraft project might have a top-level WBS similar to the following.

- 1.0 Aircraft (system)
  - 1.1 Fuselage (assembly)
  - 1.2 Tail (assembly)
  - 1.3 Wings (assembly)
  - 1.4 Engines (assembly)

This WBS could be further divided into the subassembly, component, and subcomponent levels. The division might be similar to the following for the engines.

- 1.4 Engines (assembly)
  - 1.4.1 Engine cowling (component)
  - 1.4.2 Engine mounting (component)
  - 1.4.3 Engine fueling (component)
  - 1.4.4 Engine (component)

In the above example, the engine cowling, mounting, and fueling could be manufactured in house and would need further definition of the sub-elements. The engine, however, would typically be contracted for, and the only requirement is that a specification be written for the procurement process. Again, dividing the WBS is needed only to the level of management.

Another way to depict the WBS is through a graphic form as shown in Fig. 7.18.

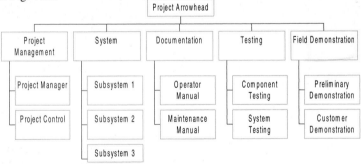

FIGURE 7.18   Work breakdown structure.

7.12.3   Putting Together the WBS

The following suggestions can facilitate the development of the WBS for any project, large or small.

- Creation of the WBS is a project team effort.
- Designers of the WBS should be people experienced in project management
- Designers should have knowledge about the project end items (deliverables) and the potential resources available for the project.
- Look at the project structure as a whole, the end items, and the means to accomplish them.
- Develop the code for the WBS numbering as the project WBS is being designed.
- Assign the code number for identifying the WBS work packages.
- Solicit assistance from project stakeholders, as needed.
- Usually develop the WBS in a top-down manner. However. the identify of individual work packages can be summarized and interrelated into the total project.
- Use brainstorming and focus groups, as required, to identify and position the work packages in the proper place in the WBS.
- Validate the WBS with knowledgeable people beyond the project team.

7.12.4   Maintaining the WBS

The WBS requires little or no maintenance for a stable project. Projects that have "discovered work" or changes to scope may dictate that significant change be made. For example, adding new technical features and functions to a product will typically drive a need for updating the WBS. This update ripples through to the budget and schedule.

Replanning a project may dictate that the entire WBS be reestablished using pieces of the original WBS. As a rule of thumb, any major change to a project should result in updating the WBS. If completion of the project is not apparent and easily envisioned, one should rework the WBS to set the baseline again.

Typically, the WBS is changed only if the customer's requirement changes. A valid WBS should be sufficient to complete the project with only minor changes. Adding elements poses a problem because of the hierarchical nature of the coding scheme. However, reducing work can be easily accommodated by merely an annotation that the element is no longer part of the project.

7.12.5   Key User Questions

1. Do the members of the project team understand how a WBS is developed and the purposes to which a WBS can be put?

2. If needed, have "brainstorming" techniques been used in the development of' the WBS?

3. Have the work packages coming out of the development of the WBS been evaluated as to the resources required, methods of obtaining resources, and the authority and responsibility assignments needed to translate those resources into the project design, development, and construction (production)?

4. Has the project team worked with relevant project stakeholders in the development of the WBS?

5. Has adequate testing of the WBS been done to help ensure its credibility and reliability?

7.12.6   Summary

Development of the WBS is a disciplined, systematic methodology to identify the work packages involved and show their relationship to other work packages and to the project as a totality. The many purposes which the WBS serves were expressed in this section. The work packages of the WBS provide the key focal point around which the management functions of planning, organizing, and control are carried out on the project. The WBS is an indispensable link between the project objectives and goals and its outcome

7.12.7   Annotated Bibliography

1. Warner, Paul J., and Paul M. Caesar, "Putting Together a Work Breakdown Structure," in David I. Cleland, ed., *Field Guide to Project Management*, 2nd ed (New York, NY: John Wiley & Sons, 2004). This chapter provides key information on how and why a WBS should be developed. The authors conclude quite properly that the WBS is a highly valuable management tool in delivering project success.

2. Milosevic, Dragan Z., *Project Management Toolbox* (New York, NY: John Wiley & Sons, 2003), pp. 152–166. The author provides a basic explanation of the WBS to include such topics as a top-down and bottom-up approach to constructing the WBS. He then suggests that advantages and disadvantages of using a WBS. Finally, in his summary, he claims that the WBS is often viewed as the single most important element in project management.

7.13   EARNED VALUE MANAGEMENT

7.13.1   Introduction

The concept of earned value management system (EVMS) has been in use for 40 years in the US Government, but under different names. It has been used, tested for validity, and improved through application to major contracts awarded by Department of Defense elements, Department of Transportation, General Services Administration, and other agencies. Government contractors have used EVMS to measure performance parameters and report them to the government agencies.

Some criticism of the EVMS has been leveled because individuals thought that the measure of historical data was not representative of the future. One individual commented, "It is like driving by looking in your rearview mirror." Despite any criticism, the earned value management system is the best method for managing large, complex contracts.

EVMS criteria span all business areas

The most recent version of the EVMS has evolved from many years of experience—both good experiences and bad experiences. One of the major contributors to the success of the system is that it forces performing contractors to plan projects through with rigor and accuracy. To be certified in the earned value management process, one must meet 32 criteria that span all business areas.

The fundamentals of the EVMS are as follows:

- Must plan all work from project beginning to end
- Must integrate the total scope of the work, schedule for the project, and the cost of the project into a baseline plan for measuring project progress
- Must be able to assess progress objectively at the work performance level
- Must be able to analyze and assess variances from the baseline plan and be able to forecast the potential impacts of variances
- Must provide the same information to all levels of management to permit decision-making and direction of management actions

7.13.2   Application of EVMS

EVMS is a tool for large projects that require extensive and detailed planning with the requisite control to avoid project overruns and schedule slippages. Small projects with visibility into the processes do not justify the time and expense of detailed planning. Further, variances from the anticipated performance parameters on small projects do not have the negative impact on the organization like those in major projects.

Variances from the plan for large projects of, say, 15 percent cost may make a difference between whether an organization stays in business or fails. Small projects, on the other hand, may have a 15 percent variance and the negative impact could be a routine matter. For example, a 10 percent negative variance on a large project of $1.5 billion could represent a loss of $150 million, whereas a small project of $500,000 would only have a loss of $50,000.

EVMS applied to large, complex projects can give indications when future variances are significant and the trend is adverse.

EVMS also provides a means to measure progress on a project for such activities as making incremental payments to contractors. Payments are often made on earned value for contract performance, less a small percentage. These progress payments provide a revenue stream for contracts that is less than the actual contract value at a point in time and reduces the need to borrow working capital.

7.13.3   EVMS Planning

The EVMS process is critical to the success of establishing and operating a project performance measurement system. There are no shortcuts, and skipping items in the process can fatally flaw the information generated through EVMS. Rigorous application of the principles and practices is needed to implement the tool for any large project.

- *Statement of work* (*SOW*). It is fundamental to the process that the project has a complete SOW to define all the work. The SOW, which describes the work scope, may be augmented by a technical specification for a product. The specification defines the technical parameters, such as form, function, and fit, that a product will possess when completed. The combination of the SOW and Specification forms the baseline parameters for the project's work.

- *Work breakdown structure* (*WBS*). A WBS is typically used to decompose the project's product and the accompanying functions into work packages. The WBS serves as a tool to define the lower echelon elements of a project—product, services, functions, and data—as work packages that can be assigned to individuals or team for accomplishment as shown in Fig. 7.19.

- *WBS dictionary*. The WBS dictionary is developed to describe all levels of the project as it is decomposed. The purpose of this document is to

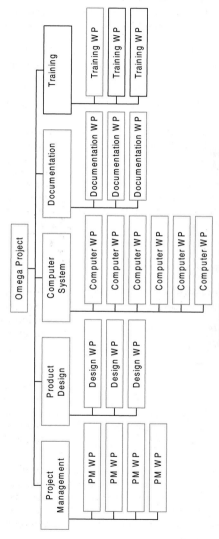

FIGURE 7.19 Work breakdown structure.

ensure complete communication between the contracting agency, the performing contractor and internal to the performing contractor's organization. Definitions of the work should rely on standard descriptions where possible, but may have unique descriptions for new or complex items.

- *Project schedule.* A project schedule is used to lay out the work activities over time and to sequence activities in the order in which they are to be accomplished. Realistic scheduling is based on several items:

  - Complete understanding of the scope of work
  - Accurate decomposition of work to task or activity level
  - Tentative assignment of qualified resources to task/activities
  - Proper sequencing of tasks/activities in schedule
  - Adequate time to perform tasks/activities
  - Accurate "facts" and "assumptions" to bridge information gaps

- Project budget. A budget is a time-phased expenditure plan that has estimates of the cost for each item in the WBS. The validity of the budget resides in the cost estimating process and accurate prices for labor, materials, and functions. Long-duration projects may also have escalated prices of work to account for inflation or other factors changing prices. Considerations for the budget would be the following:

  - All work has been identified.
  - Estimates of cost have a realistic basis.
  - Estimates of cost include all burdens (i.e., overhead costs, profit/fees, and other indirect expenses).

Realistic and accurate planning are prerequisites for any project. Lack or shortfall in project planning will be apparent when implementing and exercising an EVMS. An example of the sensitivity of the EVMS to errors occurred several years ago when a contractor showed a negative slope on a cumulative cost curve. The contractor assumed earned value for work accomplished, but the work was overpriced by nearly 10 percent. When the error was found, the easy corrective measure was to reduce the earned value by the cumulative overcharge. This reduction in earned value was represented on the charts by a negative slope.

> **There is no substitute for realistic and accurate project planning**

7.13.4 EVMS Process

EVMS uses existing systems within an organization to plan, implement, control, and correct project activities. A change to the existing systems is required only when there is a shortfall in the capability to perform at a given level or there is no current system.

EVMS uses a WBS to define the total project's work and links that with the performing element of the organization. The intersection between the integrated organizational breakdown structure and the WBS positively links work to organization. This provides a high degree of assurance that the performing work element understands and assumes responsibility for accomplishing work packages. The work to be accomplished is defined in the schedule and the price of the work is defined in the budget as shown in Fig. 7.20.

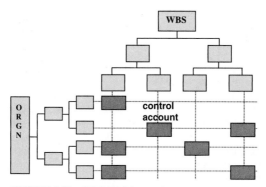

FIGURE 7.20   WBS-OBS integration.

The baseline for earned value is established by scheduling the work packages over time for accomplishment and creating a cumulative cost curve over the same period of time that assumes all work is completed on time. This concept is demonstrated in Fig. 7.21, which includes a baseline performance curve called the BCWS, or budgeted cost of work scheduled. All work accomplishment is measured against the BCWS.

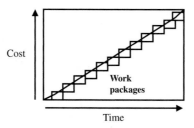

FIGURE 7.21   Basic EVMS chart.

As work is performed and valued earned, Fig. 7.22 takes on more information. ACWP, or actual cost of work performed, is the performance line that reflects any variance on cost, such as an over-expenditure to perform the amount of work completed. Variances can occur when the budgeting process is flawed, i.e., the accuracy of the cost estimate is not realistic or the prices of resources have changed.

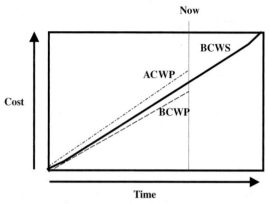

FIGURE 7.22   EVMS graph with plotted data.

BCWP, or budgeted cost of work performed, is the work complete with its price from the budget. Variances occur in the BCWP when the planned work packages are not completed or more work packages are completed. BCWP is often called "earned value" because it depicts the real amount of work accomplished and the performance toward completing the project.

There are two basic performance measures used in the EVMS—schedule variance (SV) and cost variance (CV). These fundamental measures provide the basis for identifying trends and measuring progress.

- *Schedule variance* is the difference between BCWS and BCWP. When the BCWP is depicted below BCWS, performance is less than planned. When the BCWP is above the BCWS line, performance is better than planned.
- *Cost variance* is the difference between ACWP and BCWP, the two measures related to actual performance. BCWP represents what the work should have cost and ACWP represents what the work actually cost.

It is assumed that completed work packages meet the technical specifications. EVMS does not measure technical performance, which must be validated through other means. Therefore, schedule and cost variances have little meaning unless the completed work is technically sufficient.

> **Technical performance must be validated by processes outside of EVMS**

7.13.5   How Much Variance is Tolerable?

Cost and schedule variances are relative to the project and the owning organization. High-risk research and development projects may have a high tolerance for variances but be dependent upon the degree of technical progress or technical potential. Projects that use mature technology and have a low degree of risk may have a small margin for variances—primarily in the cost area.

Experience with projects shows that the government has different metrics for variances on project. One example was that a government agency did not become concerned about cost variance until the amount exceeded 25 percent—and a new baseline was established each year for the project. Contractors enjoyed very favorable practices, where cost overruns could be as much as several millions of dollars and the government agency paid for all extra costs.

Another example is from the US Department of the Navy, where cost overruns were often sources of frustration. An upper limit of 8 percent variance was established as the critical point to trigger a review. This upper limit was for the duration of the contract rather than on an annual basis.

Changes to the scope of the project and contract are often the cause of variances. Take, for example, the case where a technical solution is failing and a new one is required. The time and money consumed on the original solution may be lost and an adjusted schedule needed to fit in the new work. Also, when there is "discovered" work that was not identified in the original plan, this needs to be considered in applying contingency reserve funds.

Legitimate changes to the scope through such actions as customer requirements growth or additional features that were not included in the original SOW will cause a change to the baseline. Typically, new work to the scope will cause a shift in the baseline (BCWS) to recognize the new effort and funds required.

7.13.6 Key User Questions

1. What is the most appropriate tool to use to divide a project into discrete parts that can be scheduled and a price assigned for budgeting purposes? Why?

2. What do you consider to be the most critical component of the EVMS process and why?

3. If there is a change to the scope of work to reduce the amount of work, how can the EVMS accommodate such change?

4. Does your organization use EVMS, and what is the typical project size and complexity for which EVMS is used?

5. Does the planning in your organization have the rigor required to implement an EVMS successfully?

7.13.7 Summary

This brief look at earned value and the requirements of the system highlights important considerations for use of this integrated performance measurement approach. To achieve the most benefit from EVMS, an organization must ensure that the planning process is adequate to support development of accurate, detailed plans. Implementation of an EVMS also needs to recognize the need to accommodate changes to project scope to include required adjustments to both schedule and cost parameters.

EVMS does not measure technical performance accomplishment, but relies on other means within the organization to ensure product quality and integrity. This is typically done through a process that validates completed product work packages—or the test, demonstration, and assembly of components.

EVMS is the best performance measurement system known for large, complex projects and promotes understanding of the entire project through decomposition, detailed planning, and cost-schedule integration. Linking the business aspects, cost, and schedule in an integrated fashion permits review and correction of variances in an incremental way.

7.13.8 Annotated Bibliography

1. Brandon, Daniel M., Jr., "Implementing Earned Value Easily and Effectively," *Project Management Journal*, June 1998, vol. 29, issue 2. Brandon gives a succinct description of the earned value system and some of the pitfalls. He provides suggestions on how to implement the system and defines the acronyms that are specific to the process.

2. Fleming, Quentin W., and Joel M. Koppelman, *Earned Value Project Management,* 2nd ed. (Newtown Square, PA: Project Management Institute, 2000), 224 pages. Fleming and Koppelman describe the earned value process and give meaning to the benefits of implementing earned value management. Both authors have significant experience in earned value management, and they provide a concise forum that lends itself to promoting understanding of otherwise complex formulas.

SECTION 8
THE PROJECT CULTURE

8.1 UNDERSTANDING THE TEAM CULTURE

8.1.1 Introduction

The project culture is also discussed in a different context in Section 2.8.

Culture is the set of refined behaviors that people have and strive toward in their society. In this section culture is defined as the synergistic set of shared ideas and beliefs that is associated with a way of life in the team-driven enterprise. Some of the likely key cultural features to be found in an organization that uses teams comes from:

• The management leadership-and-follower style practiced by key managers and professionals

• The example set by leaders of the organization

• The attitudes displayed and communicated by key managers in their leadership and management of the organization

• The assumptions held and communicated by key managers and professionals

• The organizational plans, policies, procedures, rules, and strategies

• The political, legal, social, technological, and economic systems with which the members of an organization interface

- The perceived and/or actual performance characteristics of the organization
- The quality and quantity of the resources (human and nonhuman) consumed in the pursuit of the organization's mission, objectives, goals, and strategies
- The knowledge, skills, and experiences of members of the organization
- Communication patterns
- Formal and informal roles. (Paraphrased from David I. Cleland, *Strategic Management of Teams* (New York, NY: John Wiley & Sons, 1996), p. 100.)

The project culture is essentially a way of life in the project environment

---

8.1.2 Cultural Strength

A strong working culture is like magic. The organization believes and practices a philosophy of creativity and innovation, facilitated by a participative leader and follower style. In addition, the following characteristics tend to exist:

- As employees serve on teams, they welcome the opportunity to participate in influencing the organization's purposes.
- Employees find that their opinions are valued and they feel as if they are being treated as thinking adults and key organizational members.
- A strong feeling of interdependent relationships emerges among people from different functional entities and organizational levels.
- An enlightened feeling of the acceptance of change in the organization develops. Several strategies are required to enhance the team focus in the management of change in the organization. These strategies include:
  - Enhancement of the organizational culture so that people at all levels, and in all specialties are encouraged to bring forth ideas for improvement in their areas of responsibilities.
  - Development of an organizational culture that seeks to abandon that which has been successful through the continuous improvement of existing products, services, and processes.
  - Become a learning organization through explicit recognition that all organizational members will have to retrain and relearn new technologies and professional strategies to escape obsolescence.

- Organizing the organization's resources so that explicit opportunity is available to bring a team organizational focus to the development and implementation of new organizational initiatives, that will bring forth new products, services, and processes.
- Providing a strategic management capability by which organizational leadership is proactive in providing the resources, vision, and disciplines to manage the future through the use of product and process projects. (Paraphrased from David I. Cleland, *Strategic Management of Teams* (New York, NY: Wiley, 1996), p. 102)

When the managers of an organization elect to use alternative teams in its operational and strategic strategy, significant cultural changes will ultimately come about. Empirical, practical, experience-based evidence suggests that when alternative project teams are used the before-teams and after-teams cultural characteristics look like those described in Tables 8.1 and 8.2, respectively. There are more characteristics described in Table 8.2 than in Table 8.1. The current fascination of using teams described in the literature probably accounts for this.

8.1.3 Maintenance of a Team-Driven Culture

Certain actions can help develop and maintain such a culture.

- Managers and team leaders must design and implement an ongoing, disciplined approach in planning, organizing, and controlling the team management system for the enterprise. The Project Management Systems described in Section 7.10 is an excellent model to use in this respect.
- Teams should be regularly reviewed to determine their progress, the effectiveness of the team as a contributing organizational unit, and how the cultural ambience of the team is developing and melded with the culture of the enterprise.
- Adequate descriptions of authority and responsibility should be developed and be maintained so that team members understand their individual and collective roles.
- Team members should be given ownership in the decision affecting their team.
- Team leaders should use as many strategies as possible such as brainstorming and other participatory means to get the members involved.
- Feedback must be provided on a regular basis to the team members.
- Team leaders must ensure that adequate resources are provided to the team so they can get their job done.

TABLE 8.1 Cultural Characteristics Before Teams

There are formal rules and procedures to govern individual behavior.
There is hierarchical managerial authority.
There are narrow definitions of work responsibility.
There is a bureaucratic culture.
Change comes about through top-level-directed programs
There are more organizational levels.
Leadership comes out of the organizational hierarchy.
Reorganizations come from the top.
Individual efficiency and effectiveness are paramount.
Authority and responsibility flow within the hierarchy.
There is a command-and-control mentality.
There is considerable risk of over-managing.
Organizational and system boundaries are preserved.
There is individual responsibility for decisions.
The organization is non-team oriented.
There is individual thinking.
There is blurred organizational identification.
Individuals are reluctant to assume additional responsibility.
Individuals identify primarily with individual goals.
People follow their leader.
People feel a limited degree of involvement.
People are given responsibility.
People are managed.
Managers assume responsibility for execution of the management functions
    (Planning, organizing, motivation, leadership, and control).
Managers assume responsibility for quality.
Managers are responsible for the performance of their subordinates.
Managers make the decisions.
People are reluctant to seek additional responsibility.
Managers supervise.
There is limited participation by people in the affairs of the organization that
    affect them.
The titles of superiors and subordinates reflect the culture.
There is limited sharing of organizational results.
Rewards are based on individual performance.
Problems are owned primarily by the managers.
People tend to withhold their opinions until the manager gives his or her opinion.
Responsibility for strategic decisions rests primarily with the managers.
Project management is viewed as a special case of management.
People are reluctant to change because they usually have little participation in the
    development of the rationale for the change, the selection of the strategies to
    bring about the change, or in the execution of change strategies.

Source: David I. Cleland, *Strategic Management of Teams* (New York, NY: John Wiley & Sons, 1996), p. l06.

TABLE 8.2 Cultural Characteristics After Teams

There is systems thinking.

There are blurred organizational boundaries.

There are formal and explicit interdependencies.

There are closer relationships between customers and suppliers.

There are changes in the implicit contract about what employers and employees owe to each other.

External environmental considerations (home, family, school) are put on the organization's agenda.

Problems and opportunities are seen as systems related to larger systems.

Project management is seen as an element of strategy to deal with and to facilitate change in the organization.

Organizational people are working at thinking together.

Ad hoc and other forms of teams are becoming commonplace in the strategic and operational management of the organization.

The gap between organizational potential and performance is being reduced.

People accept, even insist, on greater responsibility for their own work, for their organization as a whole.

People are truly involved in the business of the organization.

People participate in the management of the organization: the design and execution of the planning, motivation, leadership, and control of the organization.

Managers become facilitators and coaches.

People manage themselves.

Organizational managers manage the context; they provide and allocate resources; and design (with the help of the teams) and implement the systems.

Managers (with the help of the teams) provide the required organizational design.

People assume responsibility for quality.

People are responsible for their own performance.

Teams work with the customers and suppliers.

It is recognized that the people doing the job know the most about how the job should be carried out.

Teams evaluate the performance of the team as a whole and of the team members.

Team members counsel team members on individual performance: the team may recommend the release of those individuals who do not perform at the standards expected of their team.

Organizational decisions, results, and rewards are shared.

Teams take over some of the functions performed by individual managers in the hierarchy.

The number of orrganizational layers is reduced.

Teams assume the supervisor functions, and those jobs change considerably or disappear.

People perform a wide range of work in their jobs.

There is a greater sharing of information about the organization, its problems, opportunities, successes, and failures

TABLE 8.2   Cultural Characteristics After Teams (*Continued*)

Senior managers recognize their limited power in mandating organizational change and renewal from the top. They encourage a non-directive change process.

Managers work at creating a culture for change in the organization, and a culture that supports the operation of different kinds of teams.

The main thrust in change gains momentum at the periphery of the organization and spreads to the core of the organization.

Change is viewed as a learning process for all of the organizational members.

Coordination, information sharing, and teamwork become critical to making the organization function more efficiently and effectively.

Commitment, loyalty, trust, and confidence become clear characteristics of the culture.

Willingness to seek new knowledge, develop new skills, and change attitudes becomes socially acceptable in the organization.

A shared vision for the organization and its purposes and an appreciation of the importance of competitiveness become apparent to organizational members.

Words like supervisor, employee, and subordinate tend to fall into disuse; everyone is called associate, member, coordinator, or coach.

Promotions are based on results as well as the learning, teaching, coaching, and facilitating role of people.

People are committed to the organization, to individual and organizational goals, to members of the peer group, and to the continued improvement of the organization.

A permanent change in the way the organization is run has been undertaken.

People like being able to influence the affairs of the organization to which they belong.

People's expectations are driven by their own motivations and by how they see themselves fitting into the vision shared by organizational members.

Learning (by everyone in the organization) takes on new importance.

Mangement's job is to promote the vision, provide the conditions for people to share in the vision, facilitate the use of resources in the organization, and build a culture in which the individual's and the organization's best interest are served.

Change is seen not as a goal but as an endless journey from the present to the future.

Status in the organization becomes less dependent on the organizational role held and more on the results one is able to accomplish, both individually and as a member of the organizational teams.

Senior managers are concerned with the strategic management of the organization in the sense of managing and orchestrating change so that values are created for the organization and its stakeholders that did not previously exist. The oversight role of the senior managers during the change process (which is continuous) becomes a critical responsibility.

TABLE 8.2 Cultural Characteristics After Teams (*Continued*)

Change and organizational renewal come about without imposing them.

Members of the organization become more aware of the competitive pressures facing the organization.

Knowledge of the competition dictates high standards for individual and organizational performance.

Managers alter fundamentally the way of doing things—there is less chance that they will over-manage the organization.

Authority through the use of knowledge, interpersonal skills, leadership capabilities, building and maintaining alliances, and expertise—totaled in the ability to influence people—becomes more important than the authority of the formal role that an individual holds in the organization.

Leadership becomes a criterion for promotion at all levels in the organization. People's careers are directed to encourage further development of their potential leadership skills.

The organizational culture tends to facilitate creativity, motivation, prroductivity, and quality.

Consequently, there is an enhanced change for continuous improvement in the way in which the organization does business.

Participative management and consensus decision making become key characteristics of the organizational culture.

Distinctions between managers and workers become blurred.

People have more fun at work.

Policies, procedures, and rules still exist, but people have a greater understanding and respect for such guidelines to influence behavior.

There is a discuss-and-decide mentality.

People feel involved.

Everyone participates, at some level, in the organizational decision processes.

People do not wait for problems and opportunities to be assigned to them; they look for problems and opportunities that can be undertaken either individually or through an organizational team.

Leadership for change rests with everyone in the organization.

There is less resistance to change in the organization. Because people play vital roles in discovering and bringing about the needed change, they find that the change is less threatening.

People become more outspoken, questioning the existing order of things, questioning decisions and the right of people to make decisions.

More people understand the management processes, develop improved interpersonal skills, and see the larger systems context off problems and opportunities; individuals develop empathy for the people who have responsibility for improving organizational performance.

Source: David I. Cleland, *Strategic Management of Teams* (New York, NY: John Wiley & Sons, 1996), pp. 107–109.

- Managers and team leaders must recognize the key people-related cultural factors and utilize them. These factors include:
  - Rewarding useful ideas
  - Encouraging candid expression of ideas
  - Promptly following up on team and member concerns
  - Assisting in idea development
  - Accepting different ideas—listening to that team member who is marching to a different drummer
  - Encouraging risk taking
  - Providing opportunities for professional growth and broadening experiences on the project
  - Encouraging interaction with the project stakeholders so that there is an appreciation by the team members of the project's breadth and depth. (Paraphrased from David I. Cleland, *Strategic Management of Teams* (New York, NY: John Wiley & Sons, 1996), p. 114.)

8.1.4   Key User Questions

1. Do the members of the organization and the teams understand the nature of an organizational "culture" and how that culture can impact the behavior and effectiveness of the teams?
2. Has the organizational documentation (plans, policies, procedures, protocols, etc.,) been evaluated to determine what influence such documentation has had on the organizational culture?
3. Do the team members understand and appreciate the importance of the team culture in the overall efficiency and effectiveness of the organization?
4. What features (such as leader-and-follower style) have influenced the culture of the organization and the teams?
5. In a comparison of "successful" and "unsuccessful" project in the organization, has the organizational and team culture had any impact on the outcome of the projects?

8.1.5   Summary

In this section, a culture was defined as the set of refined behaviors that people have in the society to which they belong, be it a nation, a family, an organization, or a team. More specifically, a team culture is the environment of beliefs, customs, knowledge, protocols, and behavior patterns in a team. It was suggested in the section that the culture of the organization and of the teams should be mutually supportive.

8.1.6 Annotated Bibliography

1. Cleland, David I., *Strategic Management of Teams* (New York, NY: John Wiley & Sons, 1996), chap. 5, "Team Culture." This chapter puts forth the idea of a culture, how a culture is supported in an enterprise, and how the enterprise culture impacts the culture of the teams operating within the enterprise.

2. Cleland, David I. and Lewis R. Ireland, *Project Management: Strategic Design and Implementation*, 4th ed. (New York, NY: McGraw-Hill, 2002), chap. 20, "Cultural Considerations in Project Management." This chapter considers the use of project teams in maintaining continuous improvement in the management of an organization. Insight is provided into the cultural ambience in which projects are best managed.

8.2 POSITIVE AND NEGATIVE ASPECTS OF TEAMS

8.2.1 Introduction

In this section, both the positive and negative results of using teams will be presented. In Table 8.3 the positive results are shown:

The positive results are basically of two types: (1) specific and measurable direct accomplishments such as improvements in productivity, quality, reduction in cost, and so forth; and (2) indirect results such as

TABLE 8.3   Team Positive Results

- Productivity increases
- Quality improvements
- Cost reductions
- Earlier commercialization
- Improved supplier relationships
- Enhanced customer satisfaction
- Employee satisfaction
- Greater creativity and innovation
- Improved stakeholder image
- After-sales service improvements
- Development of leadership/management potential
- Improved product, service, and process development
- Ability of teams to make and execute management decisions

Source: Paraphrased from David I. Cleland, *Strategic Management of Teams* (New York, NY: John Wiley & Sons, 1996), pp. 273–274.

increased employee satisfaction, enhanced culture, and emergence of innovation and creative behavior. Overall, the positive results of using teams provide clear evidence of the wisdom of using teams as a primary organizational design alternative to support operational and strategic purposes in the enterprise. However, teams can have their problems.

> **Both positive and negative results
> arise from the use of teams**

8.2.2   The Negative Side of Teams

Project teams and alternative teams are not a panacea. There is no question that teams can produce positive and beneficial results for the organization, but there is a cost associated with the use of teams. Most of the negatives that are associated with teams can be identified and credited to "failures" on the part of responsible people in the organization, such as:

- Inadequate delegation of authority and responsibility compounded by the ambiguities that people sense in their individual roles in the organization.
- Changed relationships with "subordinates" and with peers because of the lack of clarification of what the new authority and responsibilities will be in the organization.
- Acceptance of the concept of teams but becoming disenchanted when the teams are appointed and begin working.
- In the case of a unionized company, teams are usually perceived as negative because the risk arises of the members having more fidelity to the team with less fidelity to the union.
- Belief that teams undercut the traditional manager's roles in the sense that such managers no longer have "command and control" management authority.
- A general devaluation of the role played by managers and supervisors and the sense that their status has been reduced in the organization.
- More emphasis on interpersonal skills, which can be threatening to a manager who has limited skills.
- A failure to understand the need for empathy among the managers of the organization.

- Concern over how the merit evaluation and p
  by members of the teams.

   If the following planning failures are audite
from the teams can be expected.

8.2.3   Planning Failures

- Managers take a hands-off approach, "order" the use of teams, and then leave the teams to their own destinies. This will result in a waste of everyone's time.
- Managers do not hold the teams responsible and accountable, and they avoid regular reviews of the team's progress. If team deadlines are missed, or if their work is not of high quality, managers do nothing.
- Management fails to devise a strategy for handling conflicts among team members.
- Management tells the teams that they have all of the authority needed to do their job, but fail to designate the specificity of that authority through appropriate documentation.
- Management fails to make the objectives and goals of the team clear, or fails to instruct the team to study probable and possible objectives and goals, clarify them, and seek management review of these.
- Management withholds or provides resources as a means of rewarding or punishing the team.

8.2.4   The Costs of Teams

Using teams is not without its cost. Some of the costs that are incurred include the following:

- Maintenance costs for keeping the knowledge, skills, and attitudes of the team members current.
- Training and education costs for the team members' attendance at training and education sessions. When a team member is out for training, someone on the team has to pick up the absent member's duties.
- Costs of potential interference with the creative and innovative skills of the loner.
- Costs of management's role, which is not diminished, but rather is changed as the responsibilities of facilitator, counselor, teacher, coach, and strategist take on new meaning and duties in working with teams.

, of going from negative to positive attitudes about teams.

's of properly training people to work in teams.

Managerial costs to develop an understanding of how managerial roles have changed vis-a-vis teams.

- Costs of changing reporting authority-responsibility relationships. A manager may no longer have approval authority held previously under a command-and-control culture. For example, a concurrent engineering team will conduct an ongoing design review through the workings of the team. Having an engineering manager "sign off" on the design becomes superfluous considering that an ongoing design review has been carried out by the team.

- Costs of making decisions using teams. However, even though team decision making is costlier, the decisions that are made are more thoroughly analyzed and evaluated. More people are involved in the decision process, and there are opportunities for more in-depth evaluation.

- Costs of non-support. Any organization tends to have a base of institutional knowledge and memory. If the teams are not able to access this institutional know-how, their ability to produce meaningful results can be hampered.

8.2.5 Key Ingredients of Teams

The conditions cited in Table 8.4 must exist in teams:

TABLE 8.4 Key Ingredients of Teams

- An atmosphere where trust is given.
- Team members are devoted and support each other on the team.
- There is a strong conviction that teams are the way to go.
- People are committed to the team's work and the people in the team.

8.2.6 Key User Questions

1. Have both the benefits and costs of using teams been considered in the decision to use alternative teams as key elements of organizational design?

2. In planning for the use of teams, has care been taken to avoid some planning failures that can occur in preparing for the use of teams?

3. Has consideration been given to how the merit evaluation and pay issues will be handled?

4. Are the team members loyal and committed to each other?

5. What are the major issues in the organization that might militate against the successful use of teams?

8.2.7 Summary

In considering the use of teams, managers should remember that the following considerations and conditions are known. First, each organization is different, as is each organization's culture. Careful analysis of the culture of the organization can provide insight into some of the likely negatives and the problems to be encountered in the use of teams. Second, teams are not the end; rather, they are a means for focusing the use of resources to deliver value to the oganization and its stakeholders. There are obvious as well as subtle costs to using teams. Appreciating these costs can be useful in developing and executing a meaningful and successful strategy for the use of teams.

8.2.8 Annotated Bibliography

1. Cleland, David I., *Strategic Management of Teams* (New York, NY: John Wiley & Sons, 1996), chap. 13, "The Negative Side of Teams," and 14, "Team Results." These chapters provide insight into some of the impacts that teams can have on the organizations and its stakeholders. When teams are used, there can be both positive and negative results. Positive results reflect excellent management. Negative results can usually be associated with poor management processes and practices.

2. Spiegel, Jerry, and Cresencio Torres, *Manager's Official Guide to Team Working* (San Diego, CA: Pfeiffer & Company, 1994). This book is written for managers, supervisors, and team leaders who find themselves responsible for developing effective work groups or teams. It provides development process. Some of the important subjects discussed are team formation and assessment, communication, team development, roles, decision-making, resolution of team conflict, problem-solving, and the politics of team work.

8.3   MOTIVATING THE PROJECT TEAM

8.3.1   Introduction

Motivation is a system of forces and relationships, originating both within and outside the individual, that influences behavior. It causes people to behave in certain ways. Motivation flows from a need that people have—something that they want from the organization and the project team. It is a major field of study in management thought and theory.

If you dig deeply into any organizational problem, there is a high probability that you will come to "people" problems. In the management of projects, the matter of motivation takes on special significance since the project manager, and the project team, have to deal in some way with the challenges of motivating many individuals—the stakeholders, over whom they have limited de jure or legal authority. In these circumstances, the interpersonal skills of the members of the project team take on special importance. In the material that follows, some of the basic ideas about motivation are presented.

The project manager must always be concerned about what motivates the project team

---

8.3.2   Maslow

Maslow's *Needs Hierarchy* is an hypothesis that people have five sets of needs as depicted in Fig. 8.1. In this figure, the needs of physiological well-being and the satisfaction of safety and security are called the *primary* needs of people. The third, fourth, and fifth levels are called *secondary* needs.

A brief explanation of these needs follows:

Physiological needs are at the bottom of the hierarchy. Satisfaction of these needs, food, water, and sufficient shelter, is essential to living.

Safety and Security needs include protection from the elements and harmful environments, from threats to one's life and well-being, and freedom from arbitrary and capricious management actions.

Belonging and Social Activity means that most people cannot live by bread alone. There is satisfaction of social belonging, affection, mem-

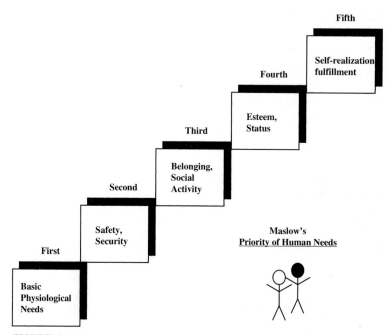

FIGURE 8.1 Maslow's order of priority of human needs. (Adapted with permission from Keith Davis, *Human Relations at Work,* 3rd ed. (New York, NY: McGraw-Hill, 1967.)

bership, or affiliation. Accepted and full membership in a family unit is important. Membership and being accepted on a project team are important as well.

Esteem and Status needs motivate people to not only seek affiliation but to become active in influencing the culture in the organization to which they belong. They enjoy belonging to a group because of the acceptance they feel, and the satisfaction they gain by contributing to that group.

Self-realization and Fulfillment needs, at the top of the hierarchy of needs, explains an individual's drive for achievement, creativity, and self-realization. It means that individuals have become what they want to be, and in part explains why people that have had much wealth and many honors continue to work hard at their profession.

The characteristics of these secondary needs include:

- They are strongly conditioned by experience.
- They vary in type and intensity among people.
- They change within any individual.
- They are often hidden from conscious recognition.
- They are nebulous feelings instead of tangible physical needs.
- They influence behavior.

Maslow ranks as the premier innovator in developing and putting forth his seminal theory of the Hierarchy of Needs. He was followed by another key innovator in the field of human behavior. Douglas McGregor postulated the concept of Theory X and Theory Y in his explanation of why people behave the way they do. He proposed that managers tend to hold a Theory X or Theory Y set of assumptions about employees.

Theory X takes a pessimistic view of human nature in their work:

- The average human being has an inherent dislike of work and will avoid it if he can—*PEOPLE ARE LAZY*
- Because of this human characteristic of dislike of work, most people must be coerced, controlled, directed, threatened with punishment to get them to put forth adequate effort toward the achievement of organizational objectives—*MOTIVATION IS ACCOMPLISHED THROUGH FEAR OF PUNISHMENT.*
- The average human being prefers to be directed, wishes to avoid responsibility, has relatively little ambition, wants security above all—*PEOPLE ARE NO GOOD.*

On the other hand, Theory Y sees human nature in the place of work, in a more positive light. It assumes:

- The expenditure of physical and mental effort in work is as natural as play—*WORK IS A NATURAL ACTIVITY.*
- External control and the threat of punishment are not the only means for bringing about effort toward organizational objectives. Man will exercise self-direction and self-control in the service of objectives to which he is committed—*PEOPLE CAN MANAGE THEMSELVES.*
- Commitment to objectives is a function of the rewards associated with their achievement—*MOTIVATE THROUGH POSITIVE REWARDS.*

- The average human being learns, under proper conditions, not only to accept but to seek responsibility—*PEOPLE ARE BASICALLY RESPONSIBLE.*

- The capacity to exercise a relatively high degree of imagination, ingenuity, and creativity in the solution of organizational problems is widely, not narrowly, distributed in the population—*EVERYONE HAS THE CAPACITY TO BE CREATIVE AND INNOVATIVE.*

- Under the conditions of modern industrial life, the intellectual potentials of the average human being are only partially utilized—*CHALLENGE YOUR PEOPLE.*

- Implications of Theory X and Y
  People respond as they are treated.
  Managers are responsible for the conduct of their people.
  Participation and cooperation are critical.
  Managers must be sensitive to needs of people.
  Work can be a source of personal satisfaction.
  State of human relations depends on quality of the leadership, and the management philosophies that are used.
  Importance of interpersonal skills in the success of management.

Notwithstanding the abundance of literature on the subject of human behavior and motivation, there are a few basic ideas that the people managing projects should recognize, and practice. Maslow's Hierarchy of Needs and McGregor's Theory X and Theory Y are two of these basic ideas. Another deals with the idea of job motivational factors.

8.3.3   Job Motivational Factors

In Table 8.5, a questionnaire is shown which can be used to gain insight into what motivates the team members, or any other stakeholders. (Cleland and Kocaoglu, *Engineering Management* (New York, NY: McGraw-Hill, 1981), p. 101.)

Administering this questionnaire to individuals can provide insight into their attitudes and the conditions under which they are best motivated. The authors have administrated this questionnaire to several thousand technical people. A pattern of typical responses from these people is reflected in Table 8.6 (*Ibid.*, Cleland and Kocaoglu, p. 102).

A review of the responses to this questionnaire indicates that good pay is important.

TABLE 8.5   Job Motivational Factors

Please indicate the five items from the list which you believe are the most important in motivating you to do your best work.

_____ 1. Steady employment
_____ 2. Respect for me as a person
_____ 3. Adequate rest periods or coffee breaks
_____ 4. Good pay
_____ 5. Good physical working conditions
_____ 6. Chance to turn out quality work
_____ 7. Getting along well with others on the job
_____ 8. Having a local house organ, employee paper, bulletin
_____ 9. Chances for promotion
_____10. Opportunity to do interesting work
_____11. Pensions and other security benefits
_____12. Having employee services such as office, recreational and social activities
_____13. Not having to work too hard
_____14. Knowing what is going on in the organization
_____15. Feeling my job is important
_____16. Having an employee council
_____17. Having a written description of the duties in my job
_____18. Being told by my boss when I do a good job
_____19. Getting a performance rating, so I know how I stand
_____20. Attending staff meetings
_____21. Agreement with agency's objectives
_____22. Large amount of freedom on the job
_____23. Opportunity for self-development and improvement
_____24. Chance to work not under direct or close supervision
_____25. Having an efficient supervisor
_____26. Fair vacation arrangements
_____27. Unique contributions
_____28. Recognition by peers
_____29. Personal satisfaction

Source: Adapted from David I. Cleland and William R. King, *Management: A Systems Approach* (New York, NY: McGraw-Hill, 1972), p. 370.

8.3.4   Key Considerations Regarding People

People are motivated for many reasons in their place of work. A few of the principal ideas behind such motivation have been put forth in this section:

- Project managers and members of the project team should be aware of the basic factors that motivate people.

TABLE 8.6 Job Motivational Factors Most
Frequent Response

Chance for promotion
Chance to turn out quality work
Feeling my job is important
Getting along well with others on the job
Good pay
Large amount of freedom on the job
Opportunity for self-development and improvement
Opportunity to do interesting work
Personal satisfaction
Recognition by peers
Respect for me as a person.

- The assumptions that you hold about people, as put forth by McGregor, are reflected in the attitudes you display when working with people, and the way that you treat them.
- The notions about the behaviors of people hold true for all of the stakeholders with which the project manager must deal.
- Most of us are as logical in our work as our emotions allow us to be.
- Most of us have a need for "belonging." A project team that is properly led can do much for satisfying our need for belonging.

There are many factors that influence our job performance. These include our physiological condition, our psychological attitudes, our political beliefs, and moral standards. In addition, our professional standards, our prejudices, and our habits also impinge on our job performance.

8.3.5 Key User Questions

1. Do the team leaders and members of the team have a basic understanding of the characteristics and forces that serve to motivate people?
2. Has everything been done by the team leader, and other leaders of the enterprise, that could be done to create a cultural ambience that helps to motivate team members to do their best work?
3. Do the reward systems of the organization serve to motivate the project team members? If not, why not?

4. Are there any forces or conditions in the organization or in the project team that serve to "demotivate" individuals or team members?

5. Are the project teams sufficiently motivated?

8.3.6 Summary

In this section, a very brief explanation was offered that describes some of the basic factors that motivate people to do their best work. It was recognized that human motivation is a large field of study in the area of human behavior. Presented in this section were a few practical tools and techniques that could be used to gain a basic understanding of the motivational environment of a project team.

8.3.7 Annotated Bibliography

1. Maslow, A. H., *Motivation and Personality* (New York, NY: Harper, 1954). This book was written by the creator of the "Hierarchy of Needs" of people. It is a classic in the field of human behavior. Even though it was published over 50 years ago, its theory and explanation of people behavior is unparalleled in its originality.

2. McGregor, Douglas, *The Human Side of Enterprise* (New York, NY: McGraw-Hill, 1960). This is the book that launched the notion of Theory X and Theory Y concepts and perceptions of people by managers. It too is a classic whose concepts and beliefs are valid today, 44 years after its publication.

8.4 PROJECT TEAM BUILDING AND DEVELOPMENT (PTBD)

8.4.1 Introduction

Team building and development are the act and process of forming, growing, and improving the knowledge, skills, and attitudes of individuals from different needs, organizational units, and professional backgrounds, into a cohesive, motivated, dedicated high-performance team. Team building should be an ongoing "way of life" in the leadership/management of a team. If a team is well managed, an effective PTBD is likely underway on a continuing basis. Basic assumptions about the typical behavior of people which facilitate their becoming a contributing team member include:

- Individuals closest to the work being done in an organization know the most about how that work should be done.
- Participating as a contributing member of a team increases an individual's commitment and loyalty, resulting in potentially high morale, work satisfaction, and quality work performance.
- Meaningful work can be a source of personal satisfaction and, given the right conditions, people will seek responsibility and accountability.
- An individual's fullest potential is best realized in work that encourages freedom of thought and action, initiative, and creativity.
- People are inherently creative and can be developed to improve continuously their technical and leadership capabilities.
- The more people are kept informed about their work and the performance of the organization to which they belong, the more dedicated and capable they will be to make and implement decisions in their work responsibilities.(Paraphrased from David I. Cleland, *Strategic Management of Teams* (New York, NY: John Wiley & Sons, 1996), p. 79.)

> **Team building can enhance the knowledge, skills, and attitudes of the project team**

8.4.2   Features of an Effective Team

A key objective of a PTBD strategy is to provide a fully integrated team that has the characteristics of a fully integrated team as noted in Table 8.7.

PTBD requires that the project leader and the members of the team work on a continuing basis to assess their effectiveness.

8.4.3   Continuous Team Development

As the team works at its responsibilities periodic self-examination is required. This self-examination can be facilitated by asking, and discussing feedback, from questions that are shown in Table 8.8.

TABLE 8.7   Characteristics of A Fully Integrated Team

- Members of the team feel that their needs for participating in meaningful activities in the organization have been satisfied through active membership on a team.
- Team members contribute to the team's culture of shared work, interests, results, and rewards.
- People on the team feel a strong sense of belonging to a worthwhile activity, take pride in the team activity, and enjoy it.
- Team members are committed to the team, its activities, and achievement of its objectives and goals.
- People trust each other, are loyal to the team's purposes, enjoy the controversy and disagreements that come out of the team's operation, and are comfortable with the interdependence of working on the team.
- There is a high degree of interaction and synergy in the team's work.
- The team's culture is results-oriented and expects high individual and team performance.

Paraphrased from David I. Cleland, *Strategic Management of Teams* (New York, NY: John Wiley & Sons, 1996), p. 80.

TABLE 8.8   Team Performance—Self Assessment

- Are we effective in achieving our purposes? If not, why?
- What is going right on our team? What might be going astray in our team's work and in the way the team is being led and managed?
- What are the strengths of our team? What are its weaknesses?
- How well are we doing in settling controversies and disagreements?
- Are we developing a distinct and supportive culture for the team? If not, why?
- Do we help each other in making our team an effective entity?
- Are there any nonparticipants on the team? If so, what are we going to do about these individuals?
- Are the team leader and the team facilitator fulfilling their roles?
- Is working on this team enjoyable, and do the team members feel that membership on this team is beneficial to their career objectives?
- Is there anything that we would do differently on this team if we were given the opportunity?

PTBD begins during the early forming of the team. The culture of a team when it is being formed includes:

- There is concern about individual and collective roles.
- There is uncertainty how the work is to be designed and carried out.

- Objectives, goals, and strategies for the team are preliminary and may be unclear.

- Insufficient time working together as a team raises questions of trust, respect, commitment and other supportive factors.

- There may be difficulties relating personal objectives to team objectives.

- Individual performance standards have not been established.

- Team members probably do not understand how the technical and managerial aspects of the project will be handled.

- Likely limited team spirit

- Team leadership and direction are unclear.

- There are opportunities for power struggles and conflict.

A frank discussion of the answers to the following questions can help:

- What questions do the team members have about the team, its purpose, and how it will operate?

- What questions do the team members have about their individual and collective roles on the team?

- What are the likely objectives and goals of the team and the expectations of team members for their role in meeting those objectives?

- What do team members expect of each other?

- How will a team member's shortcomings or nonparticipation be handled? What do team members expect of the team leader? The team members?

- How will decisions be made on the team? How will checks for consent and consensus be carried out?

- What can the team, its leader, and its members do to build and maintain a cultural ambience of trust, loyalty, respect, candor, and commitment?

- How will conflict be handled on the team?

8.4.4   Managing Team Conflict

Project managers devote a lot of psychological and social energy in dealing with conflict. Conflict can arise over ideas, technical approaches, or processes. Conflict can arise from personal behavior. In dealing with personal behavior, a few guidelines are in order:

- Some controversy is to be expected when people of different experiences, qualifications, and values work together.

- Petty or personal behavior has no place on any team. If the project leader is unable to resolve such behavior, then a counseling specialist from the Human Resources department should become involved.
- Disagreement over substantive issues should be distinguished from personal behavioral issues. Open and thoughtful discussion of the former can contribute to effective conflict resolution. There are many ways to resolve conflict on a project team. The following quote below says so much about the importance of resolving conflict:

> "We increasingly understand that psychological and social energy is tied up in suppressing conflict, that conflicts not confronted may be played out in indirect and destructive ways, and that the differences that underlie interpersonal conflict often represent diversity or complementary of significant potential value to the organization. An interpersonal or organizational system that can acknowledge and effectively confront its internal conflicts has a greater capacity to innovate and adapt."
> R. Walton, *1969 Interpersonal Peacemaking: Confrontations and Third Party Consultation* (Reading, MA: Addison-Wesley, 1969)

8.4.5 Conflict Resolution

A process that can help to resolve conflict on a project team includes the following:

- Gather and have a full understanding of the facts around which the conflict has occurred.
- Effort should be made to reach agreement on what the real issue(s) are at the base of the conflict.
- Potential impact on the project's work or the project team members should be determined.
- Careful identification of the alternatives, and their relative costs and benefits, should be considered.
- The team should come up with appropriate recommendations, identifying majority and minority opinions if required.

If the team is unavailable to resolve the conflict on its own, then a resolution should be sought from higher level managers. If the above protocol is followed, the chances are increased that the conflict resolution can be realized without appeal to higher level managers for resolution.

8.4.6 Ongoing Discussions

By having ongoing discussions along the following points regarding what it takes to achieve a participatory culture for the team, a subtle form of team building and development can result:

- Are team members truly comfortable serving on a team in which the members are expert in diverse fields and will show independence of thought and action?
- Do team members truly value the ideas and opinions of other members?
- Can individual team members accept disagreement and controversy?
- Do team members really want to know what is going on with the rest of the team members, or do they think that what they do is the most important and that things, if let alone, will fall naturally into place?
- Do team members really enjoy working on a team?
- Would some of the team members rather be working alone? If so, they can still make contributions to the organization's purpose and would be better off working by themselves. There are some people who are not team players unless they are in charge of the team.
- Do team members enjoy and value contributing to the growth of fellow team members? Are team members pleased to see a member of the team receiving special recognition for doing something well, or is there likely to be jealousy when a team member is given individual recognition?
- Do team members show true and dedicated interest in making and implementing the best possible decisions to support the team's work?

8.4.7   Team Leader Actions

If the team leader is able to assume effective responsibility for the matters listed in Table 8.9, the cohesiveness of the team will be enhanced.

TABLE 8.9   Team Leader Actions

- The final decision regarding who will perform which work assignments on the team, including deadlines and performance standards.
- The final decision regarding who serves on the team, and the right to request reassignment of a team member who is not performing adequately in spite of continuing counseling.
- The right to comment in writing to the team members' immediate supervisors about performance evaluations of team members. This right should include recommending people for special recognition, bonuses, or exceptional merit increases.
- The right to set the tone for reviewing individuals' and the team's ability to produce results. Schedules, deadlines, reporting protocol, finances, and daily oversight of the team's operation are key responsibilities of the team leader.

8.4.8   Some Further Key Ideas

In addition to what has been presented, a few additional ideas that can help the team leader.

- Seek the recommendations of team members on as many of the team strategies as possible.
- The team leader's role is to facilitate.
- Encourage the maximum participation of all team members.
- Be sure to clearly identify individual and collective roles on the team.
- Try to facilitate a relaxed cultural ambience on the team.
- Always encourage listening by all team members and set the example in doing this!
- Be receptive to all ideas, and seek maximum involvement in selecting ideas for further development.
- Adverse criticism and ridicule should be avoided at all costs.
- Provide feedback to the team—both good and bad news.
- Require team members to report on their work on a regular basis.
- Encourage team members to disclose both bad and good news, and to point out areas where conflict may happen.
- Finally, set a professional standard for the practice of strong interpersonal skills on the team.

8.4.9   Key User Questions

1. Has the project team leader facilitated a cultural ambience that sets the tone for an ongoing development of an effective program of PTBD?
2. In maintaining an effective team, has the team leader drawn on the guidance that is suggested in this section?
3. Was a PTBD initiative started when the team was being formed—and does it continue throughout the life of the project?
4. Are the team members comfortable with bringing bad news to the attention of the team leader? If not, why not?
5. If professional assistance is required in PTBD, has the assistance of a specialist in the Human Resources department been involved?

8.4.10   Summary

PTBD is an important process to be carried out in leading a project team. While both formal and informal means can be used to help a team out in this regard, much can be done through informal means and working with the team during its life cycle along the lines recommended in this section. Indeed some PTBD will happen, although it might be negative in a team that is poorly managed.

Any team will have conflict. Such conflict over substantive issues involving the team's work can be expected. Conflict that arises through the behavior of the members can weaken the team's competency if not resolved early and effectively. There should be a protocol on how conflict should be handled.

8.4.11 Annotated Bibliography

1. Cleland, David I., *Strategic Management of Teams* (New York, NY: John Wiley & Sons, 1996), chap. 5, "Team Culture." This chapter describes how effective leadership of a team can result in an effective ongoing development of the professional and interpersonal skills of the team members. A series of important questions and suggestions are presented in this chapter.

2. Thamhain, Hans J., "Working with Project Teams," chap. 18 in David I. Cleland and Lewis R. Ireland, *Project Management: Strategic Design and Implementation,* 4th ed. (New York, NY: McGraw-Hill, 2002). Professor Thamhain provides valuable guidance on how the performance of teams can be appreciably enhanced through an ongoing strategy of team building and development. He suggests several models and checklists that can help to address the right issues when launching a strategy to improve the interworkings of a project team.

8.5 ROLE AND RESPONSIBILITIES OF THE PROJECT TEAM

8.5.1 Introduction

Defining the roles and responsibilities of the project team is critical to each person understanding their duties and obligations for performance. These roles and responsibilities may also include the obligations to other teammates as well as stakeholders. Defining the duties and obligations also provides a means for selecting resources for the project work.

Undefined roles and responsibilities lead to possible gaps in coverage of vital project elements and possible overlap. Identifying roles and responsibilities for each individual in the project team provides assurance that all areas are covered. It also saves time and effort in recovery of various missed functions.

Roles and responsibilities addressed here focus on the business aspects of project management. Technical aspects such as engineering or physics, are specific to projects and roles and responsibilities are defined at the onset of the project. Further, project team member roles that relate to the discipline or technology are also omitted. There may be a need to identify the roles and responsibilities of these individuals in the project plan.

For this Section, only the roles and responsibilities of the project team will be addressed. Senior management, project sponsors, customers, and other project stakeholders typically do not have defined and documented roles. However, their roles and responsibilities for a specific project may be defined in the project charter. Stakeholders external to the project should be included in the project charter and specifically defined for each project.

> **The effective team understands its individual and collective roles**

8.5.2   Key Business Titles and Functions in Projects

Because different projects use titles that equate to the industry being served, all project titles are not appropriate here. There are common positions within projects that can be described and roles identified. These roles and responsibilities for the business areas are addressed here.

Titles of project team members may include the following: project manager or leader, project planner, and project controller. These titles vary among industries, but they include only the business functions of a project. The technical functions may be accomplished by a project engineer or technical specialist for the industry being served.

The expertise employed in each position is bounded by the skill, knowledge, and ability of the person performing the work. Generally, the positions require the items displayed in Table 8.10.

The project manager or leader is the single point of authority to direct the project's work to a point of technical convergence and delivery of the products or services. This person is appointed to plan, organize, direct, and control the efforts of all persons working on the project, as well as overall responsibility for the internal project work, coordination of interfaces with own organization's entities, and customer interface.

A project manager or leader would typically have the roles and responsibilities detailed in Table 8.11.

TABLE 8.10 Position and Expertise Requirements

Position title	Areas of Required Expertise				
	Planning	Directing	Organizing	Controlling
Project Manager or Leader	Planning concepts, principles, skills, knowledge.	Leadership, managerial, counseling, coaching, mentoring skills.	Developing work plans, assigning work to people.	Receiving reports, initiating corrective actions, follow-up on assigned actions.
Project Controller	Planning skills, writing skills, information collection skills, analytical skills.	None	Formatting reports, analyzing information.	Collecting information, reporting performance figures, tracking performance.
Project Planner	Planning skills, writing skills, information collection, project execution knowledge.	None	Formatting plans, assembling information.	Rework plans, distribute new plans, adjust plans.

TABLE 8.11   Project Manager/Leader Duties

| Project Managers/Leaders Duties | |
Task management	Project leadership
• Structure the project phases • Assign activities • Estimate activities • Schedule the phases • Review assigned work • Track status and progress, replan project • Report status and progress • Manage change for project and product	• Create and sustain a vision • Assure good public relations • Manage expectations • Communicate within team • Inspire and empower team • Build a cohesive team • Sustain a "can-do" attitude • Provide team support and rewards

The project controller is the one who maintains the status and progress of the work through tracking and collecting information. This position is critical to the completion of the project on time and reporting the progress to senior management and the customer. Typically, this person assembles the information for the project manager or leader.

The project controller must collect the information, conduct some analysis to determine the validity of information, and format information into a report. The assembled and formatted information becomes the project's documentation for progress. Accuracy and validity of collected information are typically checked by the project controller.

The project planner position requires a person who has planning knowledge and skills that provide the expertise to develop and publish a coherent project plan. This person must have in-depth knowledge of planning concepts and principles as well as possess communication skills to ensure the plan conveys the proper information to others.

Some organizations consider the project planner primarily a schedule developer. The fully qualified project planner is one who can collect information and place it in a format that describes the work to be done and how it will be done. This includes all components of the project plan such as the scope statement, risk plan, quality plan, and procurement plan.

8.5.3   Skills and Knowledge Requirements

Key positions within the project require specific skills and knowledge, which impact the project success when they are not fully developed. Any weakness in these areas will negatively impact the project and reduce the effectiveness of the project team in getting its work done.

Table 8.12 includes the basic requirements for each position and identifies areas for training or other methods of raising the proficiency of individuals selected for these project positions. The "X" indicates some degree of proficiency is required without specifying the scope of knowledge or skill.

Different organizations will have unique requirements for the level of skills and knowledge in each item. For example, the project leader is not identified as having skills in the use of project scheduling software. It is the planner's job to develop the schedule and the controller's job to main-

TABLE 8.12   General Knowledge, Skill, Ability Requirements

Task, knowledge area, or skill	Project leader	Project controller	Project planner
Proficient with tools of project management			
• Computer	X	X	X
• Scheduling software		X	X
• Word processing software	X	X	X
• Electronic spreadsheet software	X	X	X
• Graphics software	X	X	X
Communication skills (oral and written)	X	X	X
Knowledge of project life cycle and methodology	X	X	X
Interact and communication with:			
• Senior Management	X		
• Project Team	X	X	X
• Project Sponsor	X	X	X
• Functional Managers	X	X	X
• Internal Customer	X		X
• External Customer	X		X
• Industry and Trade Representatives	X		
Conducts information and decision briefings with:			
• Senior Management	X	X	X
• Project Team	X	X	X
• Project Leader	X	X	X
• Functional Managers	X		X
• Internal Customer	X		
• External Customer	X		
• Industry and Trade Representatives	X		

tain and update the schedule. Training the project leader in scheduling software diverts his or her attention away from the primary role.

Communication skills are essential if the project is to succeed. First, project team members must develop a plan that requires writing skills and then brief the plan to others through oral presentation skills. Weak communication skills are easily identifiable and are reflected in project planning and execution.

8.5.4   Key User Questions

1. Defining roles and responsibilities for key project positions has several advantages for the project and organization. Name three advantages.
2. The roles of the project planner and project controller differ in the skill sets. What is the difference between the two roles?
3. The project leader is not required to possess skills with automated project scheduling software. Why is this true?
4. The role of the project planner includes oral presentation communications. Why does the project planner need this skill?
5. What other business roles in your organization should have defined skills requirements? Why?

8.5.5   Summary

Selected project team members require knowledge, skills, and abilities that are supportive of the planning and execution processes of the project. These critical areas will be reflected in the project's documentation, communication of requirements, and progress assessments. Weak knowledge and skills will result in weak plans; strong knowledge and skills will result in solid plans.

The identification of roles and responsibilities for key members of the project team establishes boundaries for work and communication without overlap or gaps in the work. Defining roles and responsibilities provides a means for rapid startup of projects and consistent application of the resources on the right functions within the project. Weak or non-existent roles and responsibilities allow selection of work that is pleasing and avoiding work that is onerous.

The non-technical members of the project team, such as the project leader, project controller, and project planner, must possess the critical skills for the business area of the project. These critical skills should be criteria for selecting and appointing people to these positions. Other skills may be desirable, but these knowledge and skill areas will enhance the team's capability.

8.5.6 Annotated Bibliography

1. Kast, Fremont E., and James E. Rosenzweig, *Organization and Management—A Systems Approach* (New York, NY: McGraw-Hill, 1974). Although this book was published thirty years ago, it provides excellent insight into the subject of individual and collective roles in the management of an organization. By using the systems approach as a basing point, the authors prescribe how organizations and their management can improve how people can better define, understand, and carry out their job assignments.

2. Flannes, Steven W., and Ginger Levin, *People Skills for Project Managers* (Vienna, VA: Management Concepts, 2001). This is a primer on the challenge of handling people and their problems in the context of managing projects. The book is filled with methods and tools for handling people problems that involve communication, role performance, motivation, behavior skills, and much more. It is a valuable book for the individual who wishes to bring about key leadership skills in their project management work.

8.6 PROJECT MANAGER CAPABILITIES

8.6.1 Introduction

A competent project manager has certain key attributes, which provide both a conceptual and practical framework to guide his or her behavior in the management of a project. Capable and successful project managers possess the following competencies:

- The possession of certain *knowledge* (K)—the fact or condition of knowing something with a familiarity gained through learning and experience.

- The possession and demonstration of *skills* (S)—the ability to use one's knowledge effectively and efficiently in execution or performance as a project manager.

- The possession of proper *attitudes* (A)—positive feelings and an open mind regarding a fact or state. A simple expression captures the desired balance of personal competency:

$$K + S + A = \text{Competency}$$

A balance in all of these competencies is essential to being able to successfully manage a project. A shortcoming in any of these competencies can limit the project manager's ability to manage the project. For example:

- An adequate knowledge of project management theory and practice but a lack of skill in applying that knowledge.
- Possession of skills in general management but an inability to exercise those skills in the matrix organizational culture.
- An inadequate knowledge of the theory and practice of management due to lack of training or experience in the discipline.
- A negative attitude towards people—such as that described as "Theory X" by Douglas McGregor in Section 8.3 of this Handbook.
- Lack of the strong interpersonal skills required of a project manager in deliberations and interfaces with project stakeholders.

> **Competency is the ability to perform**

8.6.2   An Adequate Knowledge Base

The key *knowledge* bases required of a successful project manager include an overall understanding of the theory and practice of general management to include the fundamentals involved in:

- An appreciation of the strategic management of an organization.
- A working knowledge of project management theory and practice.
- Execution of the management functions of planning, organizing, motivation, direction, monitoring, evaluation, and control.
- The making and execution of decisions.
- The management of diverse organization stakeholders.

The *skills* of a project manager include:

- Ability to apply management concepts, processes, and techniques in a matrix organizational paradigm.
- Ability to respect and treat with dignity project stakeholders in the deliberations of the project.

- Ability to coach, facilitate, mentor, and counsel project team members as they work in the matrix context.
- The ability to build and maintain alliances with project stakeholders in their support of the project.
- The ability to communicate effectively with all project stakeholders.
- The interpersonal skills that enable the project manager to work through and with diverse people and personalities involved on the project, such that the respect of these people is gained, and they are responsive to the project manager's guidance and support.

The *attitudes* that the project manager has should include:

- Respect for all of the project stakeholders who are involved on the project.
- A perception of people as suggested in McGregor's "Theory Y" described in Section 8.3 of the handbook.
- The need to provide the intellectual and emotional criteria that project stakeholders will accept and emulate.
- Being responsible to the professional and personal needs of the project stakeholders as they support the project team members.
- Recognition that the use of a "matrix" organizational design is simply "the way that we do things around here."

Taking the above brief explanation of the *knowledge, skills,* and *attitudes* that are found in competent project managers, the project manager's success is dependent on a combined set of demonstrated personal capabilities. Figure 8.2 depicts these capabilities.

Project Manager Demonstrated Personal Capabilities

- **Understand Technology**
- **Strong Interpersonal Skills**
- **Appreciate System Perspective**
- **Understand Management Process**
- **Decision Context**
- **Produce Results**

FIGURE 8.2   Project manager competencies.

A presentation and discussion of these capabilities follows:

- *Understand the technology—The ability to understand the "technology" involved in the project.* The word "technology" is used in the sense of a method followed to achieve the project purposes, and deals with the total technological purpose that the project provides, such as a bridge, highway, aircraft, an information system, an order entry system, in providing value to the users. The project manager does not need to have an in-depth understanding of the technology, as members of the project team will provide that, but he or she should have sufficient knowledge to ask the right questions and know if the right answers are being provided.

- *Interpersonal skills*—Having that blend of interpersonal skills to build the project team, work with the team members and other stakeholders so that there is a cultural ambience of loyalty, commitment, respect, trust, and dedication to the team, the stakeholders, and the project. A project manager should be aware that the single greatest cause of failure in managers is their lack of interpersonal skills.

- *Understand the management process*—Being able to understand the management process through an appreciation of the basic functions of planning, organizing, motivation, direction, and control. This includes setting up the management systems that support these basic functions within the context of the project—particularly in the manner in which the use of resources required to fulfill the work package requirements.

- *Appreciate the systems viewpoint*—The ability to see the systems context of the project. This means that the project is viewed as a set of subsystems, the project management system, and is a subsystem of a larger system of the organization, the strategic management system.

- *Decision Context*—Knowing how to make and implement decisions within the systems context of the project. The making and implementing of decisions requires a consideration of such fundamentals as:
  - Defining the decision problem or opportunity
  - Developing the databases required to evaluate the decision
  - Considering the alternative ways of using the resources to accomplish the project purposes
  - Undertaking an explicit assessment of the risk and cost factors to be considered
  - Selecting the appropriate alternative
  - Developing an implementation strategy for the selected alternative
  - Implementing the decision

- *Produce Results*—Of course the key characteristic of a successful project manager is the ability to produce results in the management of the project.

8.6.3   Key User Questions

1. Do the project managers in your organization have the requisite blend of knowledge, skills, and attitudes to perform their project management responsibilities in an excellent manner?

2. Have training programs been set up in the organization to train the project team members on how they can better support the project needs?

3. Do the project managers understand the difference between knowledge, skills, and attitudes, and how these factors can influence how well the project is managed?

4. Does the organization encourage people who work in project management to develop a program of self-improvement to better qualify them to enter the project management career ladder in the organization?

5. Has the organization established any standards for the competency of project managers?

8.6.4   Summary

A summary description of the key elements that enhance a project manager's capability was presented in this section. It was noted that a project manager's competencies center around the knowledge, skills, and attitudes that the individual has, and is able to use to influence the project environment. The section closed with the idea that a project manager's success is dependent on a combined set of personal capabilities, which inherently contain the key elements of knowledge, skills, and attitudes.

8.6.5   Annotated Bibliography

1. Cleland, David I., and Lewis R. Ireland, *Project Management: Strategic Design and Implementation,* 4th ed. (New York, NY: McGraw-Hill, 2002), chap. 16, "Project Leadership." In this chapter, a discussion is presented on the notion of leadership as applied to the project management environment. The chapter recognizes that there have been over 5000 research studies in leadership, over a hundred definitions, and an abundance of opinions on what constitutes leadership. By drawing on some of the leading authorities in the field, a paradigm for the essential conditions for effective leadership in project management is presented.

2. Culp, Gordon, and Anne Smith, *Managing People (Including Yourself) for Project Success* (New York, NY: Van Nostrand Reinhold, 1992). This practical guidebook prepared by authors who have had 40

years of successful project management experience shows how the integration of people and technical skills can be combined in the successful project manager. The authors make the important point that consistent project success comes when project managers focus their energy on people, in particular the ones on their project team, and other stakeholders.

8.7 THE POLITICAL PROCESS IN PROJECT MANAGEMENT

8.7.1 Introduction

Politics play an important role in projects because of the meaning of politics: "serving self-interest." Self-interest is not in itself a pejorative term, but it can be used to the detriment of the interests which we are obligated to serve. Self-interest can also confuse those who have only the interest of the organization in mind when performing on the project.

Politics can mean several things. It can mean that we support some projects and not others. We may put more emphasis and energy to supporting the projects of our personal choice and little effort into projects that we do not believe are worthwhile. It can also mean that we withhold support from projects that do not serve our personal interests.

Politics also plays an important role in interpersonal relationships. This may include whether or not a person is liked or respected. Differences among people and how they respond to different situations can create perceived likes or dislikes.

Individuals sometime fail in projects because they do not understand the politics and who has the most influence. Influence does not necessarily come from a position, but it may be derived more from the relationships between individuals. It is obvious that all managers in an organization do not have equal influence.

> **Politics absorbs much of the time of the project stakeholders**

8.7.2 Project Selection and Sponsorship

Projects are often started based on subjective criteria, i.e., whether senior management likes or dislikes the person championing the project. The likes or dislikes of the champion may be for any number of reasons. Typically, senior management will believe and trust a person who has similar traits and style as they. Figure 8.3 summarizes the selection of projects because for political reasons.

FIGURE 8.3   Project selection and sponsorship.

Senior management does not have a lot of time to select projects and personally perform all the checks to determine the feasibility of the projects. Therefore, they place their trust in the project champion and follow that person's recommendations. This interpersonal relationship is the driver for many solutions.

Senior management sponsorship of an individual will cause the sponsored person to be given more latitude than an unknown individual. Sponsorship is not a bad concept; it is used to nurture a junior into a position that will benefit the organization. The sponsored individual is known for what he/she can do and also known for who is the sponsor.

Take, for example, the senior manager who has a staff member that has been serving him/her for several years. The staff member has the trust and confidence of the senior manager. This same concept is extended to project managers who have worked with the senior manager over several years to build trust and confidence in his/her ability.

8.7.3   Personal Sponsorship

Often individuals are selected by a senior manager for sponsorship because of their potential for promotion. The sponsored individual has access to the sponsor and can often ask for and receive advice on critical career issues. Why an individual is selected for sponsorship is a personal choice of the senior manager.

Sponsorship has many benefits for the sponsored individual and for the organization. Senior managers want to do well for the organization in both the near- and far-term. By selecting and mentoring a promising person, the organization can tap the potential of the best and brightest. The promotion system most often does not select the long-term potential, but is focused on the current performance at a given level.

8.7.4   Project Impacts

Projects may be allocated resources based on personality rather than priority or urgency of need for the project. Allocation may be made based on confidence in the requester more than objective criteria. This is a human trait and will typically occur when there is no priority system in an organization.

Sponsored individuals will receive preferential treatment at project reviews and may have access to information that is not given to others. The sponsor has a vested interest in making "his person" look good. This may be obvious to the person on the outside, but perhaps not as obvious to the sponsor or the sponsored person.

8.7.5   Examples of Politics in Projects

Some examples of projects affected by politics can bring the situation to light. These examples identify the situation without getting into the specific details of who and where. Examples are provided for learning without criticism:

- A major project was considered the best managed project in the nearly 50 projects in the organization. Everyone in the organization knew the project manager and senior managers had openly identified this individual as one of the organization's promising leaders. It was clear that this individual could do no wrong and that all measures of his success followed the words of his sponsorship.

  After 12 months, a new project manager was appointed. Six weeks after the new project manager assumed the job, a review of the project indicated that it was the worst of the nearly 50 projects. The best project became the worst project with the same people except for the project manager. New measures of success were applied to the new project manager.

An interview with the original project manager's college room mate gave some insight. This person's personality and interpersonal skills were very high, and the college room mate stated that everyone liked the project manager. In the 15 years since college, the project manager had earned the trust of senior managers and had been able to please them with his performance.

The new project manager did not have the confidence of senior managers and was not well known in the organization. The major difference between the two was how they were perceived. The new project manager was every bit as competent as the first project manager, but didn't have the sponsorship.

- In 1989, a project manager was friendly with the president of the company. This relationship gave the president confidence in the project manager's capability. Accidentally, this project manager was found to be manipulating project funds, i.e., taking funds from another project. When reported, the president refused to believe the manipulation of funds although the situation was documented.

  The project manager whose funds were short reported the matter and was criticized for being distrustful of the president's friend. This project manager resigned to take another position because of the conflict caused by reporting the manipulation of funds.

  Subsequently, the president's friend was promoted to vice president. However, his employment was terminated within a month. He was found to be using company information to build a business for himself. This did not correct the incident that occurred several months prior, but did work to the detriment of the organization.

- In 1998, two project managers were working in an organization on similar projects. One project manager was an employee and the other was a contract consultant. Both were equally competent and effective. However, when resources were assigned under a matrix organization, the employee received priority.

  The consultant project manager did not know the resource allocator well and was not familiar with the procedures for requesting resources. The second project suffered when the employee project manager would receive adequate resources and the consultant would always be short of resources.

  Known people in an organization will typically receive preferential treatment in the allocation of resources or other considerations. It is important for outside personnel to become known and for them to understand the organization's procedures and practices.

8.7.6 Working in a Political Environment

There is an old saying that refers to the "three ships to success." The best ship is kinship. With kinship, such as a father-in-law owning the business, one just needs to nurture that family relationship. The second ship is sponsorship. Sponsorship is having a senior manager take a personal interest in your career and spreading the word that you are being mentored for high positions. The third ship is showmanship. Showmanship is the ability to dazzle superiors, peers, and subordinates with your presence.

The story continues that it is better to have kinship and then sponsorship. If you have neither of these, you need to have showmanship. It is true that one must be known for his/her abilities and competencies. Spreading the word may be by others or through demonstrated performance.

By recognizing that there may be kinship and sponsorship, one can excel in an organization by being viewed as more competent and possessing more ability that the sponsored person. This places the burden on the individual to first recognize the situation and to be better than the sponsored person without commenting on any weaknesses of that person.

Some guides to working in a highly political environment are as follows:

- Focus on your efforts and not the efforts of the sponsored person.
- Always present yourself in the best light without comparison to the sponsored person.
- Do not use perceived difference in treatment as an excuse.
- Do not look upon the sponsored person or the sponsor as an enemy.
- Never criticize the sponsored person or the sponsor.
- Use facts and figures when requesting support, and never exaggerate the requirement or need.
- Remember that the sponsored person's weakness is not necessarily your strength.
- Treat others with respect and understand the political process.
- If new to the organization, learn the practices and procedures rapidly. Understand that all the rules will not be enforced or used.
- If an outsider, such as a consultant, demonstrate your unique knowledge and skills through work. Don't try to sell yourself with words or references to past accomplishments.
- Be responsible, be respectful, and be reasonable.

Other rules may be appropriate to supplement this list based on experiences in organizations. Each organization will have its unique environment that can be defined by such areas as:

- Publicly-owned corporation with stockholders
- Privately-owned by a few major shareholders
- Family-owned and managed by a few individuals
- Private not-for-profit organization managed by an executive director under the guidance of a board of governors
- Professional association managed by an executive director under a board of directors, which changes each year

Each of these organizations has their sources of power and tenure for management. The constituency served by each is different and the atmosphere can differ greatly depending upon the management styles and interests being served.

8.7.7 Key User Questions

1. What are the benefits of senior management sponsoring a project manager?
2. What are the challenges with senior management sponsoring an project manager?
3. How should a project manager respond to an individual who is related to the president and why?
4. How can a project manager demonstrate his/her worth when several other project managers have sponsorship from senior managers?
5. Why would the structure of an organization affect the political environment and how would two separate structures differ in politics?

8.7.8 Summary

Politics in organizations occur and different relationships develop based on sponsorship or kinship. The self-interests of politics may be detrimental to the organization and can create conflict. A person working in an organization must recognize and understand either wittingly or unwittingly, that politics will always be present.

Relationships based on personal preference and personal style similarities create the political environment. The person being sponsored will most often be relied upon to have the best and most accurate information. While this is not true, an unsponsored person must work to thrive in the environment.

Following the basic precept that good work will be recognized, one must always excel and have a presence that is acknowledged by senior management. This requires setting a few personal rules of not degrading the sponsoring person or attempting to gain sponsorship. It requires doing the best for the organization with the full expectation that this contribution will make a difference.

Different organizational structures have different political environments because of the perceived and projected values of the organization. Management serves different interests and, therefore, will have different styles to meet their obligations. Recognizing the interests served by management will help a person with the political situations.

8.7.9 Annotated Bibliography

1. Baker, Bud, "Political Strategies for Project Managers," in David I. Cleland, ed., *Field Guide to Project Management,* 2nd ed. (New York, NY: John Wiley & Sons, 2004). This chapter provides pragmatic insight into the responsibilities of project managers in managing their political constituencies. The author concludes that politics matter, and effective project leaders know that politics can spell the difference between ultimate success or failure for the project. With this knowledge, the effective project leader takes a proactive stance in working with relevant stakeholders who can influence the outcome of the project.

2. Pinto, Jeffrey K., *Power and Politics in Project Management* (Sylva, NC: Project Management Institute Publications, 1996), 159 pp. This book gives insight with regard for successful management of projects in politically sensitive areas.

SECTION 9
PROJECT COMMUNICATIONS

9.1 The Project Management Information System
9.2 Project Communications
9.3 Communication in Project Meetings
9.4 Negotiations

9.1 THE PROJECT MANAGEMENT INFORMATION SYSTEM

9.1.1 Introduction

The Project Management Information System (PMIS) contains the intelligence essential to planning and control of the use of resources on the project. Also, the PMIS provides the basis for determining where the project is with respect to its cost, schedule, and technical performance objectives, and where the project fits in the overall context of the organization strategy.

Informal information arises from interactions with project stakeholders, through the "informal organization" and from many miscellaneous sources. An important value of informal information is that it can give an important indication of how people outside of the project team actually feel about the project. A model of the linkages in supporting project and organizational purposes is portrayed in Fig. 9.1.

The purpose of a PMIS is to provide for the design and development of a Project Management System (PMS) for the Project (Section 7.10). A PMS in turn provides a model against which resources can be monitored, evaluated, and controlled on the project. This leads to a determination of what the probabilities are regarding the project results. These results are then evaluated by senior managers/users/sponsors to determine opera-

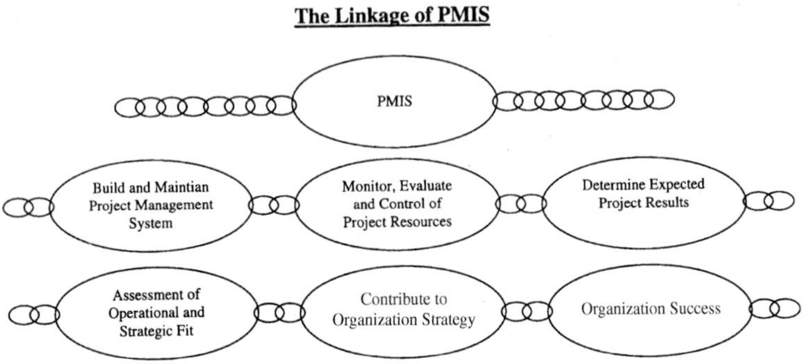

The Linkage of PMIS

FIGURE 9.1   The linkages of PMIS.

tional and strategic fit. This in turn provides for an assessment and realization of how the project contributes to organization strategy, and organization success.

9.1.2   Key Questions regarding PMIS

Project managers should consider the following questions when determining their needs for a PMIS:

- What information is required for the adequate planning, organization, direction, and control of the project?
- What information is needed to keep the project stakeholders informed and managed?
- What information is needed to keep the organization's key managers informed about the status of the project?
- What other projects or programs in the organizational interface with a particular project about which project stakeholders need to be informed?
- What information is required about any project that would enable senior managers to assess its operational and strategic fit in the organization?
- Is information available that is required to make and implement decisions regarding the project?
- Is there too much information about the project?

- What is the cost of not having adequate and relevant information about the project?
- Does the existing PMIS add value to that project?

9.1.3   Expectations Arising from a PMIS

A project manager should have high expectations for the PMIS on a project. These expectations include:

- Understand where the project stands relative to cost, schedule, technical performance objectives, and its likely operational and strategic fit.
- Provide the intelligence needed to plan, organize, direct, and control the project.
- Keep the project stakeholders informed on the status of the project.
- Allow the planned and controlled use of resources to support the project.
- Facilitate communication among the project stakeholders to include the transfer of both "good" and "bad" news about the project status.
- Predict the likely future outcomes of the use of resources on the project.
- Help to recognize project "success" and project "failure."
- Test strategies in the use of resources on the project.
- Comprehend the need for and the likely outcomes of changes on the project.
- Finally, the PMIS can be used to ascertain past, present, and expected future status of the project.

9.1.4   The "Best" PMIS

The "best" PMIS provides a disciplined basis to:

- Identify and isolate significant variances and possibly the reason why a project deviated from the plan.
- Emphasize, where possible, the quantitative and the specific qualitative factors likely to impact the project.
- Provide insight into the specific corrective actions that can be planned and executed to include the assignment of appropriate authority and responsibility.
- Indicate likely effects on the project baselines, to include insight into what revisions are needed, when, and why.

- Provide insight into the specific corrective actions that can be planned and executed to include the assignment of appropriate authority and responsibility.
- Provide intelligent, relevant, and timely information available to facilitate the making and execution of decisions on the project.
- Build around the work breakdown structure (WBS) to provide the capability of ascertaining the status of the work packages at all times on the project to include identification of the work package, its associated cost code and schedule, and the individual responsible for the work.

9.1.5 Principles of a PMIS

In Table 9.1, a summary of PMIS principles is given and more fully described below.

TABLE 9.1 Summary Principles of PMIS

- Improve project management quality.
- Stakeholder awareness.
- Reflect work breakdown structure.
- Life cycle coverage.
- Formal/informal sources.
- Facilitates decision making.
- Supports organization information system.
- Reduces project "surprises."
- Focus on critical project areas.

- The quality of the management of a project is likely a reflection of the quality of the PMIS.
- All project stakeholders need information on the status of the project.
- The work breakdown structure provides the common denominator for information for the management of the project.
- Information is required on the project preceding and during its entire life cycle.
- Information for the management of the project comes from a variety of sources, both formal and informal.
- Information provides the basis for informed decision making and execution on the project.
- The PMIS should interface and be compatible with the larger information systems of the organization.

- The PMIS should minimize the chances of the project manager and other key stakeholders being surprised on project developments.
- Be exception-oriented, focusing on critical areas needing attention rather than simply reporting on all areas of the project.

9.1.6   Key User Questions

1. Does an adequate PMIS exist along the lines suggested in this section for all of the organization's projects?
2. Do the project team members understand the importance of a PMIS in committing and using resources on the project?
3. Is the PMIS for the project "user friendly" in the sense of providing and facilitating the effective and efficient use of resources on the project?
4. Is the PMIS used as an essential part in the process of planning and controlling the project?
5. Do the principal project stakeholders trust the PMIS to provide accurate and timely information on the project?

9.1.7   Summary

In this section, several key elements in the design and implementation of a PMIS were presented. The point was made that relevant, accurate, and timely information is needed to manage a project. Several expectations and principles were suggested as having value in guiding the thinking and action of project stakeholders.

9.1.8   Annotated Bibliography

1. Cleland, David I., and Lewis R. Ireland, *Project Management: Strategic Design and Implementation,* 4th ed. (New York, NY: McGraw-Hill, 2002), chap. 12, "Project Management Information System." This chapter is a comprehensive explanation of a PMIS that includes information for assessing, designing, and operating a system that supports the management of projects in an organization.
2. Tuman, John, Jr., "Development and Implementation of Project Management Information and Control Systems," in David I. Cleland and William R. King, eds., *Project Management Handbook* (New York, NY: Van Nostrand Reinhold, 1983), p. 500. This chapter provides both a

conceptual framework and practical guidance on how to establish and use a PMIS to support project purposes.

9.2   PROJECT COMMUNICATIONS

9.2.1   Introduction

Communications plays an important role in projects to integrate and align the efforts of the stakeholders. Understanding the fundamentals of communication and the project environment is important to the success of enhancing the exchange of information to positively affect the project.

The role of communications in the management of projects is presented in this section. Guides to improve the effectiveness of project communications are offered and considerations are given for communicating with participants in the project.

9.2.2   What is Communications?

Communications is the process by which information is exchanged between individuals through a common system of symbols, signs, or behavior. Some key considerations about communications include:

- May be the most important skill required of a project manager and members of the project team.
- Information is exchanged during a communication process.
- Some means by which communication occurs include plans, policies, procedures, objectives, goals, strategies, organizational structure, linear responsibility charts, leader and follower style, meetings, letters, e-mail, telephone calls, team interaction, and the examples set by the project manager and members of the project team.
- People communicate with each other through actual touch, by visible movements of their body (non-verbal communications), and through the use of written or spoken symbols.

In any communication effort a few basic concepts apply:

- Be as specific as possible about the information to be exchanged.
- Know something about the expectations of the sender and the receiver.
- Consider the perceptions of both the sender and receiver.
- Select the means or medium of exchange for the communication.

- Consider the timing of the communication effort.
- Consider how a misunderstanding of the message might happen.

A few additional guidelines about communication between people include:

- Have the interest and motivation to listen actively and carefully to the message being sent.
- Be sensitive to the sender's purposes in sending the message.
- Consider fully the ways in which a message can be sent, whether verbally, in writing, or by nonverbal means.
- Plan for and provide timely feedback in response to the message being received.
- Ask for clarification of the message, or its intent, if needed.

9.2.3   A Communication Model

Figure 9.3 is a model of the communication process.

.... = Noise
Source -- the originator of the message
Encoding -- the oral or written symbols used to transmit the message
Message -- what the source hopes to communicate
Channel -- the medium used to transmit the message
Decoding -- interpretation of the message by the receiver
Receiver -- recipient for whom the message is intended
Feedback -- information used to determine the fidelity of the message
Noise -- anything that distorts, distracts, misunderstands, or
interferes with the communication process

FIGURE 9.3   Communication model. (***Source:***   James L. Gibson, John M. Ivancevich, and James H. Donnelly, Jr. *Organizations, Structures, Processes, Behavior* (Dallas, TX: Business Publications, 1973, p. 166.)

9.2.4 Role of Informal Communication

People join informal groups at their place of work for social contact, companionship, emotional support, and so forth. The project manager's interpersonal style will influence the informal communications on the project team. To facilitate the value of informal communication the project manager should:

- Accept the fact of informal communication and how it can help or hinder the project team and other stakeholders.
- Find ways to get feedback through the informal organization.
- Use the informal leaders as a source for information about the project and its stakeholders.
- Recognize that much of the cultural ambience of the project is reflected in the attitudes and behavior of the people in the informal organization.

9.2.5 Listening is a Difficult Part of Communication

The ability to listen is an important skill. There are some emotionally-based reasons for ignoring the need to develop better listening skills:

- Listening may uncover some unknown or unexpected problems.
- Project team members may withhold bad news about the project or their work.
- Bad news in an organization does not flow upward easily. But good news flows upward quickly in an organization.
- It is a human characteristic for people to not want to listen to anything that is contrary to their preconceived ideas or prejudices.
- People can think faster than they can listen. This often causes people to react too soon to a message.
- People tend to listen for only the "facts" and ignore the hidden meanings that are often behind the facts.
- We tend to emotionally turn off what we do not want to hear, yet we are "all ears" when something is being said that we want to hear.
- In listening, all too often people are planning a "rebuttal" or response before the message is completely sent by the originator of the message.
- It is easier to "talk down" to subordinates than it is to "talk up" to superiors, or to "talk horizontally" to peer groups and associates.
- In listening, the full potential of effective communication is not realized because:

- Without good listening, people do not talk freely.
- Only one bad listener is needed to impair the flow of communication.
- Messages can be distorted because of "noise" along the communication network.

9.2.6   Emotional Considerations

The most influential issues in communication are the emotional barriers such as ethics, morals, beliefs, prejudices, politics, biases, and beliefs. People on the project team should keep the following tendencies in mind when sending or receiving messages. Table 9.2 contains some of the major emotional factors that impact communication.

TABLE 9.2   Emotional Considerations in Communication

- Hearing only what you want to hear, rather than what the sender intended.
- Undue emotional involvement regarding the subject matter being transmitted.
- Ignoring the contents of a message that runs counter to what we want to hear and believe.
- A preconceived image of the receiver perhaps tainted by past experiences.
- Different means of symbols and cultures, such as in the global marketplace.
- A preconceived evaluation of the source to include positive, negative, or indifferent feeling as to the credibility of the source.
- An ego that gets in the way of the sender or receiver sending or receiving a meaningful message.

9.2.7   Project Management Communication Difficulties

- People will withhold information on a problem in the hope that the problem will go away.

- Team members may not want to share information that they believe is critical to the success of the project.

- The project manager speaks but does not listen. This one-way communication fails to use needed feedback to see how things are going on the project.

- The project review meetings, which should have maximum two-way communication, turn out to be one-person shows.

- People do not understand the basic communication process depicted in Fig. 9.3, thus reducing the chances of effective communication.

9.2.8   Non-Verbal Communication

Subtle hidden messages can be contained in various forms of non-verbal communication such as:

- Facial expressions.
- Body movements, such as nodding, eye movements.
- Actions, such as putting one's feet up on the desk.

  Non-verbal communication may be divided into four categories:

- Physical—such as facial expressions, tone of voice, sense of touch, sense of smell, and body motions.
- Aesthetic—such as creative expressions such as playing instrumental music, dancing, painting, and sculpturing.
- Signs—mechanical communication, such as signal flags, the 21-gun salute, horns and sirens.
- Symbolic—such as religious, status, or ego-building symbols.

9.2.9   Written Communications

In project management, written communications include proposals, reports, plans, policies, letters, memoranda, and other means of transmitting information. Effective writing is an art. The best writing is reflected when the message is simple, clear, and direct. Writing is also a science, with proven methodologies that consider audience, writer, reader objectives, ideal design for different information types, etc. Writing is to be practiced at all times by a manager and team members.

  The field of writing is so huge that we have not the space to describe what needs to be done. A few general guidelines can be useful:

- There are ample article and book readings on writing that can be useful in improving writing skills.
- Always ask the question: Have we written our message clearly?
- Any written message should be informative and easy to understand.
- A proposal or report which uses simple, understandable language and uses tables, bar charts, pie charts, and graphs effectively will have a better chance of being understood than one filled with technical jargon, vague concepts, and ambiguous narrative language.
- Effective writing depends on adequate preparation. Build a "model" of what you want to say before you start writing.
- Establish the basic purpose of the intended message.

- Take time to collect and analyze the quantitative and qualitative data bases that would enhance the message.
- Organize the material into topics and subtopics as appropriate.
- Prepare an initial draft of the message, and if possible, have someone review it in terms of:
  - Is it objective?
  - Is it logical?
  - Are there any fallacies in the reasoning behind the message?
  - Does the message say what was intended?
  - Is there too much (not enough) detail?
  - Are acceptable rules of grammar, punctuation, format, numbering, and abbreviations used?

9.2.10 Key User Questions

1. Do the members of the project team understand the basic considerations involved in effective communication on the project affairs?
2. Considering the criteria suggested in this section, how effective is the communication on the project with all of the project stakeholders?
3. Are the members of the project team able to deal realistically and adequately with the emotional issues that would be present in maintaining communication within the project team, and with outside stakeholders.
4. Have past or present experiences regarding communication with project stakeholders contained any specific "success" or "failure" examples that could be used in acquiring the project team with how to communicate better in the future?
5. Is the project manager able to listen effectively when dealing with the project stakeholders?

9.2.11 Summary

In this section, the matter of communication was presented to include its importance in the management of the project team. Some of the basic concepts and processes of communication were presented. A communications model was presented as a conceptual guide to how project stakeholders can think about the communication processes. Listening was presented as an important, and difficult part of communications. A few suggestions were presented on how better listening can be carried out on the project. The section ended with a brief outline of how better written communications can be carried out in the management of the project.

9.2.12 Annotated Bibliography

1. Cleland, David I., and Lewis R. Ireland, *Project Management: Strategic Design and Implementation,* 4th ed. (New York, NY: McGraw-Hill, 2002), chap. 17, "Project Communications." This chapter provides basic, philosophical insight into the process of communications and how communications with and between the project stakeholders can be improved.

2. Kuprenas, John A., "Impact of Communication on Success of Engineering Design Projects," *Proceedings of PMI Research Conference 2000 21–24 June 2000, Paris France,* Project Management Institute. This paper, based on research within an engineering design project, is divided into two broad types—information distribution and human understanding. The hypothesis of this research project was that each element is essential to success. This hypotheses was found to be partially true. More importantly, the research set a preliminary strategy on how to improve communications in an engineering project.

9.3 COMMUNICATION IN PROJECT MEETINGS

9.3.1 Introduction

A project manager spends most of his or her time in some aspect of communicating: writing, reading, listening, or speaking—much of it done at meetings.

There are three cardinal rules for speaking at meetings:

- Tell them what you are going to tell them!
- Tell them!
- Tell them what you told them!

Project meetings can be a waste of time—and some are! Project meetings are usually called for the purpose of:

- Telling—passing on information about the project.
- Selling a concept or proposal—for example, a new strategy in the management of the project.
- Solving—working with the people in the meeting to come up with a solution to some problem or opportunity, such as participating in a design review to select product design.

- Education and training—instruction to upgrade the knowledge, skills, and attitudes of the project team members.

9.3.2 Planning a Meeting

In Fig. 9.2 a few of the basic purposes of project meetings are depicted. A project manager who is planning a meeting should clearly understand the purpose for which a meeting is being held!

Planning a meeting involves making decisions on agenda, material to use, time, place, purpose, expected outcome (deliverables) and the information that is needed by the participants. Key planning questions include:

- Is the meeting really necessary?
- What is the meeting intended to achieve?
- What is the issue for which the meeting should be held?
- What are the facts needed for the meeting?
- What are the potential alternatives or solutions?
- What recommendations might come out of the meeting?
- What would happen if the meeting were not held?

FIGURE 9.2   Basic purposes of project meetings.

9.3.3 Organizing the Meeting

Some key considerations in organizing a meeting include:

- An agenda
- A suitable location
- Identification and notification of the participants
- Designating the chairperson
- Selection and dissemination of the information to the meeting participants
- How the meeting will be conducted

9.3.4 Effective and Ineffective Meetings

Douglas McGregor (*The Human Side of Enterprise* (New York, NY: McGraw-Hill, 1970), pp. 32–34) describes in everyday common sense terms effective and ineffective meetings. His description of such meetings can provide a project manager with standards against which planning and execution of a meeting can be carried out. According to him, effective meetings are characterized by:

- An informal, comfortable atmosphere
- Participation by everyone in the discussions
- The objective of the meeting being well understood
- The members listen to each other!
- There is comfortable disagreement
- Most decisions reached by a consensus
- Criticism which is frank and comfortable
- People who are free in expressing their feelings
- Taking action so that clear assignments are given
- The chairperson of the group not dominating it
- The group is self-conscious about its own operation

   Conversely, McGregor describes an ineffective meeting:

- The atmosphere reflects indifference or boredom.
- A few people dominate the discussions.
- It is difficult to discern the objective of the meeting.
- People really do not listen to each other.

- Disagreements are generally not dealt with effectively by the group.
- Decisions and actions taken tend to be premature.
- Action decisions are unclear.
- The chairperson is always "at the head of the table."
- Criticism may be present, but it is embarrassing and tension-producing.
- Personal feelings remain hidden.
- The members fail to discuss their own "maintenance."

9.3.5 Controlling a Meeting

Control of a meeting means to make sure that the meeting accomplishes its purpose. The following elements should be emphasized in controlling a meeting:

- Set the time limits and adhere to those limits.
- Start with a statement about the purpose of the meeting.
- Limit the discussion.
- Summarize progress, or lack of progress.
- Encourage and control disagreements.
- Take time to assess how well the meeting is going and what might be done to improve its operation.
- Use McGregor's description of an effective and ineffective meeting as a standard for having quality meetings.

9.3.6 Some Key Guideposts for Any Meeting

- Have a meeting for a definite purpose, and only when it is likely to achieve desired results.
- Be thoroughly prepared before the meeting to include agenda and materials to be distributed for the participants to be able to participate in a meaningful fashion in the meeting.
- Facilitate lively participation—tell people it is OK to become excited, but don't get angry.
- Summarize the meeting's progress from time to time, and bring the meeting to a definite conclusion when it has accomplished its objectives, or when further discussions would not provide any value.

9.3.7   Key User Questions

1. What importance does the project manager and members of the project team place in having effective meetings of the stakeholders of a project?

2. Has any training or reading material been provided to the project team members to sharpen their ability to communicate with stakeholders during meetings about the project?

3. Have there been any "failures" on the project that could be attributed in part to poorly planned and executed project meetings?

4. Are the project meetings planned for and conducted in an efficient and effective manner?

5. Are the project meetings in the organization carried out in the context of McGregor's description of "effective" and "ineffective" meetings?

9.3.8   Summary

In this section, a brief philosophy and the mechanics of planning and conducting project meetings have been presented. Project meetings, properly planned and executed, can facilitate the successful management of the project activities. Meetings are like any other management issue. Time must be taken to develop plans, protocol, arrangements, agenda, and the materials that the people need to prepare themselves for active and meaningful participation in the meeting. Then the meeting has to be led and controlled.

9.3.9   Annotated Bibliography

1. Cleland, David I., and Lewis R. Ireland, *Project Management: Strategic Design And Implementation,* 4th ed. (New York, NY: McGraw-Hill, 2002), pp. 493–497, "Project Meetings." This presents key basic guidelines to use in planning and conducting project meetings.

2. Jay, Antony, "How to Run A Meeting," *Harvard Business Review,* March-April 1976, pp. 120–134. This classic article provides valuable guidance on how to conduct better meetings.

9.4 NEGOTIATIONS

9.4.1 Introduction

An important part of a project manager's responsibility is the ability to negotiate with project stakeholders concerning the project's goals and objectives carried out through a process of obtaining and using resources to support the project's purposes.

Negotiation is the process of arranging support for the project's requirements through discussion, conferences, and appeals to individual members of the stakeholder's community. A few examples of the circumstances under which the project manager and other members of the project team carry out their responsibilities through a process of negotiation include:

- Gaining the concurrence of the functional managers to provide resources to support the project.

- Contracting with vendors for the provisioning of equipment, materials, and services to sustain the project.

- Contracting with the project owner for the project deliverables, and the conditions under which the project's costs, schedule, and technical performance objectives will be met.

- Making agreements with team members to include work package managers and other professionals in the organization to which the project belongs concerning their individual and collective roles on the project work.

- Making provisions for briefing senior management on the progress being made on the project to include recommendations for decisions which have to be made by senior management for the support of the project.

- Gaining approval and cooperation with varied stakeholder groups to include but not be limited to, such stakeholders as unions, government agencies, local community officials, professional associations, media, environmentalists, political/social organizations, educational/training institutions, intervenor groups, and consumer groups.

- Interacting with anyone or agency/institution that believes they have a stake in the project.

9.4.2 Types of Negotiation

A project manager encounters situations in which both formal and informal negotiations are required. The negotiations involved in contract negotiations and contract administration are typically done in a formal manner, although laying the ground work for such negotiations would involve informal interactions. The negotiations to resolve a conflict over the assignment of people from the functional manager's organization would tend to be informal. To further clarify the nature of negotiations:

- *Formal* negotiations are involved where an expected outcome would be a contract or commitment in which the negotiating parties consummate a contract through which certain goods and services are exchanged for some consideration—financial or otherwise.
- *Informal* negotiations are carried out where the contracting parties seek agreement on something of value to be given for some consideration. The negotiation leading to a member of the project team accepting a specific role on a project would be informal.

In any negotiation, the parties are expected to have adequate knowledge of the matter under evaluation, the skill in being able to come to agreement, and an attitude that will respect and perform on the agreed to conditions. In any negotiation, adequate knowledge of the technical matter under negotiation, plus acceptable interpersonal skills are essential.

9.4.3 Conflict

All project matters to be negotiated contain the potential opportunity for a contradiction over the matters in question. For example:

- The scope of the project—what is to be accomplished
- The schedule for the completion of the work packages and the total project itself
- Costing considerations
- The quality of the project and its services that are to be delivered
- Utilization of the people on the project
- Communication effectiveness
- The project uncertainties and risk likely to be involved
- Management of the procurement strategies with the project vendors
- Manner in which the project subsystems are to be integrated into the total project deliverables
- The operational and strategic fit of the project when it is completed
- Perception of the project's success or failure from the perspective of the stakeholders

- Resolution of the potential conflict in the use of resources to accomplish project ends involves the development of certain key strategies:
  - Preparation for the negotiation to include consultation with other members of the project team to identify interests
  - Identification and priority of the "issues" around which the negotiation is to be carried out
  - Development of proposals to integrate the issues and desired outcome of the matter(s) under review
  - Learning about the "other side" to include an assessment of the strengths/weaknesses and probable strategies of the other party with which the negotiations are being carried out
  - Organization and indoctrination of the individuals who will be conducting the negotiations

9.4.4 Power in Negotiating

Like any interaction among people, the possession of control, authority, or some influence is important, and can result in power over the people with whom the negotiations is being carried out. The sources of power available to the negotiators include:

- De jure and de facto authority of the negotiators. (See Section 2.3.)

- Ability to reward to gain compliance. This can include tangible rewards such as money, promotions, and other emoluments. Other rewards can be intrinsic, such as compliments, praise, recognition, or a written praise of a person's performance on a project team.

- Negative rewards such as a form of punishment for example, delaying payments to vendors, disqualifying a subcontractor, or recommending the removal of a team member for cause. Caution should be used in exercising negative rewards, which can be counterproductive.

- Specialized unusual knowledge of the circumstances and factors around which the negotiation is being carried out, such as a research engineer's superior knowledge of a technical area being considered in the contract negotiations.

- Possession of, and ability to use information that is available to a person involved in the negotiations.

- Expert knowledge buttressed by previous demonstrated competence in which project outcomes have been gained by some member of the project team based on their image in the matters under consideration.

- Political competency to include the building of partnership alliances with stakeholders.

• The parties undertaking the negotiation should remember and respect the following guidance suggested in Table 9.3 in which power can be used:

TABLE 9.3   Power in Negotiation

• The power of the other side should be considered—it may be as great or greater than yours.
• The use of power always exposes the user to risks and costs.
• A perception of power can be as effective as the actual power of a contending party.
• Power should consider the perceived expectations of the other party.
• Negotiators should remember their objectives, goals, and planned strategies for the negotiation underway.
• Always go to the negotiations armed with alternative positions and fall-back strategies.
• The efficiency and effectiveness with which decisions are made during the negotiations can enhance the power of the negotiator.

9.4.5   Some Common Dangers of Negotiation

In developing and executing a strategy for negotiation, the contending parties should recognize some of the potential dangers and problems as cited in Table 9.4.

TABLE 9.4   Negotiation Dangers

• Failure to consider the competency of the other party.
• Being inflexible in the bargaining position, to include a lack of willingness to compromise.
• Expecting too much concession on the part of the other party.
• Perceiving negotiation process as a contest of wills between parties each of which wants a win-lose objective on their part.
• Unwillingness to negotiate to a win-win outcome.
• Failure to develop strategies for the negotiation in which adequate preparation has been undertaken by the negotiating parties.
• Trying to negotiate with too many other parties rather than identifying those individuals in the other party, who have the competency and authority to speak for the other team.
• Unable to separate the problems/opportunities from the symptoms.
• Failure to think through the alternative strategies that deal with alternative means, if possible, to reach agreement with the other party.
• Failure to remember that people negotiate, and they have much the same values, past experiences, biases, prejudices, and motivation to gain an upper hand in the process of negotiations that you would have.

9.4.6 Key User Questions

1. Do the members of the project team understand the basic mechanisms and processes involved in the design and execution of successful negotiation strategies?

2. Have training and indoctrination programs been carried out for those individuals who are members of key negotiating groups in the organization?

3. Do the members of the project team understand the knowledge, skills, and attitudes that are precursors to the ability to design and execute successful negotiation strategies? Do such members understand the basics of both formal and informal negotiations?

4. Do the members of the project team understand their responsibilities for negotiation on behalf of the project, as well as the limits of authority that they have in such negotiations?

5. Is the ability to negotiate considered to be a key ability of the project management in the organization?

9.4.7 Summary

In this section, some of the basic information on the matter of negotiation in the management of projects was presented. The reader is cautioned that the ability to negotiate is a key and desirable attribute of the project manager, and that it is a field of study, about which much has been written. The purpose of this section is to present some of the basic notions about the theory and process of negotiation, and to encourage the project team members to seek additional help in the project management literature, and through consultants to upgrade their abilities in this important activity.

9.4.8 Annotated Bibliography

1. Magenau, John M., "Negotiation Skills," in Jeffrey K. Pinto, *Project Management Handbook* (San Francisco, CA: Jossey-Bass, 1998), chap. 21. This chapter was written by an individual who has done research and published about negotiation and dispute resolution. The chapter provides an overview of the process and typical conditions under which negotiation is carried out.

2. Owens, Stephen D., and Francis M. Webster, Jr., "Negotiating Skills for Project Managers," in David I. Cleland, ed., *Field Guide to Project Management* (New York, NY: John Wiley & Sons, 2004). This chapter provides basic and essential information regarding the advantage that

negotiating skills provide the project manager. Recognizing that there is conflict involving a determination of what and how many resources are going to be available to the project and where they will be used, the authors make a compelling case that negotiating is one of the fundamental methods that can be used to resolve conflict.

SECTION 10
IMPROVING PROJECT MANAGEMENT

10.1 THE "NEW MANAGERS"

10.1.1 Introduction

In this section, a description will be given of the characteristics of a "new manager." These new managers, who have emerged in recent years, occupy positions of responsibility in contemporary organizations; however, their style is far from the traditional "Command and Control" managers of yesteryear. In essence, these "new managers" must earn the right to manage other people in changing organizational environs.

The emergence of the project manager in the early 1950s helped to pave the way for the coming of these new managers. In the growing team-driven organizations in contemporary organizations, these new managers are starting to look like today's project/team managers.

10.1.2 Challenges Facing New Managers

The new managers are leaders, mentors, facilitators, coaches, sponsors, advocates, chaplains, comforters, trainers, teachers, team players, entrepreneurs—who must work to provide an environment for the people in the organization to work together with economic, psychological, and social satisfaction. Some of the major challenges that these managers face include the following:

- Lifetime learning to maintain competitive knowledge, skills, and attitudes in an ever-changing environment
- Openness in providing the members of the organization with full information on the organization's performance in the marketplace, and in other activities of the organization that impact both organizational and personal performance
- A cultural environment that abandons traditional manager behavior such as watchdog, controller, bureaucrat, "supervisor," or other characteristics that convey the message that "I'm in charge"
- Growing changes leading to more outsourcing, contract services, and temporary employees
- Recognition that the persons doing the work know the most about how to do the work. Therefore, the new manager explicitly recognizes this and works hard to gain the respect and trust of the employees, and takes extraordinary measures to provide maximum empowerment to these individuals
- Maximizing the participation of people in all of the affairs of the organization
- Abandonment of the single-focus job as we know it today
- Living with strategic initiatives in the organization that constantly executes downsizing, restructuring, and retrenchments to maintain competitiveness
- Living in a cultural ambience in which organizational hierarchy has lost its traditional meaning through a growing design where both internal and external stakeholders play major roles in the influencing of organizational strategies

10.1.3 The Team-Driven Enterprise

Use of teams has accelerated in the last few years, which will operate in a cross-functional and cross-organizational operating environment in which individuals and teams are the key organizational design to accomplish change. Supervision of these teams in the traditional sense is giving way to the following characteristics described in Table 10.1.

TABLE 10.1   Characteristics of "New Manager" Responsibilities

- Provide resources, and work with the teams in judging whether or not the resources are being used judiciously.
- Work with team leaders as well as managers in the limited hierarchy that exists.
- Counsel people rather than "supervise" them.
- Evaluate organizational performance through close interface with the teams in the enterprise.
- Provide broadened opportunity for creativity and innovation to flourish at all levels in the organization, from the senior manager level to the individual team members.
- Provide broad freedom for the teams to operate, essentially working in an environment in which organizational and territorial borders have lost their traditional meaning.

10.1.4   Role Changes

For traditional managers and supervisors the transition to the "new manager" management philosophy can be a real threat because of:

- A sense of loss of status or power, or even the job
- A lack of being able to understand the reverberations set in motion by the broadening use of teams in organizational strategy. This lack of understanding causes a fear of the unknown
- A poorly defined or misunderstanding of the role of the teams
- Fear of personal obsolescence and self-motivated changes that require new knowledge, skills, and attitudes

10.1.5   Some Comparisons

Table 10.2 shows an empirical comparison of changes in management/ leadership philosophies done in the context of the traditional "Command and Control" and the "New Manager" "Consensus and Consent" mode of operating.

10.1.6   The Project Management Influence

The strategic pathway for the emergence of the new managers was influenced by the emergence of the project and team leaders, who had to learn

TABLE 10.2   Changes in Management/Leadership Philosophy

The Old World Command and Control	The New World Consensus and Consent
Believes "I'm in charge."	Believes "I facilitate."
Believes "I make decisions."	Believes in maximum decentralization of decisions.
Delegates authority.	Empowers people.
Executes management functions	Believes that teams also execute management functions.
Believes leadership should be hierarchical.	Believes that leadership should be widely dispersed.
Believes in "Theory X."	Believes in "Theory Y."
Exercises de jure (legal) authority.	Exercises de facto (influential) authority.
Believes in hierarchical structure.	Believes in teams/matrix organizations.
Believes that organizations should be organized around function	Believes that organizations should be organized around processes.
Follows an autocratic management style	Follows a participative management style.
Emphasizes individual manager's roles	Emphasizes collective roles.
Believes that a manager motivates people.	Believes in self-motivation.
Stability.	Change.
Believes in single-skill tasks.	Believes in multiple-skill tasks.
Believes "I direct."	Allows team to make decisions.
	Believes that a manager leads, as opposed to directs.
Distrusts people.	Trusts people.

Source: David I. Cleland, *The Strategic Management of Teams* (New York, NY: John Wiley & Sons, 1996), p. 249. This table appears in Section 2.8 in a slightly different context.

to operate in a cultural ambience in which they did not have traditional command-and-control authority. The change has been profound, and will continue as teams become more influential in both the strategic and operational management of the organization.

10.1.7   Key User Questions

1. Do the existing managers, at all levels, subscribe to the "new manager" philosophy, or are these managers still stuck in the "command-and-control" mode of operation.

2. Have strategies been developed and executed to provide for the suitable training and indoctrination of contemporary managers in the "new manager" environs?

3. Have project managers in the organization been brought in to work in a cultural ambience in which the management philosophies of the "new managers" predominate?

4. Have provisions been designed and executed for the maximum education and empowerment of project managers, team managers, and other managers in the "new manager" organization?

5. Have the team members been afforded the opportunity to work with their counterparts in the suppliers and customers of the enterprise, and through so doing broadened their authority and responsibility.

10.1.8 Summary

In this section, the "new manager" was described as an individual who has emerged in part through the influence of project and team managers. Such an individual, when suitably empowered, provides the leadership and managership in the organization that draws on the competencies of people at all levels—even extending out to suppliers and customers. Some traditionally biased managers cannot, and will not, adapt to the "consensus and consent" style that is characteristic of the "new managers". Such individuals cannot survive in contemporary organization where a prevailing belief exists that the person doing the job knows the most about how to do it. Such persons should be empowered to the maximum to carry out their work responsibilities with the highest degree of participation in the affairs and strategies of the organization.

10.1.9 Annotated Bibliography

1. Cleland, David I., *Strategic Management of Teams* (New York, NY: John Wiley & Sons, 1996), chap. 12. This reference provides insight into how the traditional role of managers is changing from a "supervisory" philosophy to one in which facilitation, mentoring, and coaching are needed to enforce the "consensus and consent" ambience of the organization. The chapter contains several tables that compare the style of new managers and traditional managers.

2. Kiechel III, Walter, "A Manager's Career in the New Economy," *Fortune,* April 4, 1994, pp. 68–72. This watershed article provides provocative insight into how contemporary management philosophies are changing, and how successful managers must adjust to the new philosophies to cope with the growing influence of the use of teams as

elements of operational and strategic initiatives in today's organization. The article is particularly important because *Fortune* magazine is considered to be one of the more sophisticated and visionary publications in the field.

10.2 PROJECT MANAGEMENT MATURITY

10.2.1 Introduction

Organizations are seeking new methods to assess their maturity in project management to ensure they have the most capable process for their businesses. A project management maturity model (PMMM) will give the framework for comparing the degree of development of a project management capability. A standard model may be used to guide internal development of the capability as well as provide a means of assessing one's capability in a comparison with competition.

The PMMM is a guide that characterizes the project management capability of an organization at several levels—typically five levels. The levels are numbered 1 through 5, with level 1 being the first state evaluated. Level 1 is the ad hoc state of project management and is typically the result of random development of a project management capability. Level 5 is the state of optimizing the process for continuous improvement on the capability.

Efforts to develop the model for project management capability centers around the efforts of Carnegie Mellon University's Software Engineering Institute (SEI). SEI has developed a Capability Maturity Model (CMM™) that is being used to improve the capabilities of companies to develop software. The SEI CMM uses project management within its framework to achieve a "repeatable process" and attempts to obtain predictable results from the work efforts.

Several Project Management Maturity Models (PMMM) have been attempted by authors to put forth a framework for companies to use as a guide. These models have replicated much of the CMM™ and typically follow the same format. There has been no official endorsement of any PMMM that would give it a basis for formalizing the process.

The concept of a formal model for project management capability is valid and there is a need to continue with the development of better models for organizations. Project Management Maturity or Capability Models can provide a standard or baseline from which organizations can consistently measure initial capabilities and progress to improving them.

10.2.2 Software Maturity Model

The Carnegie Mellon University Software Engineering Institute has a model for software development that has five levels of maturity. Maturity is equated to a level of capability in this model and gives a measure of the capability to perform software development. This model is defined as levels shown in Table 10.3.

The SEI CMM is an excellent springboard for structuring a project management maturity model. It has the basic requirements for a model that can be tailored to meet the needs of project management both within the software industry and other industries. The rigor of the model can be transported to an organization for use in project management.

TABLE 10.3 SEI CMM Levels for Software

Level	Definition
1—Initial	Process is characterized as ad hoc. Stability of the process is uncertain and may be chaotic. Few processes are defined and success depends upon individual effort.
2—Repeatable	Basic project management processes are established to track cost, time, and functionality. The necessary process discipline is in place to repeat earlier successes on similar projects.
3—Defined	Process for both project management and engineering activities is documented, standardized, and integrated into a standard software process for developing and maintaining software.
4—Managed	Detailed measures of the software process and product are collected. Both the software process and products are quantitatively understood and controlled.
5—Optimized	Continuous process improvement is enabled by quantitative feedback from process and from piloting innovative ideas and technologies.

10.2.3 Quality Management Maturity Grid

Phillip Crosby has a similar maturity model with a different name, the Quality Management Maturity Grid. This grid is useful in understanding the progression of quality (and any other process) within the context of US businesses. The grid, shown in Table 10.4, has five levels that show growth of capability.

TABLE 10.4   Crosby's Quality Management Maturity Grid

Stage	Definition
I—Uncertainty	The lack of understanding and an appreciation for the problem at hand.
II—Awakening	Recognizing the problem and the value of processes to the business.
III—Enlightenment	Initiating improvements and learning more about ways to make additional advances within the work efforts. Being supportive of change and helpful in making change.
IV—Wisdom	Personal participation in the process and engaging in improvements. Placing continuing emphasis on improvements to the process.
V—Certainty	Consider management of the process an essential part of the organization's system.

This quality maturity grid may be the basis for the SEI CMM and has the structure characterized at five levels, but defines the people involvement in the quality management process. It emphasizes management recognition of the need for quality in an organization and progressively more management involvement as one moves up the scale. This is also an excellent basis for developing the project management maturity model.

10.2.4   Project Management Maturity Model

As the profession is recognized and grows, there is a need to define one or more models for implementation into organizations. This implementation depends upon the size of the organization, the need for more formal processes, the needs of the business, the methodology used for developing and delivering products, and the cost of establishing a model within an organization. Models used in an organization provide a means to measure the effectiveness of project management, or the means to determine the project management capability.

Fincher and Levin define a Project Management Maturity Model in five levels, shown in Table 10.5, which is similar to the description of the SEI/CMM. It is helpful to understand their approach to a maturity model.

The graph in Figure 10.1 depicts the overall approach to the growth and maturity of project management in an organization. The level of sophistication raises the maturity and the assumed capability of an organization to perform in the project management arena with a greater degree of effectiveness.

TABLE 10.5   Fincher-Levin Project Maturity Model

Level	Definition
1—Initial	Process is missing and work is accomplished on an ad hoc basis. The success of the project is dependent upon individuals. There is no formal project management methodology.
2—Repeatable	Project personnel are trained in the fundamental elements of project management and related skill or knowledge areas. A methodology is in place and being used. Emphasis is made on replicating the process to ensure the work outcome is repeated.
3—Defined	All areas of project management are defined and processes documented. Project management practices are collected and uses to enhance project efficiency and effectiveness.
4—Managed	The project management process is measured and controlled. Project difficulties are anticipated by management and solutions found before they have a great impact on the project.
5—Optimizing	Focuses on process improvements and fine tuning the methodology to maintain pace with changes to technology. Processes are in place and being used properly. All personnel are trained and performing at a high level of competency.

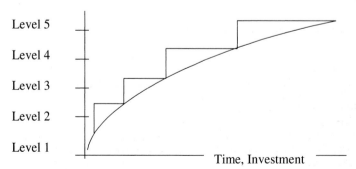

FIGURE 10.1   Conceptual model for maturity.

Another depiction of the project management process maturity model is shown in Table 10.6. It is similar to the Fincher-Levin model, but has different terminology to identify the five steps. Again, it is a five-level process that is designed to achieve maturity in project management.

While these project management maturity models accurately display step-by-step approaches to gaining capability, it may not be the most advantageous attempt to improve one's business. The traditional approach to building on a project management capability is flawed with misperceptions. Project management is often introduced into a company by the use of tools. The typical approach is shown on the left while the more effective approach is shown on the right in Table 10.7.

The most effective project management capability maturity model may be one that does not have steps, but is on a continuum. The organization selects those features of a maturity model that most directly affects its business line and gives the most gain for the cost. Recognizing that some features will be required before others to achieve a valid capability for predictable project results, some features may be deferred.

TABLE 10.6 Generic Project Maturity Model

Level	Definition
1—Ad Hoc	Basic project management processes in place.
2—Planned	Individuals perform project planning.
3—Managed	Systematic project planning and control.
4—Integrated	Integrated multi-project planning and control.
5—Sustained	Continuous project management process improvement.

TABLE 10.7 Project Management Development Order

Typical	Preferred
Tools	Methodology
Skills Training	Standards
Methodology	Techniques
Techniques	Tools
Standards	Skills Training

A case can be made that some companies would not benefit from a Level 5 under any of the preceding models and that it would be an excessive expense to pursue and maintain that model. Maintaining a level 5 capability with the requisite training of personnel may only be cost-effective if all personnel are in projects. If they rotate in and out of projects, one may find the proper balance with less project management capability at less cost.

Using Table 10.8, a linear model that combines the features of each component in order from top to bottom would be a tailored project management maturity model. This tailored model should give greater return because it fits the organization rather than the organization trying to fit a generic model. The increased capability is measured against the needed features and not a standard that attempts to meet all criteria.

A model that uses the features of project management can be depicted as an increase in capability in each area. A cost of achieving each intersection can be computed and the benefits of achieving and maintaining that capability can be weighed against business requirements. The model in Table 10.8 provides flexibility and a tailored approach to constructing a model to follow.

Because the above model is to be tailored to the requirements of the organization, criteria would be needed for each intersection and what level of capability is needed. Only the capability required to perform the project management tasks would be included. No expenditures would be made to "flesh out" the model without some derived benefit.

TABLE 10.8 Component Integration at PMMM Levels

Area	Level 1	Level 2	Level 3	Level 4	Level 5
Methodology	Ad hoc	Partial	Capable	Capable	Capable
Standards	Few	Partial	Capable	Capable	Capable
Techniques	Ad hoc	Ad hoc	Partial	Capable	Capable
Tools	Random	Standard	Standard	Capable	Capable
Skills Training	Random	Partial	Capable	Capable	Capable
Personnel Competency	Unknown	Partial	Partial	Partial	Capable
Project Management Capability	Ad Hoc	Partial	Integrated	Capable	New Processes
Other	To be identified	To be identified	To be identified	To be identified	To be identified

10.2.5 Key User Questions

1. At what level is your organization's capability to perform project management?

2. In what area could you make the most gain in improving your organization's project management capability?

3. What benefits do you see in improving your project management capability according to a maturity model?

4. What improvements have been made in the past for improving your project management capability in a formal and structured way?

5. How long will it take for your organization to reach the next level of maturity?

10.2.6 Summary

There are several capability maturity models for project management. Each has its advantages and disadvantages. It may be desirable to use a model that is tailored to meet the business requirements of the organization rather than have expenditures for features of unknown value. The model should also consider the order in which different features are developed.

A project management maturity model will result from the combined efforts of several authors and will leverage the SEI CMM. This new PMMM will provide the basis for measuring an organization's capability to deliver products using project management processes. The PMMM will provide the framework for continuous improvement of processes so project management becomes a core competency.

There are benefits to having a project management maturity model from which to measure the capability of an organization. A model provides the structure and a baseline from which to measure progress in advancing project management capabilities. It takes the measurement from random process based on subjective estimates to a consistent approach to measuring and building the capability.

10.2.7 Annotated Bibliography

1. CMU/SEI Technical Report 24-93, *Capability Maturity Model for Software, Version 1.1,* Software Engineering Institute, Carnegie Mellon University, Pittsburgh, PA, February 1993, chap. 1 and 2. This Report describes the SEI CMM framework and gives an overview of the purpose of the maturity model. Chapter 1 focuses on the framework and Chapter 2 describes the maturity levels for software.

2. CMU/SEI Technical Report 25-93, Key Practices of the Capability Maturity Model, Version 1.1, Software Engineering Institute, Carnegie Mellon University, Pittsburgh, PA, February 1993, chap. 2 and 4. This Report describes key practice areas of the different levels of the maturity model. Chapter 2 gives an overview of the CMM and Chapter 4 guides a person through the interpretation of the CMM.

10.3 PROJECT RECOVERY FOR THE PROJECT TEAM

10.3.1 Introduction

Many projects become overwhelmed during implementation and it becomes obvious that the project is doomed to failure if it continues on the current course. There is an identified lack of convergence on the technical solution that denies benefits to the customer. Current plans are not working and the project is unstable.

This realization that the project is not fulfilling its promise of benefits for the customer seldom recognizes the toll taken on humans. The project leader is typically stressed and the team members have most often worked far in excess of the number of hours expected. The human resource is depleted of spirit and energy because of heroic efforts to do the impossible.

Senior managers focus on getting the technical effort back on track while the project team's poor state is only recognized as an incapable group to perform the tasks. The project team leaders are often replaced because of a perceived ineffectiveness in completing the project. Perhaps this replacement of the project leader is to energize the team into a more effective group.

10.3.2 Challenged Project Environments

Many projects become difficult to manage because of weak planning or discovered new requirements. Unrealistic completion dates and inflexible scope requirements, while there is a lack of human resources for the work, also contribute to unfriendly project environments. New requirements cause the workflow to change and the project team members get changes to work assignments on a continuous basis. Changes upon changes materially reduce productivity and progress, often to the point that no progress is achieved from a significant expenditure of effort.

Team members become frustrated because each task seems to be a new start without recognition of the dedicated efforts made to advance the project. Progress is measured by senior management in terms of the number of hours expended. Realistic measures of progress are forgotten and the assumption is often taken that "given enough hours of effort on the project, it will meet the objectives."

10.3.3   Degradation of Team Member Capability

Team members dedicated to the project will put forth extraordinary efforts. These efforts may or may not advance the project. Loyalty to the project team leader, the performing organization, and the customer will cause team members to make extra efforts to ensure project success. Dedication and loyalty have the following detrimental effects.

Figure 10.2 summarizes the factors that contribute to project team capability degradation.

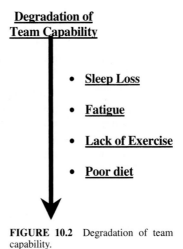

**Degradation of
Team Capability**

- **Sleep Loss**

- **Fatigue**

- **Lack of Exercise**

- **Poor diet**

FIGURE 10.2   Degradation of team capability.

Degradation factors and the causes are described below. Each factor is described as a single element, however, they all interact on the project team members. These are the major contributors to project team performance declines.

- Sleep loss through extra work either before normal work hours or staying late to complete work. The sleep loss is typically from working extra hours each day, and sometime around the clock to meet artificial deadlines. Team members are not trained to work 12 to 18 hours a day six and seven days a week.

• Fatigue from continuous work without a rest break. When work weeks become 60 to 80 hours over several weeks, an individual's efficiency declines severely and the productivity equates to perhaps only 40 hours or less of true effort. The stamina for individuals working overtime typically diminishes significantly in six to eight weeks.

• Stress induced by family members can significantly reduce the productivity level of an individual. Extra time spent away from home and excessive fatigue when the team member is home can generate an undesirable family situation.

• Lack of exercise contributes to reduced capacity to think and the physical stamina required to work many hours without rest. Many tasks on projects require individuals to sit at desks or attend meetings where the limited opportunity to obtain even the minimum exercise inhibits both mental and physical capacity.

• Poor diet from food purchased from vending machines or fast-food restaurants will affect the stamina of individuals. Over a long term, a diet of high-calorie, low-nutrition food will affect the body functions and result in diminished capacity to perform at even moderate levels of efficiency.

This list of major contributors to reduced efficiency and effectiveness can also generate health problems. Team members may experience more frequent colds and respiratory problems. Headaches, chest pains, aches in joints, and difficulty breathing can be symptoms of the fatigue and generally reduced state of health.

Long periods of work without proper rest can also lead to safety issues as well as mistakes in work. Safety issues can include bodily injury from inattention to work around equipment, driving while fatigued in heavy traffic or reduced visibility weather, and general inattention to routine safety procedures because of the inability to recognize potential hazards. Mistakes in work increase with fatigue and reduced mental alertness. Mistakes result from both physical and mental fatigue.

10.3.4 Project Team Leader Responsibilities

The project team leader is responsible for the health and welfare of the individuals on the team. When the health and welfare of the team declines, the team leader will be unable to effectively complete the project. It takes a delicate balance between keeping the team focused on performing the work and placing the health and welfare of the team in jeopardy. Unfortunately, the team leader often believes and acts as though the individual team members will not damage their health for the sake of the project.

Team leaders need to be aware of the signs of fatigue or adverse changes to individuals. These signs can often be detected before there is a major decline in a person's ability to function. First, team leaders must know the individuals on the teams and understand how they function in a usual situation. Deviations from the usual behavior give indication of reduced capacity. Some indicators are:

- Lethargic or unusually slow performance that would indicate fatigue or poor health.
- Slurred speech or a poor pronunciation of words that would indicate slow mental responses to situations.
- Staggering or inability to move about in a controlled fashion that would indicate an inability to physically perform routine activities.
- Laughing or loud talking when the individual is usually quiet and reserved.
- Quiet and withdrawn from the team when the individual is usually outgoing and gregarious.
- Change in wearing apparel from conservative to extravagant or the reverse.
- Any radical change in mood, motivation, or initiative that would indicate a mental shift.
- Bickering or arguing over trivial matters.

The number of hours that a team member works is the responsibility of the team leader. Excessive hours can deplete the energy of the team members and reduce their effectiveness. Individuals will respond differently, but some areas that show the adverse effects of too many hours without sleep are:

- Studies show that individuals deprived of sleep tend to fall asleep more frequently between 4 am and 7 am. Their capacity for reasoning and problem solving is diminished to a low point. Alertness is diminished to minimum levels.
- Individuals functioning with four to five hours of sleep each night will decline in alertness and performance after five to seven days. The alertness and performance degradation may be difficult to identify if the individuals are not under continuous observation.
- Individuals become totally ineffective after 48 to 72 hours of sleep deprivation. The recovery time for individuals depends upon the physical

and mental health of each person. Recovery is not on a linear basis, but may require 12 hours sleep to recover to some marginally effective state.

10.3.5 Project Personnel Recovery

Prevention is the best method of avoiding recovery situations for stressed, fatigued project team members. First, the project team leader must ensure that he/she is capable of thinking through the recovery process. A project team leader who has exceeded the capability to continue to function at an adequate level of effectiveness may be the wrong person to direct rehabilitation for the team.

Effective project leaders can assess the situation to determine whether the entire team requires a period of recovery or whether there are only selected individuals. Following the assessment and identification of stress and fatigue in individuals, the team leader should use a combination of items, as suggested in Fig. 10.3, that would cause recovery over time.

Project Personnel Recovery

- **Sleep 10-12 hrs, Shower**

- **Exercise, Social contact**

- **Eliminate Stress**

- **Proper Training, Reassignment**

FIGURE 10.3 Project personal recovery.

- Sleep deprivation—have team member sleep 10 to 12 hours and perform light exercises upon waking. A hot shower and fresh clothes will make the individual feel better about him or herself. Eating light meals and social contact will speed recovery. Note that an initial period of sleep may only provide some immediate relief and that additional sleep may be required over several days.

- Fatigue from working excessive number of hours—have team member perform light exercise, such as walking or climbing stairs. A hot shower and fresh clothes will change the individual's feeling about him or herself. Eating light meals and social contact that does not discuss work is also helpful. It may also be necessary to change the individual's tasks to ensure a sense of something new and different to achieve a fresh mental outlook.

- Stress—identify the causes of stress and eliminate them to the maximum extent. If the stress is caused, in whole or part by personal problems, one may seek counseling by a healthcare professional. If the project is causing the stress, it may be best to transfer the individual to another project or job area.

- Inability to perform tasks—individual performance issues can be caused by sleep deprivation, fatigue, or stress. Team leaders should also determine whether the individual has the skills and knowledge to perform the tasks. Often, individuals are assigned to high-pressure projects without the requisite capability to perform. They will not admit to not having the skills and knowledge to perform the tasks. It is either training or reassignment as the alternatives for these individuals.

Other factors may be identified that affect individual performance and will need to be addressed. Some items may require assistance from individuals external to the project. Project leaders should always seek assistance from senior management, human resource department, and healthcare professionals at every opportunity.

10.3.6   Prevention Measures

Prevention of degradation of the human resources assigned to the project is perhaps the best method of project recovery for the project team. Identifying and addressing issues early on is the best recovery approach. Early identification limits the scope of the problem and allows repair of the situation before major problems are encountered.

There are several preventive actions that the project team leader may use to avoid team member "burnout" during intense efforts on projects. Figure 10.4 summarizes activities that prevent project team member burnout.

The summarized items are explained in the following paragraphs with some limited detail. Prevention is always better than recovery, especially when the recovery period is dictated by the states of nature.

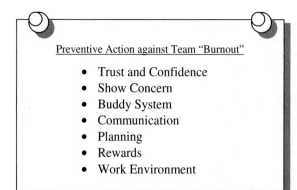

FIGURE 10.4   Burnout prevention actions.

- Trust and Confidence—team leaders must have trust and confidence in the team leader. Consistent, fair treatment of team members will build on this trust and confidence. Trust is on loan to the project leader until he/she proves that the trust is warranted. Trust is built based on "a promise made and a promise kept." Broken promises and intentional untruths destroy the trust in leadership.

- Show Concern—project leaders must show genuine concern for the individual as well as the team. Project leaders must understand the issues that face team members, collectively and individually, to identify opportunities to improve the environment. A project leader who has only his or her interests in mind and demonstrates this through actions will find that the individuals have no alternative but to serve their individual interests.

- Buddy System—when new members are assigned to the project team, it is helpful to have a person team up to support their orientation. It also serves to be a support system between two colleagues, where one can help the other. This buddy system works well when one person identifies the induction of stress, fatigue, or sleep deprivation during intense periods.

- Communication—communicate frequently and on a personal level about the need for pacing one's self and asking for help when needed. Brief team members on signs of fatigue, stress, and sleep deprivation that are early indicators of reduced capacity. Discuss the need for proper sleep and diet with breaks from intense activity. Inform team members that their personal well-being is important to the success of the project and the team's ability to meet challenges.

- Planning—if planning is weak or the project is unstable, the team leader must take action to prevent switching priorities or making changes to the work. Stabilize the work and avoid doing work because it "shows progress." The team members must know that all work being done is important or they will not be concerned with the quality of work.

- Rewards—do not reward team members for working extra hours. Reward achievement. Working numerous hours and being rewarded for not accomplishing the work sets expectations that only time on the job is important. Talk to individuals working more than the usual schedule and determine why they are taking so much time on the job. Do not penalize team members for the dedication and loyalty that they are demonstrating, just change the criteria for rewards to achievement rather than time expended.

- Work Environment—ensure the work environment is conducive to the type of work being performed. Temperatures that are vary frequently, temperatures extremes, and temperatures that are different from those previously encountered by the project team have debilitating effects on performance. Project leaders need to consider all factors of the work environment to make it conducive to a well-functioning team.

It is the responsibility of the project team leader to care for the needs of team members to reduce fatigue, avoid illness, reduce stress and avoid sleep deprivation. A team that is "maintained" by the team leader should perform in a superior way to one whose needs are ignored. Using expert judgment in dealing with the issues of individual and team performance degradation is a major responsibility of the team leader.

10.3.7 Key User Questions

1. What methods of preventing team member burnout during intense implementation have you experienced?
2. How can the project team leader's action prevent team member burnout on projects?
3. How can a buddy system prevent degradation in the quality of work by supporting team members?
4. Who is responsible for preventing team member burnout during intense implementation?
5. What are some of the measures for recovery from sleep deprivation over extended periods?

10.3.8   Summary

Project recovery requires project leaders and senior managers to consider the rejuvenation of the project team. A challenged project typically has team leader and team member burnout. This burnout comes from intensive efforts that often result from sleep deprivation, stress, and fatigue. The project work environment often contributes to the depletion of energy and motivation of team members.

Project leaders are responsible for the team member capability through preventive and recovery actions. Prevention is the most effective because it does not allow the human resource to become severely degraded before corrective actions. Recovery actions take time and effort that may be required for project technical recovery. Both preventive and recovery actions are needed for challenged projects.

10.3.9   Annotated Bibliography

1. Department of the Army, Washington, DC, Field Manual 22-9, *Soldier Performance in Continuous Operations,* December 12, 1991, chap. 1–4, Appendix A. This document describes situations that degrade individual performance through stress, fatigue, loss of sleep, and irregular work hours. It provides suggested methods of prevention and recovery from stress and the leader's role in managing the situation.

2. Verma, Vijay K., and Hans J. Thamhain, *The Human Aspects of Project Management* (Upper Darby, PA: Project Management Institute Press, 1996), 268 pp. This book is one of the few comprehensive treatments of human resources in the project environment and management techniques for those resources. It provides an understanding of the requirements for managing people in dynamic environments.

10.4   PROJECT RECOVERY FOR THE CHALLENGED PROJECT

10.4.1   Introduction

Projects often become unstable and indications are that the promised benefits will not be delivered to meet expectations of the customer. Convergence on the technical solution is frustrated by numerous problems. Current plans, if any, are clearly not providing the guidance to the project leader or project team to successfully move ahead on project work. Progress is either at a standstill or there is loss of progress.

Senior management realizes that something must be done to deliver the benefits of the project to the customer. Typically, the first action is to place more resources on the project. This addition of resources assumes that more is better and that more people doing the same thing will find the solution. Next, there is a realization that more is not better and the process must be flawed. The second action will often be the relief of the project team leader.

Senior managers tend to focus on the desired results without attempting to understand the reason for project failures. Senior managers look to the project team leader as the single source of information—often a person is so overwhelmed with recovery issues that he or she does not have time to fix process problems. The project team leader may also be severely fatigued from working extra hours.

10.4.2   Challenged Project Environments

Projects typically get into trouble because of weak or incomplete planning. New requirements may indicate that the scope of the project was not fully understood prior to implementation. Continual changes to the product definition is a sign of project instability. Weak or incomplete planning is evident when there is only the most basic of schedules for a project plan. This combination of poor scoping and poor planning is a signpost on the road to failure.

Signs of challenged projects are shown in Table 10.9.

TABLE 10.9   Signs of a Challenged Project

- Lack of clear understanding of the work to be accomplished to complete the project
- Reports that assume a higher productivity rate for the future than is really being accomplished now.
- Overly complex plans that cannot be understood by the entire project team
- Overly simple plans that do not give adequate guidance to meeting the project's objectives and delivering the benefits of the project
- High rate of change requests with low rate of closure on changes
- Excessive rework being performed and little actual progress on planned work.
- Surprises with new work that indicates scope is expanding or is not understood.
- Project team members are uncertain as to their work to be performed in the future.
- Project team members waiting for instructions on what to do next.
- Project team members stressed and fatigued from working an excessive number of hours.

10.4.3 Recovery Actions

Recovery from a failing project requires smart thinking. The situation must be assessed to determine the real reasons that there is a failure happening. Both the project team leader and senior managers are reluctant to stop the project activity, regardless of the ineffectiveness of current work, to assess the project's process. There must be a "snapshot" of the project in time and activities to make an informed decision on corrective actions.

Steps that need to be taken to assess the project prior to initiating recovery actions are shown in Fig. 10.5. These top level activities provide the direction for improving the project.

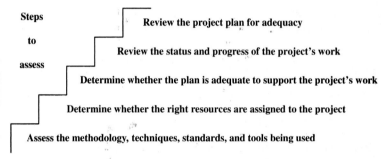

Steps

to

assess

Review the project plan for adequacy

Review the status and progress of the project's work

Determine whether the plan is adequate to support the project's work

Determine whether the right resources are assigned to the project

Assess the methodology, techniques, standards, and tools being used

FIGURE 10.5   Steps to assess project.

Within the steps for assessing the challenged project, there are detailed checkpoints. These detailed checkpoints provide a sharp focus on the project and may include other areas for consideration, including:

- Review the project plan for adequacy
- Level of detail in plan
- Number of assumptions in plan
- Number of issues in plan that have not been resolved
- Comprehensiveness of plan
- Completeness of plan
- Understandability of plan
- Use of plan by those performing the work
- Review the status and progress of the project's work
- Progress and status on schedule
- Productivity compared to planned progress
- Actual expenditures versus budgeted expenditures

- Percent of product complete as compare to the plan
- Total number of changes initiated for the project or product—
  - Number of changes in process
  - Number of changes completed
  - Degree of difficulty to make the changes
- Determine whether the plan is adequate to support the project's work.
- Plan is adequate to guide project to successful completion.
- Plan and work are similar for completed work.
- Plan contains description of work to be completed.
- Determine whether the right resources are assigned to the project.
- Compare plan to actual resources to identify differences.
- Compare assigned personnel to requirements to determine fit of skills.
- Assess whether resources are knowledgeable of the project's work and whether they understand their role in the project.
- Assess the methodology, techniques, standards, and tools being used.
- Is the methodology capable of providing a product?
- Is the methodology being followed by the project team?
- Are best practices being employed?
- Are project management standards being employed?
- Are the tools being used adequate for the project?

10.4.4   Areas for Capability Assessments

Recovery from any of the listed flaws is difficult to perform. If the process is flawed, then the probability of attaining the technical solution is low. One needs to assess the capability of the project leader, project team, and project's process to determine if the project has a high or low probability of completion. Table 10.10 is a matrix that gives an approximation of six combinations of capabilities.

The range of capabilities can vary within elements. However, it is the balance between areas that makes the difference. All elements need at least a partial capability if the project is to be successful. This matrix may serve as a guide in assessing failing projects to determine where the emphasis should be placed.

The most important element is the project's process. If the process is flawed, neither leadership nor hard work will bring the right solution. The process must be fully capable to achieve high-performance on projects. The process includes the planning and the implementation of the plan.

TABLE 10.10 Capability Assessment Matrix

	Capability assessment					
Element	Capable	Capable	Capable	Capable	Capable	Capable
Project's Process	Yes	Yes	Yes	No	No	No
Project Leader	Yes	No	No	Yes	Yes	No
Project Team	Yes	Yes	No	Yes	No	No
Ranked Desirability	1	2	3	4	5	6

Note: While it is recognized that the capability of the project leader, project team, and process may be rated in increments between fully capable and incapable, this simple solution demonstrates the need for assessing the different areas. Further, it tends to weight the capabilities as being more important for the process, then the project team, and last by the project leader.

Flawed processes typically are overcome by brute force and deviation from plans to give a less capable product.

A project leader needs both the business aspects, i.e., project management capabilities, as well as an understanding of the technology. The fully qualified project leader has demonstrated the capability to deliver the products of the project to customers in a manner that meets their expectations. These deliveries are made with a capable process and capable project team.

Project team members should possess the skills and knowledge associated with their responsibilities. This match may not always be perfect, but it should approximate the requirements. Mismatched skill and knowledge requirements will typically result in flawed outcomes for the product.

10.4.5 Project Recovery

Identification and ranking of the important aspects of the project provides the solution to recovery. The recovery solution should range from delivery of a fully capable product to the customer to cancellation of the project. Any solution must result in satisfying the cutomer's requirements, either the original or negotiated requirements.

Some project recovery methods have been identified as general approaches. The following list of actions is a start and may be supplemented by innovative methods. The only caution is that solutions should be simple and easy to implement.

- Correct flaws in the process to ensure a capability to complete the project. These corrective actions may result in a modified process with more or fewer steps. Major changes to the process may result in a complete new start for the project. Thus, only the changes necessary to complete the project should be attempted during recovery. Future projects, however, will benefit from an improved process capability.

- Revise the project plan to reflect what can be done within the constraints identified. If the challenge is technical, address the technical solution first and fit the time and cost around the solution. If there is no feasible project plan, prepare one that guides the project team to the right solution.

- Replace a project leader who is either incapable of performing the work or who is stressed through fatigue. Retain the project leader as a source of information if a new one is assigned to the project.

- Add resources to the project team only after assessing the need for new resources. More resources may have a detrimental effect on the project if there is work being performed and the new resources must learn the work. New resources should be fully informed on the new solution to completing the project as compared to just placing them on the project.

- Resolve issues and limit assumptions in the project plan to simplify the recovery solution. If major assumptions must be used, then assess the risk associated with each assumption not coming true. Avoid high-risk areas where possible and have contingency plans for those that must be taken.

- After the recovery solutions are formulated into a new plan, assemble all the project team and brief them on the new guidance. Existing members of the project team will have individual solutions while new members may have new solutions. The team must function as one and have alignment on the recovery plan. During this critical stage, all team members must follow the plan.

10.4.6   Prevention Measures

Prevention and early detection of unstable or challenged projects gives the best chance for recovery. Prevention is often the result of knowing when the project starts to vary from a good plan. Measuring the key indicators gives warning and allows senior management to assert influence on the projects direction.

Unstable projects result from taking too many opportunities to improve the project's product or through "discovered requirements" that must be included in the work. Unstable projects can be prevented through better planning prior to starting work. New opportunities inject instability into projects because it is assumed that the change is either required for a complete product or there is an enhancement to the product.

The best prevention is to plan the project and follow that plan until it is complete, making only minor modifications as required. A plan that reflects the work to be accomplished and provides a clear picture to the delivery of a technically qualified product is best. Uncertainty injected in the plan through many assumptions and open issues creates the foundation for failure.

10.4.7 Key User Questions

1. What is one of the first indicators of a project that is challenged?
2. What is the most important part of project planning to ensure success?
3. What corrective measures would you recommend for a three-month project with 22 assumptions?
4. How many changes to the product's technical specifications should be allowed on a six-month project?
5. What are some prevention actions that you would take on a three-year project that has 75,000 hours of effort?

10.4.8 Summary

Projects become challenged and indicators of failure emerge in the early stages of implementation. These indicators are usually not recognized until a significant incident or event clearly shows the project is deviating from the desired results. Often the deviation requires a major effort to align the work to product the desired results.

The causes of the impending failure must be identified and assessed as to each one's impact on the project. Process flaws probably have the greatest impact on the project, which is closely followed by weak or inadequate planning.

Prevention of errors in projects is the simplest and least costly form of corrective action. However, when there is a project that has all the indicators of becoming a failure, the corrrective action follows a set plan to fix those items which have the greatest imnpact on project completion. These corrective actions range from minor revisions to the plan to a complete replanning effort.

10.4.9 Annotated Bibliography

1. Glass, Robert L., *Computing Calamities: Lessons Learned from Products, Projects, and Companies That Failed* (Upper Saddle River, NJ:

Prentice Hall, 1999), 250 pp. This book describes 30 of the worst failures in the computer industry. It show readers how to prevent disaster from happening.

2. Glass, Robert L., "Universal Elixir and Other Computing Projects Which Failed," *Computing Trends,* December 1997. This book contains fictionalized tales about real failed computing projects. The book is a classic in its field; it has been reprinted three times, and its lessons learned are as fresh today as they were when it was originally written.

10.5 PROJECT STABILITY

10.5.1 Introduction

Stability of projects is often directly related to the planning phase when items that resolve uncertainty and define the project's outcome are either omitted or not recognized. Project stability is directly related to the level and detail of the planning as well as the availability of valid information at the time of planning. Project instability is the result of either unknown future states of the process or the lack of measures to adequately plan the project.

Planning is a weak area for many projects. In the face of uncertainty, there is optimism that the project will be executed with clarity. Critical items are left to chance and issues are not resolved prior to execution. Planning may often be incomplete.

The project leader and project team are typically technically qualified within their fields and disciplines, but have little training or experience in planning concepts. Highly technical people may ignore planning and consider it an unimportant part of the project. Planning requires a person to envision the future and commit to that course of action. Many people have difficulty with this.

10.5.2 Project Stability by Elements

Planning elements often give advanced warning of the degree of uncertainty and probable instability in the project. It takes experience in identifying the planning elements and how much detail is needed. One must also have the ability to determine when and where critical information is missing.

There is a balance between the degree of project planning and the stability of a project. Figure 10.6 depicts that balance. More project planning weight pushes the project's stability up.

**Project
Planning**

**Project
Stability**

FIGURE 10.6   Project planning—project stability balance.

Major elements for consideration during planning are describe below. These elements are first tier and different organizations will add to this list for their respective project planning.

- Project goals are broad in that they only indicate the end result of the desired outcome. However, this end result may be interpreted differently by several people. The only real way is to have measures of success that describe the components of the end result. For example, if the goal is to build a transportation system, the end result may be described in technical detail as the combination of a bus and train system that moves people over a given geographic area. Time and cost goals are much easier if the scope of the project is defined. Time is a date or several dates. Cost is a price for the project, either a lump sum or a phased expenditure.

  Use measures of success to focus efforts on the end product or the outcome for a process. Each goal should have one or more measures of success.

- Constraints are often listed as a part of the project planning. Constraints are negative statements of limitations on the project, but do not tell the positive side of the project. The missing element is the positive side of the planning. This would be more accurately described as facts. Facts can address both the positive and negative sides to set a foundation for planning.

  Identify constraints under the category of facts. Insufficient facts at the start of a project is an indication of instability in the planning. Address both the positive and negative aspects of the facts.

- Assumptions must always be positive future states in a project. Assumptions can never be negative because if one assumes a negative situation, then there must be action to correct this assumed outcome. Negative assumptions takes on the characteristics of risk events.

  Assumptions supplement facts. When there are facts available, then there is no need for an assumption. Too many assumptions indicate instability in the project. A rule of thumb is that if there are three times as many assumptions as facts, one should review the project for stability.

A second rule of thumb is that there are more than one assumption for each week of project duration, the project stability should be reviewed.

- Risk events are adverse activities that may happen to disrupt the project. A risk event may vary in the degree to which it disrupts the project from some minor cost or delay in the project to total failure of the project.

  Risks need to be addressed early in the project's life cycle. All major risks need to be mitigated or a contingency plan in place. Minor risks should be reviewed for mitigation or acceptance of the risk.

- Product description and completeness of the technical or functional description should be assessed. Unclear requirements will most likely lead to the wrong solution. The customer's definition of the requirement and the one assumed by technical people may vary significantly. This variance or unspecified requirement by the customer may lead to differennt solutions for the project.

  Product descriptions should always be reviewed with the customer early in the planning cycle. There is a need to have complete agreement as to the product prior to starting work. Only research projects or projects that are iterative to develop a project may have a "soft" product description.

10.5.3   Indications of Project Instability

During execution, projects will have indicators of problems. Single indicators may be symptoms of poor information. Multiple indicators probably suggest a downward spiral of the project.

Signs of unstable project execution are listed below. Recognizing the warning signs early is important for recovery:

- Project leader lacks a clear understanding of the work remaining to complete the project. This indicates an unclear plan for completing the work, and may also indicate that there is uncertainty as to the work actually completed.

- Project leader renders reports that assume a higher productivity rate for the future than is being accomplished now. While there is usually a learning curve on projects, it is highly improbable that significant increases in productivity will be achieved in a short period of time. Optimistic increases in productivity rates need to be explained.

- Overly complex plans that cannot be understood by the entire project team. Complex plans typically are not developed for the performing team. These plans are for show or to conceal the true course of action. Plans need to be simple and easy to understand.

- Overly simple plans that do not give adequate guidance to meeting the project's objectives and delivering the benefits of the project. Plans that do not address all areas and include assumptions that "everyone knows that" should be rejected. All work must be planned and the performing individual(s) agree to the work.

- High rate of change requests with low rate of closure on changes. Many change requests and few closures indicate the project is becoming more unstable with each new change. Typically, changes will be compounded and overlaid on one another. Changes to the project need to be tightly controlled or the project halted until a new baseline is established.

- Excessive rework being performed and little actual progress on planned work. Rework is waste in labor and money. Rework also demonstrates the fact that there is inadequate planning or too much uncertainty. Replanning and aligning the project with its goals is required.

- Surprises with new work that indicates the scope is expanding. Unanticipated new work is a sign of inadequate planning or agreement with the customer. New work is discovered work because the product description is weak or there was not a good understanding with the customer as to the end product.

- Project team members are uncertain as to their work to be performed in the future. Uncertainty among project team members indicates either the project leader is not keeping them informed or the planning is weak. The plan may not give adequate details. If the project leader is not meeting his/her commitments to communicate the project plan, then corrective action is needed. If the plan is weak, replanning is required.

- Project team members waiting for instructions on what to do next. Project team members should be provided work instructions for future tasks so they may, to the extend possible, perform in a continuous flow. This situation is indicative of a poorly motivated project team and reflects on the project leader. Corrective action is required to ensure team members are performing at their highest level of efficiency.

- Project team members stressed and fatigued from working an excessive number of hours. The project leader should be controlling the number of hours worked over long periods of time. Surge efforts may be required for short durations, but excessive overtime will diminish the team's ability to perform at the quality and efficiency levels expected.

10.5.4 Prevention Measures

The best prevention is to have well thought through plans that remove as much risk and uncertainty as possible in the time available for planning.

Knowledgeable and skilled planners can support completeness and thoroughness of planning. However, senior managers and the project manager must be supportive of the planning efforts.

Prevention and early detection of unstable projects give the best chance for recovery. Recognizing the indicators of a project that lack stability may be done at any project review or during execution.

Prevention is often the result of knowing when the project starts to vary from a good plan. Measuring the key indicators gives warning and allows senior management to assert influence on the project's direction.

Unstable projects can result from taking too many opportunities to improve the project's product or through "discovered requirements" that must be included in the work. New opportunities inject instability into projects because it is assumed that the chance is either required for a complete product or it is an enhancement to the product.

The best prevention is to plan the project and follow that plan until it is complete, making only minor modifications as required. A plan that reflects the work to be accomplished and provides a clear picture to the delivery of a technically qualified product is best. Uncertainty injected in the plan through many assumptions and open issues creates the foundation for failure.

10.5.5 Key User Queestions

1. What are the first indicators of an unstable project in your organization?
2. What is the most important part of project planning to ensure success?
3. What corrective measures would you recommend for a project than has more change requests than change closures during execution?
4. How many changes to the product's technical specifications are permitted on a project in your organization?
5. What are some prevention actions to avoid instability that you would take on a five-year project to avoid instability during execution?

10.5.6 Summary

Projects may be unstable from the start with poor planning through insufficient information or excessive assumptions. Project instability may also result from an excessive number of changes during implementation that causes new planning, rework, lost productivity, and impacts on the project team.

Proper planning is the best prevention of project instability. Instability indicators can be helpful in early recognition of problems and allows corrective measures to be initiated. Stabilizing an unstable project and bringing corrective actions to bear is the best method of ensuring project success.

10.5.7 Annotated Bibliography

Project stability is to a large degree the result of excellent planning for and execution of the resources used on the project. To manage a project effectively, the management functions of planning, organizing, motivation, directing, and control must be carried out. These functions must be dealt with in the context of project leadership and the strategic management of the enterprise. In this spirit we offer the following key references:

1. Cleland, David I., and Lewis R. Ireland, *Project Management: Strategic Design and Implemention,* 4th ed. (New York, NY: McGraw-Hill, 2002). This edition places emphasis on strategic management for the organization and provides leading-edge information and techniques for managing the project team—and other teams—in the design and execution of improved or new products, services, and organizational processes.

2. Lewis, James P., *Project Planning and Control,* 3rd ed., (New York, NY: McGraw-Hill, 2001). The author's purpose in this text is to translate topics relative to project planning into understandable bite-size pieces that the reader can understand. The book emphasizes the project planning as its main focus. Lewis has provided a standardized methodology that can be used for the planning and execution of a project. He also provides detailed instruction in the tools of project management, and combines management and leadership issues and strategies that complement and reinforce his basic message of project planning, scheduling, and control.

INDEX

ABOUT THE AUTHORS

David I. Cleland, Ph.D. is Professor Emeritus in the School of Engineering at the University of Pittsburgh. He is the author/editor of 36 books in the field of Project Management and Engineering Management. Dr. Cleland was elected a Fellow of the Project Management Institute (PMI®) in 1988. He has received PMI's Distinguished Contribution to Project Management Award three times. In 1997, he was honored with the establishment of the "David I. Cleland Excellence in Project Management Literature Award," sponsored by PMI®. Dr. Cleland does consulting and conducts training programs in project management around the world.

Lewis R. Ireland, Ph.D. is an Executive Project Management Consultant based in Tennessee and serving both US and international clients. He is past President and Chair of the Project Management Institute and a 20-year veteran of PMI®. He has been recognized by PMI® for his contributions by the Distinguished Contribution Award, Person of the Year, and elected a Fellow of the Institute. He currently serves as President of the American Society for the Advancement of Project Management (asapm), a professional society dedicated to better project management practices.

Coventry University